AN ANATOMY
OF THOUGHT

AN ANATOMY OF THOUGHT

The Origin and Machinery of the Mind

Ian Glynn

OXFORD
UNIVERSITY PRESS

For JMG and GJM

OXFORD
UNIVERSITY PRESS

Oxford   New York
Auckland   Bangkok   Buenos Aires
Cape Town   Chennai   Dar es Salaam   Delhi   Hong Kong   Istanbul
Karachi   Kolkata   Kuala Lumpur   Madrid   Melbourne   Mexico City   Mumbai
Nairobi   São Paulo   Shanghai   Taipei   Tokyo   Toronto

First published by Weidenfeld & Nicolson, Ltd., London, 1999

First published by Oxford University Press, Inc., 1999
First issued as an Oxford University Press paperback, 2003
198 Madison Avenue, New York, New York 10016
www.oup.com

Oxford is a registered trademark of Oxford University Press

Library of Congress Cataloging-in-Publication Data
Glynn, Ian
An anatomy of thought : the origin and machinery of the mind / by Ian Glynn
p. cm.
Includes bibliographical references and index.
ISBN 0-19-513696-9 (cloth) ISBN 0-19-515803-2 (pbk.)
1. Neuropsychology.
2. Philosophy of mind.
3. Brain.
4. Cognition.
I. Title.
QP360.G59a5   2000
612.8'2—dc21   99-41218

1 3 5 7 9 10 8 6 4 2
Printed in the United States of America
on acid-free paper

Contents

Acknowledgements

One of the pleasantest features of the writing of this book has been the extra-ordinary generosity of the many friends and colleagues without whose expert advice in disparate fields the task would have been impossible. Graeme Mitchison, mathematician, neuroscientist and molecular biologist, has been a constant source of encouragement and helpful criticism, with an eye equally alert to nuances of meaning and of style. Among my colleagues in the Physiological Laboratory, I am particularly grateful to Andrew Huxley for encouragement and appreciation at an early stage of the book; to Horace Barlow, Andrew Crawford and Trevor Lamb, for advice on vision and hearing; and to Richard Keynes and John Rogers for advice on evolution and on the origin of life. I am also grateful to Janet Allen, John Brown, James Fawcett, Peter Lewis, Hugh Matthews and Hugh Robinson. Robert Foley, of the Department of Biological Anthropology, has been a ready source of advice and help in the area of hominid evolution, and Anthony Edwards, of the Department of Community Medicine, has been similarly helpful about the genetic aspects of evolution. In the Department of Experimental Psychology, Rosaleen McCarthy has guided me in the labyrinthine paths of cognitive neuropsychology, and John Mollon not only criticized my chapters on vision but also provided a private demonstration of Edwin Land's generation of colours by superimposing black and white transparencies projected with a red light and a white light. I am grateful for helpful criticism of my discussion of language in monkeys from Nick Mackintosh, and of my chapter about emotions, from Tony Dickinson. Simon Baron-Cohen has given both encouragement and information and has made me familiar with the concepts of mindblindness and of synaesthesia. John Duncan, of the Medical Research Council's Unit of Cognition and Brain Science, has been particularly helpful in criticizing the chapter on Planning and Attention. For advice in the arcane area of the dopamine hypothesis of schizophrenia I am grateful to Douglas Ferguson of the Department of Pharmacology. The philosophical chapters owe much to the constructive (and sometimes destructive) criticism of Peter Lipton, Hugh Mellor and Gisela Striker; and at a much earlier stage to help from Myles Burnyeat and Nicholas Denyer. Among my colleagues at Trinity not already mentioned, I should like to thank Sidney Allen and Simon Keynes for help with linguistics, Roger Dawe for comments about Sophocles and the Oedipus complex, Hugh Hunt for bringing me up to date about timbre, Tony Jolowicz for a discussion about the relation between intent and liability, and Martin Roth for being a source of information and anecdote in all matters psychiatric. And I want to thank Semir Zeki, of University College, London, and Patrick Bateson, of King's College, Cambridge, for including me in informal discussion groups whose meetings have been as stimulating as they were enjoyable. For the mistakes that remain despite this wealth of advice, I accept full responsibility.

Besides experts, the writing of this book has needed the inexpert – readers unfamiliar with, yet interested in, the topics discussed. And here I owe an enormous debt to my wife and family. Though they cannot all claim to be inexpert over the whole of the field, together they have been able to provide invaluable lay criticism for every chapter – and criticism not only of substance but also of style. I am also grateful to my daughter Sarah for turning a photograph into the line drawing used in Figure 6.2.

Finally I want to thank Felicity Bryan, my agent, and Toby Mundy, my editor, for their encouragement, advice and understanding during the successive stages of the book's production.

Ian Glynn
Trinity College, Cambridge

The author and publishers either have sought, or are grateful to the copyright holders for permission to reproduce the following illustrations and quotations: 4.1, 4.2a & c I. Tattersall ('95) *The Fossil Trail*, OUP; 4.2b A. Walker & R. E. Leakey ('78) *Scientific American*, 239, Aug., 44–56, Nelson H. Prentiss; 5.1 & 5.2 J. Watson et al ('82) *Recombinant DNA: A Short Course*, W. H. Freeman & Co.; 5.3 L. Stryer ('75) *Biochemistry*, W.H. Freeman & Co.; 6.2 Redrawn from a photograph in M. Brazier ('59) *American Handbook of Physiology*, 1st edn, ed. J. Field, Ch.1, American Physiological Society, 6.4 is from the same article; 7.2 A.L. Hodgkin ('39) *J. Physiol., Lond.*, 94, 560–70; 7.3 A.L. Hodgkin ('64) *The Conduction of the Nervous Impulse*, Liverpool University Press; 9.2 N. Unwin; 10.1 C.M. Hackney & D.N. Furness ('95) *Am. J. Physiol.*, 268, C1–C13; 10.3 J.E. Dowling & B.B. Boycott ('66) *Proc. Roy. Soc. Lond., B.*, 166, 80–111, Fig.23; 10.4 Slightly modified from H.B. Barlow & W.R. Levick ('65) *J. Physiol., Lond.*, 178, 477–504; 11.3 W. Penfield & T. Rasmussen ('50) *The Cerebral Cortex of Man*, Macmillan; 12.2a W.H. Ittelson ('52) *The Ames Demonstrations in Perception*, Princeton University Press; 12.3 V.S. Ramachandran ('88) *Nature*, 331, 163–6; 12.4 O.J. Braddick & J. Atkinson ('82) *The Senses*, eds H.B. Barlow & J.D. Mollon, pp. 212–238, CUP; 12.7c R.L. Gregory (ed.) ('87) *The Oxford Companion to the Mind*, p.289, OUP; 13.1 A.D. Milner & M.A. Goodale ('95) *The Visual Brain in Action*, OUP; 13.2 A.B. Rubens & D.F. Benson ('71) *Archives of Neurology*, 24, 305–16, American Medical Association; 13.3 E.R. Kandel et al ('91) *Principles of Neural Science*, Appleton & Lange; 14.1 D.H. Hubel ('88) *Eye, Brain & Vision*, Scientific American Library, p.74, modified from *J. Physiol. Lond.*, 160, 106–154 ('62), Fig.19, 14.2 also from Hubel; 14.3 C. Blakemore ('73) *Illusion in Nature & Art* (eds R.L. Gregory & E.H. Gombrich), p.35, Fig.27, G. Duckworth & Co.; 15.2 D.E. Rumelhart et al ('86) *Parallel Distributed Processing: Explorations in the Microstructure of Cognition*, Vol.1, p.320, Fig.1, MIT Press; 16.1 N. Geschwind ('79) *Scientific American*, 241, Sept., 180–199, permission from C. Donner; 19.1 F. Vargha-Khadem et al ('77) *Science*, 277, 376–80; 21.1a Bloom & Lazerson ('85) *Brain, Mind & Behaviour*, Educational Broadcasting System, W.H. Freeman & Co.; 21.1b R.A. McCarthy & E.K. Warrington ('90) *Cognitive Neuropsychology*, p.79, Fig.4.5, Academic Press; 21.2 C.M. Gray et al ('89) *Nature*, 338, 334–7; quotations reprinted from W. Penfield & T. Rasmussen ('50) *The Cerebral Cortex of Man*, Macmillan; E. De Renzi, 'Prosopagnosia in two patients with CT scan evidence of damage to the right hemisphere' ('86) *Neuropsychologia*, Vol.24, Elsevir Science; J. Zihl et al ('83) *Brain*, 106, 313–40, OUP; J. Riddoch ('17) *Brain*, 40, 15–57, OUP; D. Hubel ('88) *Eye, Brain & Vision*, pp.69–70; H. Gardner ('77) *The Shattered Mind*, pp.61 & 68, Knopf; J. Wallman ('92) *Aping Language*, p.106, CUP; L. Wittgenstein ('58), trans. G. Anscombe *Philosophical Investigations*, 2nd edn, Pt1, para.258, Blackwell; J. Searle, *The Rediscovery of the Mind*, pp.1 & 14, MIT Press.

I CLEARING THE GROUND

1 What this book is about

The mind is its own place, and in itself
Can make a Heav'n of Hell, a Hell of Heav'n.
— Satan, in Milton's *Paradise Lost*

Out of sight, out of mind; times out of mind; put it out of your mind; you must be out of your mind; mind out! It's all in the mind; it comes to mind; it's on my mind; at the back of my mind; in my mind's eye. Of sound mind; of like mind; all of a mind. Absence of mind; presence of mind; strength of mind. Make up your mind; know your own mind; speak your mind; change your mind; never mind! Do you mind?

From time to time, most of us use most of these phrases, or phrases like them. We use them spontaneously and without anxiety, and they appear to convey information without ambiguity or confusion. So we seem to be able to understand what we mean by the word *mind*. Of course, what we mean depends on context, for both as a noun and as a verb *mind* has a family of related meanings. The mind that I change when I change my mind is not quite the same as the mind that is absent when I am suffering from absence of mind – what is changed is (usually) my *intention*, what is absent is my *attention*. What is expected of the window cleaner with the ladder, when he is told to mind the baby, is different from what is expected of the baby-minder. Minding a dog is much like minding a baby, but *minding dogs* is quite different. If you want a friend to mind your dog for an afternoon, you had better choose a friend who doesn't mind dogs.

Our competence in using the word mind in its various meanings in everyday conversation might lead us to suppose that we have a thorough understanding of all of these meanings. Such a supposition would not be justified, and for two reasons. In the first place, competence in word usage does not imply any deep understanding of the concepts referred to by the words used. We need to know very little of harmony to be able to say, appropriately, that something is harmonious. The second reason is less obvious but not less important. Intelligible conversation requires that conventions be shared between those conversing; it does not require that those conventions be based on fact or on correct interpretations of fact. Provided that we are confused in the same way, our conversation is unlikely to draw attention to our confusion.

In this book, I want to discuss a number of interconnected questions that have to do with our minds. What kind of thing is a mind? What is the relation between our minds and our bodies and, more specifically, what is the relation between what goes on in our minds and what goes on in our brains? How did brains and minds originate? Can our brains be regarded as nothing more than exceedingly complicated machines? Can minds exist without brains? Can machines have minds? Do animals have minds? None of these questions is new, and some of them are extremely old. None of them can yet be answered in a way that is wholly satisfactory, though in the last few centuries, and more particularly in the last few decades, there have been some worthwhile partial answers. Such progress as there has been has come from work in a wide range of disciplines, and this itself creates problems, for experts in one discipline may be ignorant of or unsympathetic to the conclusions (and even the problems) of experts in another. Relevant disciplines include neurophysiology (the study of normal nervous systems), neurology (the study of diseases of the nervous system), psychology, psychiatry, philosophy, and artificial intelligence (the use of computers to do things that, when done by us, seem to need intelligent behaviour). But the questions are not only – or even mainly – the concern of neurophysiologists or neurologists or psychologists or psychiatrists or philosophers or computer experts. Because the answers to them inevitably affect the way in which we see ourselves, the way in which we treat others, our attitude to animals, our interpretation of the past and our expectations for the future, they are of interest to all of us.

In the chapters that follow, I shall try to show how, and to what extent, advances in various areas of knowledge have increased our understanding of the nature of minds in general and of our own minds in particular. I shall try to ensure that the arguments are accessible to the non-scientist, either by giving the necessary background or, where that is impossible, by making clear what the reader is being asked to take on trust. But before we start to consider the contributions made by individual areas of knowledge, it will be necessary to clear the ground, for we each approach these problems supported (or burdened) with a host of assumptions and preconceptions, which, whether they are right or wrong, have an influence that is all the more insidious because we take them for granted.

The first bit of ground-clearing will be to look at what might be called the common-sense response to questions about the mind. Though common sense can be wrong – it would, after all, suggest that the earth is flat, or that white paper seen by moonlight reflects more light than black paper seen by sunlight – to reject it without good cause is to embrace gullibility and court the seductions of every philosophical, religious and scientific mountebank. And there is no doubt that the common-sense view of the mind serves us very well in ordinary life – it is the view even philosophers adopt when they are not actually engaged in philosophy. Yet we shall see that to accept it forces us to choose between two unattractive alternatives: *either* that the laws of physics that seem so universally effective in other spheres are incapable of explaining the apparent effects of our mental activities, *or* that our thoughts are without physical effects.

Anyone writing a book about the mind and rejecting the common-sense view has an obligation to establish what might be called a starting position. My own starting position can be summed up in three statements: first, that the only minds whose existence we can be confident of are associated with the complex brains of humans and some other animals; second, that we (and other animals with minds) are the products of evolution by natural selection; and, third, that neither in the origin of life nor in its subsequent evolution has there been any supernatural interference – that is, anything happening contrary to the laws of physics. (This last is, if you like, a confession of biological *uniformitarianism* – a belief that, so far as possible, it is sensible to try to explain the past in terms of the kinds of processes that occur in the present.) The first of these three statements seems to me self-evident. But if minds exist only in association with complex brains, then the origin of those brains is crucial; in which case the truth of the second and third statements is crucial. And no one can claim that those statements are self-evident. I therefore want to spend the greater part of this section in justifying them.

The notion that we are the products of evolution by natural selection is widely believed; but it is believed in the curious way that was foreseen by Darwin's friend Joseph Hooker in 1860. Writing at the height of the Darwinian controversy, he said, 'There will be before long a great revulsion in favour of Darwin to match the senseless howl that is now raised, and...as many converts of no principle will fall in, as there are now antagonists on no principle.' A belief in evolution by natural selection is widely held, but even among scientists it is often held more as a matter of faith than of reason. At a time when the British Treasury feels that Charles Darwin is too controversial a figure to appear on a banknote, a potential American Presidential candidate advocates the compulsory teaching of creationism in American schools, and more than a quarter of first-year medical students at a major Australian university reject the theory of evolution by natural selection, it seems worth reviewing the evidence and showing that recent work, so far from casting doubt on Darwin's theory, provides impressive further support for it.[1]

To the majority of educated people in the western world, my starting position about the origin of life is probably less acceptable than the idea of evolution by natural selection. Yet as a basis for trying to understand the nature of minds it is just as important. For if there is a supernatural element in the origin of life, a clue to the understanding of the mind might lie in that earlier mystery. If the origin of life can be explained without invoking any supernatural processes, it seems more profitable to look elsewhere for clues to an understanding of the mind. For this reason, I want to discuss the evidence that bears on the way in which life originated, a subject which, without much attention from the public, has been quietly transformed by work over the last three decades from an area of flighty speculation to an area in which plausible scenarios are carefully assessed.

Often in this book I have taken a historical approach. This is sometimes because the best way to grasp an intellectual position is to understand how it was reached, but there is another reason. Our understanding of the origin and machinery of brains and minds, in so far as we do understand them, is the product of a long history of investigation by people of different backgrounds working in many fields and in many different parts of the world. And the picture that has emerged, however incomplete, represents a major triumph in the history of thought. I hope, by talking not only about what we now believe but also about how we came to believe it, to convey something of the excitement and grandeur of that achievement, and also something of the problems and pitfalls. It is fashionable at the moment to decry the anthropocentric attitudes of our ancestors, to emphasize that we are one kind of animal among many, and to remind ourselves that our world is not the centre of the universe and that it was not created for our dominion. Such comments may be salutary, but the fact remains that the human brain is the most complex object we are aware of in the universe, and not the least of its many remarkable achievements is the partial elucidation of its own working.

2 The Failure of the Common-Sense View

Common sense is the most widely shared commodity in the world, for every man is convinced that he is well supplied with it.
—Descartes

Consider the relation between our minds and our bodies. The common-sense view would be something like this: we all have bodies and we all have minds. It is true that introspection, which seems to give each of us such confidence in the existence of our own mind, is not available for examining the minds of others but, given the similarities in our bodies and in our overt behaviour, it would be perverse – as well as arrogant – to maintain that only one mind – our own – exists. It also seems obvious that our minds and our bodies can interact, and in both directions. We blush with embarrassment, we are sick with fear, and we shake with rage; Shakespeare, speaking through Shylock, tells us that there are men who, 'when the bagpipe sings i' the nose, Cannot contain their urine'. And in the reverse direction, when we bark our shins or our teeth rot we feel pain; when we eat strawberries or look at dappled sunshine in a wood we feel pleasure. Smells are particularly evocative. The smell of a particular furniture polish transports me vividly into the hall of my grandfather's house in London in the late 1940s; and we have all had experiences of that kind.

But when we look more closely at this mutual interaction between our minds and our bodies we run into a fundamental difficulty. By the 1870s, the successes of classical physics had made it increasingly awkward to suppose that interaction *could* be mutual. It seemed axiomatic that the explanation of each physical event lay in an unbroken series of antecedent physical events, yet the commonsense interactionist view made it necessary to believe that in humans, and probably in some of the higher animals, some physical events were explicable only in terms of conscious mental events. And that is a notion that is wholly at variance with the development of the physical sciences during the last three centuries. With the development of physics, mutual interaction between the world of physical states and events and the private world of mental states and events became, as the Oxford philosopher Geoffrey Warnock paradoxically puts it, 'unbelievable and also undeniable'.[1]

Attempts to escape from this uncomfortable position were suggested by a number of people around 1870, perhaps most clearly by an Englishman, Shadworth Hodgson; but before describing that attempt I want to make two other points.

The first is that the dilemma created by the successes of classical physics is not helped in any obvious way by the indeterminacy of quantum physics, which introduces a random element in the coupling between very small-scale physical events.

The second point is that, though the dilemma was made acute by the successes of classical physics, there were earlier intimations of it. What we now regard as the commonsense interactionist dualist view was first codified in the seventeenth century by Descartes, who thought that animals were automata, but that in humans the automatic machinery was influenced by the 'rational soul'.[2] In 1643 Descartes received a letter from Princess Elizabeth of Bohemia saying that she could not understand how the soul, being only a *thinking* substance, could determine the actions of the body.[3] Descartes' reply was to ask her to consider three things: 'as regards body', he said, 'we have only the notion of extension [that is, the property of occupying space] ... as regards the soul ... we have only the notion of thought ... as regards the soul and the body together, we have only the notion of their union'.[4] And, he argued, 'we go wrong if we try to explain one of these notions by another, for since they are primitive notions, each of them can be understood only through itself'. Not surprisingly, this unholy and divided trinity failed to satisfy the intelligent princess, who knew a fudge when she saw one, and who in subsequent correspondence with Descartes had to be content with flattery and the discussion of less difficult problems.

It is time to return to Shadworth Hodgson. In a book with the delphic title *The Theory of Practice*, he proposed that mental events – by which he meant conscious mental events (and I shall use the term in the same sense) – were caused by physical changes in the nervous system, but could not themselves cause physical changes.[5] Like the whistle of a railway engine (which does not affect the engine), or the chime of a clock (which does not affect the clock),* they were caused by (and accompanied) physical events, but they did not themselves act as causal agents. In a slightly later terminology, they were *epiphenomena* – a word taken over from the pathologists who had used it to refer to secondary symptoms of a disease.

The idea that a sensation might be regarded as a by-product of physical events in the brain – events which could perhaps achieve their effects on the body (and its behaviour) without the mental accompaniment – had been suggested by J. M. Schiff, in Berne, in a textbook of human physiology published in 1858;[6] but Hodgson went further. Not only did he suppose that mental events could not cause physical events; he also supposed that they could not cause other mental

---

*Both metaphors were suggested later by T. H. Huxley.

events. Where one mental event appeared to cause another, what was really happening, he said, was that the physical events in the nervous system, associated with the first mental event, were causing the physical events in the nervous system, associated with the second mental event. A succession of nervous events, he argued, can give rise to a succession of mental events just as arranging coloured stones can create a mosaic picture. The stones, he compared to states of the brain; the colours, to 'feelings' or 'states of consciousness'. His analogy is peculiarly unsatisfactory, since apart from their colour the individual stones in a mosaic are often not distinguishable, but better analogies were supplied by William James (who did not accept Hodgson's theory) writing twenty years later.[7] If Hodgson is right, says James, the 'mind-history' of each man runs alongside his 'body-history', each point in the first corresponding to but not reacting on a point in the second; 'so the melody floats from the harp-string, but neither checks nor quickens its vibrations; so the shadow runs alongside the pedestrian, but in no way influences his steps'. (It was not without good reason that William and Henry James were described, respectively, as a philosopher who wrote like a novelist, and a novelist who wrote like a philosopher.)

How helpful is this theory that mind is an epiphenomenon? It is of course crucial to the theory, that mental events never occur in the absence of nervous events. Proving a universal negative is notoriously difficult, but the evidence on this point is strongly in Hodgson's favour. Claims for the existence of disembodied minds are made from time to time, not always fraudulently, but I am not aware of any that has been competently investigated and found to be convincing. So far as we can tell, mental activity is always associated with nervous activity, and later on we shall see that particular kinds of mental activity are associated with nervous activity in particular parts of the brain.

But though the epiphenomenon theory passes this first test, it is not satisfactory.

The most obvious, but least serious difficulty, is that the notion that mental events are epiphenomena – that they are caused by, but incapable of causing, physical changes in the brain – does not readily square with the results of our own introspection. Thomas Henry Huxley, who was a strong supporter of the epiphenomenon theory, gave a famous talk to the British Association meeting in Belfast, in 1874, in which he coined the phrase 'conscious automata' to describe what we would have to be if the theory were right.[8] But our immediate reaction to this phrase is to say that we do not feel ourselves to be conscious automata.

Consider this situation. I am driving home from the Laboratory when I remember that I have not posted the letter that my wife particularly asked me to post when I left home in the morning. I know that I will shortly pass a public postbox but I realize that there will be no further collections from it today. It then occurs to me that one of the Trinity College porters takes the last collection from the College post-boxes to the main Cambridge sorting office a little before seven o'clock. I reckon that I probably have time enough to get to Trinity and catch the porter, and as the car approaches the back entrance to the College I indicate that I intend to turn into it. The chain of private mental events has culminated in a public physical event. If at that moment I was stopped by a policeman and asked to explain in detail why I had indicated an intention to turn into the entrance, I should have no hesitation in recounting the links of the mental chain; indeed I should have no alternative explanation to offer.

Or consider Thurber's short story *The Secret Life of Walter Mitty*. In the course of five pages, the middle-aged, henpecked hero drives too fast, almost backs into a Buick, startles a woman in the street by saying 'Puppy biscuit', mutters 'Things are closing in!' when his wife finds him curled in a large armchair in the hotel lobby, and stands to attention in the sleet with his back to the wall of a drugstore. Most of these public physical events turn out to be the culmination of a sequence of dramatic but private mental events, the whole story being a succession of sequences in which the fantasist Mitty escapes from his wife's badgering by imagining himself, in turn, the commander of an eight-engined navy hydroplane, a great surgeon, a crack shot accused of murder, an intrepid bomber pilot leaving his Second World War dugout for a suicidal mission, and a man who, facing a firing squad, scorns both blindfold and cigarette. In the course of a few pages we have five examples of a chain of mental events leading to a physical event.

Even in those situations in which 'conscious automaton' seems only too apt a description for us – when, having thoughtlessly made a gaffe, we go beetroot – we still feel that it is our embarrassment that causes our colour; that our mental state is responsible for our physical state.

But however discouraging the results of introspection may be for Hodgson's theory, they are not fatal. Introspection – the examination of one's own thought and feelings – is the wrong tool for investigating the theory, because the physical events accompanying mental events are not open to introspection; they are not the kind of thing that we introspect about. If it is these hidden events that supply the causal connections, then an understanding of those connections cannot be expected from introspection. That the results of introspection do not support the theory is just what the theory predicts.

A much more serious difficulty, first pointed out by William James in an essay entitled 'Are we automata?' (a deliberate reference to Huxley's 'conscious automata'), is that, if mental events are epiphenomena, they cannot have any survival value.[9] Darwin's struggle for existence is a struggle in the physical world, and if mental events cannot cause physical effects they cannot affect the outcome in that struggle. But if they cannot affect the outcome – if they have no survival value – why should we have evolved brains that make them possible? You can, of course, make the *ad hoc* assumption that the physical events of which the mental events are the epiphenomena are themselves advantageous (or linked to events that are advantageous), but you then have to explain what it is about physical events in this class that makes them advantageous (or links them to events that are advantageous). That they make conscious thought possible is not relevant, for thought that merely accompanies behaviour without influencing it will be ignored by natural selection. It is astonishing that Huxley, who must have thought more about the implications of natural selection than almost any of his contemporaries, should not have commented on this difficulty in his talk to the British Association; but he did not.*

And, as James pointed out, the position is even more awkward than appears at first sight. It is not just that we have evolved brains that allow us to experience conscious sensations; we must also explain the correlation between the nature of those sensations and the survival value of the activities associated with them. By and large, pleasant sensations are associated with behaviour that promotes survival (eating, drinking, making love); unpleasant sensations are associated with behaviour that is harmful. There are, of course, exceptions: a surfeit of lampreys did not promote the survival of King John, and three great institutionary dinners in a week killed off poor Dr Grant in the last chapter of Jane Austen's *Mansfield Park*. 'Oh, show us the way to the next whisky-bar!', sings Jenny, in Brecht and Weill's *The Rise and Fall of the City of Mahagonny*, and though Jenny links her desire explicitly to survival – 'For if we don't find the next whisky-bar, I tell you we must die!' – the message of the opera is that the pursuit of pleasure is a poor policy for living. But surfeits of lampreys, institutionary dinners, whisky-bars, and even whisky itself, arrived late in evolution, and cannot be expected to have had a significant influence. Despite such exceptions, a correlation between pleasant sensations and survival-promoting events, and unpleasant sensations and harmful events, remains to be explained; and if sensations are incapable of causing either physical or mental effects, the obvious explanation is ruled out. 'The burnt child fears the fire' says the proverb, and we normally suppose that he fears the fire because the burn was painful, and that fearing the fire increased his ancestors' chances of survival. But if Hodgson is right, the feeling of pain and the feeling of fear are irrelevant. 'Pain', says Hodgson, 'must be held to be no warning to abstain from the thing which has caused pain; pleasure no motive to seek the thing which has caused pleasure; pain no check, pleasure no spur to action.' Sensations could make us aware of events, he believed, but only nervous (as distinct from mental) activity was effective in causing both physical and mental events.

And there is yet another difficulty. Even if the notion that mental events are epiphenomena is true, it leaves unexplained what most needs explaining. Why should particular physical changes in our nervous systems cause feelings or thoughts? Even epiphenomena need to be accounted for. The smoke from the

---

*Huxley's failure to see this weakness of epiphenomenalism is so surprising that James's point has sometimes been misattributed to Huxley.

engine may not move the train but its presence is not a mystery. There's no smoke without fire, we are told, and we are confident of locating the fire in the engine's firebox. The shadow running alongside James's pedestrian may not influence his steps but there is no mystery about the shadow either. Feelings, though we are all familiar enough with them, seem altogether more elusive. Even if we were in a position to say 'There's no feeling of embarrassment without such and such events in such and such bits of the brain', we would still think that to demonstrate the specified events in the specified bits was an inadequate explanation of embarrassment.

Smoke and shadows and feelings of embarrassment are in different categories. Smoke is a mixture of substances arranged in a particular way, and chemists can tell us why fires smoke; shadows, though not made up of substances, form part of the public physical world, and schoolchildren who have had their first lesson in optics can tell us why they occur; feelings of embarrassment are not made up of substances, they exist only in the private mental world, and no one can tell us – or, at any rate, no one has yet been able to tell us – why they should be associated with a particular pattern of nervous activity.

So, despite its promising start, the notion that mental events are epiphenomena has not got us out of the difficulties that a combination of commonsense and a proper respect for physics got us into. We shall return to this problem later.

I want to consider the question: can our brains be regarded as nothing more than exceedingly complicated machines? If we put this question to someone of robust common sense but with no special knowledge of physiology or psychology or philosophy, we can be fairly sure that the answer will be 'no'; and this denial will probably be justified by a string of assertions such as: 'a machine can never be intelligent', or 'machines aren't conscious', or 'machines don't feel pain', or 'machines can't love or hate or be sad or jealous or over-the-moon', or 'machines don't have free will'. And the tone of voice in which this string of assertions is delivered will betray just a hint of irritation, as if we had said something not actually improper but not quite right. This irritation may reflect nothing more than a disinclination to spend time answering Tom-fool questions, but it may also reflect a sense, however slight, of being threatened, a *frisson* of vulnerability.

The cause is probably twofold. First, all the machines with which we are familiar are manufactured; there is therefore a latent suggestion that, however remote the possibility, our brains (and presumably the rest of us) might be manufactured. And to be something that could conceivably be manufactured is to be well downmarket of something created by God in his own image. Secondly, because the machines with which we are familiar are indeed unintelligent, unconscious, unimaginative, unfeeling, incapable of passion and without free will, we suspect that the suggestion that our brains are machines might imply that the same list of unflattering terms could be used to describe us. And this seems to threaten not only our self esteem but the validity of our deepest feelings. To suggest that there is something mechanical about the sound of lamentation in Rama – that 'Rachel weeping for her children' and refusing to 'be comforted, because they are not' represents something less than the intolerable anguish of a bereft mother – is to attack us in a way that we bitterly and rightly resent. But, although, and understandably, this makes us anxious, our anxiety can be allayed. The disturbing suggestion should not be made. The implication that we are unintelligent, unconscious, unimaginative and the rest is not justified. For a machine capable of doing what the brain does must, by definition, be different from familiar humdrum machines and must be as competent as the brain in enabling us to display all the qualities that we do display – the love and the hate, the kindness and the cruelty, the serenity and the passion. To be told by a physiologist that one's brain is a machine that is different from familiar machines, and has all the properties we associate with brains, ought not to be threatening.

Except perhaps to one's patience. For the meaning of the statement is not easy to extract, as the following dialogue shows:

SCEPTIC: When you say the brain is a machine, you seem to be using machine in an abnormal sense. All machines, as the word is normally used, are contrived by someone for some purpose. Indeed the word 'machine' is derived from the Greek word *mekhos* meaning contrivance. But no one supposes that brains have been contrived by anyone. So is machine merely a metaphor, as you might say 'Little Jemima is a ray of sunshine', without implying that you could get a row of coloured Jemimas by passing her through a prism? Or do you mean merely that the brain is like a machine, or (what no one would dispute) is in some ways like a machine?

PHYSIOLOGIST: I am not speaking metaphorically, and, I am, of course, assuming that being contrived by someone is not part of the definition of a machine.

SCEPTIC: Despite the derivation of the word, and its normal usage?

PHYSIOLOGIST: Derivations don't define meanings. The word 'mechanism' has the same derivation as the word 'machine', yet we habitually talk about the mechanisms of biological processes – the pollination of orchids or the clotting of blood – mechanisms which no one has contrived. Usage is a better guide to meaning, but it does not follow that because all the machines that one usually talks about have been contrived by someone, being contrived by someone is a universal feature. Everyday cats have tails, but a Manx cat is still a cat; a zebra without stripes – an albino zebra – would still be a zebra.

SCEPTIC: So even if we accept your assertion that the brain is a machine, it lacks at least one feature – having been contrived by someone – that is common to everything that would be called a machine in everyday life. And in order to behave in the way we know it behaves, the brain must have various features that no other machine has. In insisting that the brain is a machine, you must be thinking of features that the brain has *in common with other machines* and that you consider justify the label 'machine'. It is these features (or some of them) that you must wish to draw attention to when you say that the brain is a machine. What are these features?

PHYSIOLOGIST: There are two. The first is suggested by the dictionary synonyms for machine: *apparatus, appliance, instrument*. It is the existence of a function, or purpose, or role. For machines that are contrived, the role will usually be the same as that for which the machine was contrived, though occasionally it will not be. A fire extinguisher may be used as a doorstop, and a house (which Le Corbusier tells us is a machine for living in) may be used as a defence against inflation. The role of the brain is to control the behaviour of the organism of which the brain forms part, in such a way as to promote the survival, well-being and reproduction of that organism. Even if there is no contriver and therefore no conscious purpose, there can still be a role.

SCEPTIC: What is the second feature that you feel justifies the label 'machine'?

PHYSIOLOGIST: The second feature is suggested by an early special use of the word machine: *an apparatus for applying mechanical power, consisting of a number of parts, each having a definite function*. We no longer think of machines as necessarily mechanical, but we do still think of them as functioning in accordance with the laws of physics. With all ordinary machines, i.e. machines that have been contrived by someone, the working of the machine – its mechanism – will necessarily be understood, though not perhaps by all of the machine's users, and not always in as much detail as we should like. With machines that form part of biological organisms, the extent of our understanding is very variable. We have no difficulty in understanding how the elbow joint acts as a hinge, or the hip joint acts as a ball-and-socket joint, though our knowledge of the characteristics of the lubricating fluid, and of the membranes that cover the bearing surfaces, is not complete enough for us always to be able to prevent inflammation and arthritis. We understand how the contraction of the walls of the heart's chambers, and the opening and closing of the various valves enables the heart to act as a pump; and we know how the rhythm is maintained, and a good deal about the way in which the heart's muscle converts chemical energy into mechanical work. Because it is natural to think of hinges, ball-and-socket joints and pumps as machines, no one is disturbed if elbow joints, hip joints and hearts are referred to as machines; and when Mr Charnley introduced a very successful operation for replacing the arthritic hip with a contraption of steel and plastic, no one felt threatened, and those with arthritis were particularly pleased.

SCEPTIC: And brains?

PHYSIOLOGIST: With brains, of course, the position is rather different. What brains do is so complicated, and their structure is so complicated, that, although a great deal has been discovered about the way in which brains work, we are scarcely more than scratching at the surface.

SCEPTIC: If you have not plumbed the depths, how do you know that there are not lurking leviathans whose behaviour is not in accordance with the laws of physics, as we know them. Might not intelligence, consciousness and free-will be such leviathans?

PHYSIOLOGIST: We can't be sure. I don't think intelligence or free-will are strong candidates, though consciousness is more worrying. But even if we can't be sure, the best working hypothesis is to assume that the known laws of physics will be sufficient. A great deal has been discovered about what goes on in the nerve cells and synapses of different parts of the brain, and there is no suggestion of anything incompatible with physical laws. And some of the brain's overall behaviour is remarkably machine-like. For example, Professor Merton has shown that, in controlling voluntary movements, the nervous system acts as a servo-mechanism with variable gain – to use engineering jargon.[10]

SCEPTIC: Are you suggesting that, in the remote future, it is possible that a machine could be designed capable of doing what the human brain does?

PHYSIOLOGIST: In principle, yes.

SCEPTIC: What do you mean by 'in principle'?

PHYSIOLOGIST: I mean that to design such a machine is not impossible, *either* in the sense that it is impossible to find a four-sided triangle – that task is impossible by definition – or in the sense that it is impossible to straighten the leaning tower of Pisa by stroking it with a teaspoon – that task, though not one that has been attempted, is incompatible with the behaviour of towers and teaspoons predicted by the laws of physics, as we know them.

SCEPTIC: So 'Yes, in principle' means 'No, in practice'.

PHYSIOLOGIST: It depends on what you want to do. If you want to design a machine that does everything a human brain does, then I agree that, for all practical purposes, the task is impossible, and likely to remain so. But we already have machines that do some of the things that the brain does, including – and this is the crucial point – things that until recently would have been thought quite impossible without the intervention of a mind. In *The Physical Basis of Mind,* published in 1877, George Henry Lewes – now remembered mostly as George Eliot's lover* – discusses the performance of Kempelen's chess-playing automaton. The public were surprised, he tells us, but 'every instructed physiologist present knew that in some way or other its

*But so well known in the nineteenth century that almost the first thing Dostoevsky tells us about the heroine of *Crime and Punishment* is that she had 'read with great interest . . . Lewes's *Physiology*'.

movements were directed by a human mind; *simply because no machine could possibly have responded to the unforeseen fluctuations of the human mind opposed to it* [italics added]'. Yet now we have chess-playing computers that can win against all but the very best players, and sometimes even against them. It is true that the methods used by these computers differ from the methods used by the brain, but the task performed is the task that common sense in 1877 held to be impossible without the intervention of a human mind.

SCEPTIC: You are saying that common sense is a perishable commodity.

PHYSIOLOGIST: Sometimes. And it doesn't come with a 'use by' date, so you don't know in advance. A moment ago you conjured up three 'leviathans' – intelligence, consciousness and free-will – and you implied that they could not be possessed by any machine. Yet already we have computers that can properly be described as intelligent, even if their intelligence is limited to a restricted field and not comparable with our own. Is it not possible that, in time, the other two will also lose their terrors? And wasn't the biblical leviathan a mythical beast, anyway?

SCEPTIC: Its status is uncertain. But what we do know is that it was not possible to 'draw out leviathan with an hook' – not even a hook baited by a physiologist.

3 Evolution by Natural Selection

When Charles Darwin published On the Origin of Species *in 1859, the Bishop of Worcester's wife was most distressed. 'Let us hope it is not true', she is said to have remarked. 'But if it is, let us pray that it does not become generally known!'*
—Opening of a lecture by R. D. Keynes[1]

In the days before pocket calculators, few people who used the arithmetical procedure for extracting square roots could justify that procedure, though they got the right answers. But the procedure for extracting square roots does not affect the way in which we see ourselves and our place in the world; the theory of evolution by natural selection does. Yet, although Darwin's theory is believed by a substantial majority of educated people in the western world, it is doubtful whether more than a small fraction of those who do believe could, if asked, produce much evidence to justify their belief.

That would not matter if the theory were not being continually questioned, or if it were less influential on our thinking. Indeed, Freud thought that the three most severe blows that scientific research had inflicted on 'man's craving for grandiosity' were those associated with the work of Copernicus (which displaced the earth from the centre of the universe), the work of Darwin (which displaced humans from the privileged position conferred by the first two chapters of Genesis), and his own work.[2] Not many people now would agree with Freud's assumption of the third place in the trio, but there can be no doubt about the claims of Copernicus and Darwin. And though, unlike Copernicus's *De revolutionibus orbium coelestium*, Darwin's writings were never placed on the Index,* and Darwin himself was to be buried in Westminster Abbey with a brace of dukes and an earl among his pall bearers, it is his work that necessitated the more profound readjustments of orthodox religious beliefs.

It is because of the effect that our attitude to evolutionary theory has on the way we see ourselves, our minds and the world about us, that I want to spend this chapter and the next considering evolution in general and the evolution of humans in particular.

*The *Index librorum prohibitorum*, the list of books forbidden to Roman Catholics.

THE EVOLUTION OF EVOLUTION

When Darwin published *On the Origin of Species* in 1859, the idea that species were mutable, and might have arisen by transformation of other species rather than by creation, had been discussed by naturalists for over a century, and by philosophers for even longer. But, although the notion that animals had evolved was a respectable, and at times even a fashionable, subject for discussion, it failed to be generally accepted even by those prepared to cope with its religious implications. This was partly for lack of sufficient detailed evidence that species had been transformed, but more for lack of any adequate theory to account for the transformation.

The most promising hypothesis was that changes induced in individual animals by adaptation to their surroundings could be inherited – a hypothesis particularly associated with Lamarck but also favoured by Buffon, Cabanis, and Erasmus Darwin (Charles' grandfather), and accepted, though with less emphasis, by Charles Darwin himself. Its attraction was that it could account not only for change but also for adaptive change. Giraffes, it was argued, had developed long necks by browsing on the leaves of tall trees; the kangaroo's habit of sitting upright with young in its pouch would lead to the dwarfing of its fore limbs, and its habit of leaping would develop its hind limbs and tail. We now know that the mechanisms of inheritance make it extremely unlikely that characteristics acquired in this way can be inherited, but that knowledge was not to be available until around the time of Darwin's death.

In 1830, a prolonged debate in the French Académie des Sciences between the evolutionist Geoffroy Saint-Hilaire and the anti-evolutionist Georges Cuvier caused so much excitement that, thirty years on, Mrs Gaskell, writing *Wives and Daughters*, had her Darwinian hero reading bits of Cuvier's *Le Règne Animal* to the heroine and being sought by an admiring 'M. Geoffroi St H.'. But that was exceptional; for most of the first half of the nineteenth century controversy about the evolution of species was mainly the concern of theologians and 'men of science', as they were still called. (Even in 1879, William James put the word 'scientists' between quotation marks and followed it by 'loathly word!' in brackets.) In 1844, though, a book appeared that was deliberately designed to bring the controversy to the notice of a much wider public. Called *Vestiges of Creation*, and written (anonymously) and published by Robert Chambers – an Edinburgh publisher later to found *Chambers' Encyclopaedia* – it was dramatically successful, eventually reaching twenty-one editions. Disraeli had fun mocking it. In *Tancred*, the young Lady Constance Rawleigh, who has 'guanoed her mind by reading French novels' tries to persuade the hero, Tancred, to read the startling new work *Revelations of Chaos*:

'You must read the "Revelations"; it is all explained. But what is most interesting, is the way in which man has been developed. You know, all is development. The principle is perpetually going on. First, there was nothing, then there was something; then, I forget the next, I think there were shells, then fishes; then we came, let me see, did we come next? Never mind that; we came at last. And the next change there will be something very superior to us, something with wings. Ah! that's it: we were fishes, and I believe we shall be crows, but you must read it.'

'I do not believe I ever was a fish,' said Tancred.

'Oh! but it is all proved . . . You understand, it is all science; it is not like those books in which one says one thing and another the contrary, and both may be wrong. Everything is proved: by geology you know.'[3]

Vestiges of Creation was less successful among the religiously orthodox and among those best placed to judge its scientific worth. The geologist, Adam Sedgwick, who qualified on both counts, wrote a damning review including the suggestion that the book was written by a woman because of 'its ready boundings over the tree of knowledge, and its utter neglect of the narrow and thorny entrance by which we may lawfully approach it'.[4] Darwin dismissed its geology as 'bad' and its zoology as 'far worse'[5] and Huxley wrote a review of such severity that he later had 'qualms of conscience . . . on the ground of needless savagery'.[6] What particularly disturbed Darwin was not Chambers' ignorance of geology or zoology, or even his naivety about spontaneous generation – he believed that the passage of an electric current through a solution of silicate of potash had generated insects of a hitherto unknown species – but his ideas about the cause of change. Chambers held that although living organisms normally reproduced their own kind, they had the inherent ability occasionally to produce the 'next higher' type, 'the stage of advance being in all cases very small – namely from one species to another', for example, from oats to rye. To Darwin, to postulate such sudden jumps seemed 'monstrous'. And to any critical reader, explanation by the assumption of an 'inherent ability' was reminiscent of Molière's mock explanation of the soporific effects of opium – that it contained a 'dormitive principle'.

THE ORIGIN OF THE 'ORIGIN'

To be convincing, the theory that humans had evolved from more primitive creatures needed a plausible mechanism. This, of course, was supplied by the theory of natural selection – the theory that the selective survival of the fittest among a large progeny in each generation would gradually cause the species to change, and to change in an adaptive way.* William Charles Wells and Patrick

*The phrase *survival of the fittest* was first used by Herbert Spencer.

Matthew had both, independently, thought of the principle of natural selection, but their work remained virtually unknown until after the publication of the *Origin*. Wells had described his idea in a paper to the Royal Society in 1813, but his application of the principle was restricted to the development of human races and it caused no stir. Matthew's ideas were buried in the appendix to an obscure work on Naval Timber and Arboriculture, published in 1831, and it was only during discussion of the *Origin* that he himself disinterred them in a long letter in the *Gardeners' Chronicle* in 1860.

Very near the end of the five year voyage of the *Beagle*, Darwin became deeply impressed by facts that could be explained only on the supposition that species gradually become modified, and the subject haunted him.[7] But in the absence of any plausible mechanism for modification – and especially for producing beautiful adaptations for particular habits of life, such as those found, for example, in the woodpecker or tree frog – he felt it 'almost useless to endeavour to prove by indirect evidence that species have been modified'. Despite this feeling, in July 1837, less than a year after his return to England, and in the middle of what he later described as the busiest period of his life, he started his first notebook for facts relating to the origin of species.[8]

According to his autobiography,[9] he 'soon perceived that selection – artificial selection – was the keystone of man's success in making useful races of animals and plants. But how selection could be applied to organisms living in a state of nature remained for some time a mystery'. Then, in October 1838, he happened to read 'for amusement' Malthus's *Essay on Population*, and 'being well prepared to appreciate the struggle for existence which everywhere goes on . . . it at once struck [him] that under these circumstances favourable variations would tend to be preserved, and unfavourable ones to be destroyed. The result of this would be the formation of new species'. Historians who have looked at Darwin's notebooks for this period point out that the development of his views was, in fact, more complicated.[10] Whatever the route, by 1839 he had reached a clear view of the role of natural selection.

Darwin was not impetuous. He had at last got 'a theory by which to work' and he proceeded to collect evidence to support it and – what was more troublesome – to consider the very real difficulties: how could the accumulation of innumerable slight variations account for the development of the eye, or the ossicles of the ear, or the instincts that allowed the honey bee to form its comb? How could 'survival of the fittest' explain the existence of altruism, or of insect communities consisting mainly of sterile females? How was it that some species were found in widely separated areas? If all species were ultimately related, why were there great gaps in the fossil record? And if, as was then generally believed, the characters of parents were simply blended in the next generation, how was it that any small favourable variation did not get diluted away in successive generations before natural selection had had time to act? It was not until 1842 that Darwin wrote a brief abstract of his theory, and this had only grown to 230 pages by the autumn of 1844. Twelve years later the theory was still unpublished.[11]

When Darwin was preparing to set out on the *Beagle* in 1831, his friend and former teacher John Henslow had advised him to read, but on no account to accept the views in, the first volume of *Principles of Geology*, recently published by a young lawyer and geologist Charles Lyell. Lyell's vigorous advocacy of *uniformitarianism* – the notion that the events of the past should, so far as possible, be explained by processes of the kind that can be observed in the present – had greatly influenced Darwin, who had proceeded to ignore the second part of Henslow's advice. In the intervening quarter of a century, Lyell had become a close friend of Darwin, and it was now Darwin's turn to produce disturbing views. Though still not converted to those views, Lyell urged Darwin to write them out fully, with the supporting evidence. This Darwin began to do, and on a grand scale; but in June 1858, before he had completed more than half the work, his plans were disrupted by the arrival of a letter from Alfred Russel Wallace.

Wallace was in the Malay archipelago, on the volcanic island of Ternate, from whose sultan, nearly three centuries earlier, Francis Drake had bought six tons of cloves. Lying in his bamboo house, unable to work because of an intermittent fever (probably malaria), he remembered the essay by Malthus that he had read long ago. As with Darwin nearly twenty years earlier, Malthus's argument triggered, in Wallace, the solution to the origin of species: 'Then it suddenly flashed upon me that this self-action process would necessarily *improve the race* because in every generation the inferior would inevitably be killed off and the superior would remain ... For the next hour I thought over the deficiencies in the theories of Lamarck and of the author of the "Vestiges", and I saw that my new theory supplemented these views and obviated every important difficulty.'[12]

Wallace was less cautious than Darwin – he had already swallowed phrenology and would later swallow spiritualism – and as soon as his fever subsided he summarized his thoughts in a short essay, which he sent off to Darwin, with whom he had been in friendly correspondence about an earlier article. Accompanying the essay was a letter asking Darwin to send the essay on to Lyell if he thought it worthwhile. Darwin received the bombshell on 18 June and immediately wrote to Lyell. His anguish is best conveyed in his own words:

> He has today sent me the enclosed, and asked me to forward it to you ... I never saw a more striking coincidence; if Wallace had my MS. sketch written out in 1842, he could not have made a better short abstract! ... Please return me the MS., which he does not say he wishes me to publish, but I shall, of course, at once write and offer to send to any journal. So all my originality, whatever it may amount to, will be smashed, though my book, if it will ever have any value, will not be deteriorated; as all the labour consists in the application of the theory.[13]

A week later, Darwin wrote again wondering whether, as there was nothing in

Wallace's sketch that had not been written out more fully in his own sketch of 1844, it might not be dishonourable for him to publish immediately a brief outline of his own views. But was this fair to Wallace? 'I would far rather burn my whole book, than that he or any other man should think that I had behaved in a paltry spirit.'

In the event, both he and Wallace behaved admirably. Lyell and Hooker persuaded Darwin that the right thing to do was to give a joint paper to the Linnean Society, consisting of extracts from his 1844 sketch, an abstract of a letter from Darwin to the American botanist Asa Gray (written in 1857), and Wallace's essay. The Linnean Society met on 1 July, but on 28 June Darwin's youngest child died of scarlet fever, and it was in a state of great distress that he sent Hooker a copy of his 1844 sketch on the following day. The paper was therefore presented by Lyell and Hooker. There was no discussion, and though Hooker, writing long afterwards, said that the interest excited was intense, the President of the Society – the surgeon and zoologist Thomas Bell – in his end-of-year review found 1858 uneventful: 'The year which has passed . . . has not, indeed, been marked by any of those striking discoveries which at once revolutionize, so to speak, the department of science on which they bear. . .'.

Before the end of July, Darwin had set to work to make a shorter 'Abstract' of the great mass of material that he had been accumulating over more than twenty years. His first intention was to publish it as a series of papers in the Journal of the Linnean Society, but it soon became clear that it would have to be a separate book. By this time, Hooker had become persuaded of the rightness of Darwin's (and Wallace's) views and, in early 1859, Lyell, although still not quite converted – or 'perverted' as he mockingly called it – was enthusiastic enough to approach John Murray, on Darwin's behalf, about publication. Having seen only the chapter headings, Murray agreed to publish, though he insisted that the word 'abstract' be deleted from the title. *On the Origin of Species by Means of Natural Selection* was published on 24 November 1859; a few days earlier the whole edition of 1250 copies had been sold to the trade. A second edition of 3000 copies was published six weeks later.

If Darwin and Wallace's paper scarcely disturbed the calm of the Linnean Society, publication of the *Origin* caused a storm on a wider stage. It was as if the initial small dose of Darwinism had sensitized the bodies of both scientific and religious orthodoxies so that a second exposure caused a violent reaction. Yet, within little more than a decade, Darwin's views were on the way to becoming the new orthodoxy. In 1864 the Royal Society awarded Darwin its highest honour, the Copley Medal (though the President infuriated Darwin's proposers by pointing out that the Council had 'expressly omitted' the *Origin* from the 'grounds of our award').[14] In 1867 Lyell, having at last accepted Darwin's views on evolution (though not the crucial role of natural selection), incorporated them into the tenth edition of his *Principles of Geology*, and two years later this edition was reviewed by Wallace in the *Quarterly Review* – the journal that had published the infamous (and anonymous) review of the *Origin* by the Bishop of Oxford. By 1871, Darwin, who to avoid controversy had deliberately not discussed the evolutionary origin of man in the *Origin*, was willing to publish his *The Descent of Man*. And in 1872, in the concluding chapter of the sixth edition of the *Origin*, Darwin felt it necessary to apologize for having retained sentences suggesting that most naturalists still believed in the separate creation of individual species.

THE SUCCESS OF THE '*ORIGIN*'

One reason for this success was, of course, the plausibility and power of the idea of natural selection. 'How extremely stupid not to have thought of that!', was Huxley's reflection when he first grasped the idea, for, as he pointed out, the facts of variability, of the struggle for existence, and of adaptation to conditions, were familiar enough.[15] Natural selection, even better than the ideas of Lamarck, could explain adaptive change. It abolished design without abolishing teleology – the explanation of developments in terms of their purpose. The most beautiful adaptations required neither the directing hand of God, nor the inheritance of acquired characters, nor any mysterious inherent individual propensities to develop in particular directions. All that was necessary was that those individuals with small variations that, in the prevailing circumstances, proved beneficial, should have a slightly greater chance of surviving to reproduce themselves. The innumerable instances of bodily adaptation that William Paley, in his *Evidences for the Existence and Attributions of the Deity* (1802), had regarded as the product of an intelligent and designing mind could, alternatively, be regarded as the product of natural selection.

The success of the *Origin* depended on much more, though, than a clear exposition of the plausibility and power of natural selection.[16] Without a plausible mechanism, all the evidence in favour of evolution was liable to remain unconvincing. With such a mechanism, evidence for evolution was still necessary. In the five hundred or so closely argued pages of the *Origin*, Darwin brought together a vast mass of widely scattered and previously unconnected facts which made sense only if evolution had occurred – and often made sense only if it had occurred by natural selection.

It is impossible, in a short space, to do justice to Darwin's arguments; but the assumption that we are the product of evolution by natural selection is so crucial to any enquiry into the origin and nature of our minds, that it is worth taking a brief look at the kinds of evidence he adduced, and at the way that that evidence has worn, or been added to, in the explosion of knowledge that has occurred since 1859. The evidence rests mainly on four large groups of facts: the geographical distribution of animals and plants; the comparative anatomy of animals and plants; the stages by which individual animals grow from the fertilized egg; and the nature and distribution of fossils.

Evidence from geographical distribution

A striking feature of the geographical distribution of living things is the way in which similar habitats in widely separated regions support very different animals and plants, whereas very different habitats in contiguous regions support related animals and plants. This is precisely what would be expected if existing species arose by modification of preceding species in the same or contiguous areas, but seems quite arbitrary if existing species were individually created suitable for particular climates.

If new species do arise by modification of preceding species, what should be crucial in determining geographical distribution is not distance but barriers to migration; and that is what Darwin found. The east and west coasts of Central America, for example, are separated only by the narrow and (in Darwin's time) impassable isthmus of Panama, but Darwin found that the two coasts had hardly a fish, shell or crab in common. Great differences between faunas and floras could also be found on opposite sides of lofty and continuous mountain ranges, or great deserts or even large rivers. In contrast, Darwin pointed out, the naturalist travelling great distances where there happen to be no severe barriers finds a succession of *related* animals and plants, adapted to the succession of habitats.

Particularly striking evidence came from studies of isolated oceanic islands. Such islands generally contained far fewer species than equal areas of mainland, but of these species a far higher proportion were found nowhere else. Whole groups of animals unable to survive long sea voyages without food – amphibians and terrestrial mammals, for example – were often absent, but reptiles tended to be well represented, and plants belonging to groups which on the mainland were wholly herbaceous sometimes seemed to have developed into trees. All these features could be explained by supposing that difficulties of access ensured that few species arrived on the island, and that those that did arrive were therefore able to evolve with much less competition from other species.

The most dramatic examples came from the Galapagos archipelago, between 500 and 600 miles off the coast of Ecuador, whose original fauna and flora may have arrived on masses of floating vegetation swept along by the Humboldt current. Here Darwin found giant tortoises and two species of iguana lizard not found anywhere else. There were only twenty-six species of land bird, but, of these, twenty-five were unique to the archipelago. All of these birds, however, as well as most of the other animals and plants, were closely related to South American species, and very different from those of the Cape Verde Islands, 300 miles off the west coast of Africa, which resemble the Galapagos in both climate and geology. And within the archipelago, the individual islands had species of bird peculiar to themselves yet related to species on the other islands. In particular, among the birds sent back to England by Darwin, there were thirteen new species of finch, falling into six groups, each group possessing a beak adapted to a particular kind of food. Members of one species climbed trees like a woodpecker, and bored holes in the bark but, lacking the long tongue of a woodpecker, used cactus spines to extract the insects; members of another resembled a warbler in taking insects on the wing; members of a third picked parasites off the skin of the iguanas and tortoises. Since the mainland finches were all of one species, and were seed eaters, it looked as though mainland finches that had somehow managed to reach the archipelago were able to take advantage of the absence of competition to evolve and fill various ecological niches that on the mainland were filled by birds other than finches. (Darwin himself had been thoroughly muddled by these birds; the muddle was sorted out, and the new species identified, by the ornithologist and artist, John Gould.[17])

Darwin's findings in the Galapagos were dramatic but, as he himself pointed out, the Galapagos were not unique. The animals and plants of the Cape Verde Islands were related to those of Africa just as the animals and plants of the Galapagos were related to those of America.

Darwin also considered a number of exceptions to the pattern. For some of these he could produce plausible hypotheses, but there was one curious observation that he could not explain. Nearly twenty years earlier, J. D. Hooker had noticed a resemblance between plants in the south-western corner of Australia and plants in the Cape of Good Hope, some 5000 miles distant. This puzzling resemblance, and the similarly puzzling fact that marsupials are found in Australia and America, but not anywhere else, had to wait until the following century for an explanation – an explanation which turned out to be even more startling than the biological observations themselves.

In 1912, the German meteorologist Alfred Wegener suggested that the jigsaw-like fit between the west coast of Africa and the east coast of South America, first noticed by Francis Bacon in 1620, was the result of the two continents having originally been contiguous. Extending this argument he further suggested that all of the continents had at one time formed a single land mass, and had only later drifted apart. This seemingly unlikely hypothesis became increasingly attractive, first because of the identification of similar arrangements of rock strata – the Gondwana system – in India, South America, South Africa and Antarctica, and secondly because of observations on the magnetism of rocks of the same age in the different continents, which gave incompatible positions for the earth's magnetic south pole. This incompatibility corresponded precisely with the difference between the present relative orientation of the two continents and the orientation they would have if the two pieces of the jigsaw puzzle were reassembled, with the east coast of South America applied to the west coast of Africa. From geological evidence of this and other kinds, it seems that about 43 million years ago Australia separated from a single land mass which also included Antarctica and South America. Marsupials from South America are therefore thought to have reached Australia via Antarctica, which was then not covered with ice. Following the separation of Australia they seem to have behaved analogously to Darwin's finches, evolving to fill a variety of ecological niches in the absence of competition from more advanced mammals.

Evidence from comparative anatomy

The second large group of facts on which Darwin based his theory were the facts of comparative anatomy. To compare the structures of living animals and plants, and to deduce possible evolutionary pathways, became the life work of many nineteenth-century zoologists and botanists. Darwin himself spent eight years studying the extraordinarily variable structure and the complicated life cycles of barnacles; indeed, for one of Darwin's young sons, that seemed so much the normal occupation for a father, that when visiting a friend he is said to have asked, 'Where does your father do his barnacles?'

One of the most striking pieces of evidence for evolution is the common pattern of the arrangement of bones in vertebrate limbs. 'What can be more curious,' asked Darwin, 'than that the hand of a man, formed for grasping, that of a mole for digging, the leg of the horse, the paddle of the porpoise, and the wing of the bat, should all be constructed on the same pattern, and should include similar bones, in the same relative positions?' The man, the mole, the horse, the porpoise and the bat are all mammals, but Darwin might have added terrestrial amphibia, reptiles and birds to the list. Evidence of a slightly different kind came from the study of vestigial organs. Pythons have no limbs, and whales no hind limbs, but both have vestigial hind limb bones. We ourselves, in the coccyx, have a vestigial bony tail, and our caecum has a blind outgrowth, the appendix, which serves no purpose other than giving employment to surgeons. In contrast to the twitching ears of a cat, our outer ears are fixed and fairly flat, so they are almost ineffective either for collecting sound or for helping to local-ize its source. If evolution is the result of natural selection, the existence of such vestigial structures is to be expected. As animals adapt to new environments, some organs may become unnecessary, or may even get in the way. Any reduc-tion in their size will therefore be beneficial if only by saving the resources needed to make and maintain them.

The tracing of evolutionary pathways by comparing the anatomy of plants and animals seemed to be largely complete by early in the 20th century, but within the last three decades there has been an extraordinary resurgence. This is because it is now possible to look in detail at the anatomy of the molecules that make up living organisms, and in particular at the anatomy of proteins, whose detailed structure tends to be unique for each species.

Despite their diverse properties, all proteins consist of one or more chains of small molecules – the amino-acids – of which there are only twenty kinds. And although the three-dimensional structure of proteins is extremely complicated, it seems that it is wholly determined by the order in which the different amino-acids are arranged in the chains. That order is controlled genetically: the instructions for synthesizing each protein in an animal or plant are stored as a sequence of small subunits called nucleotides in immensely long molecules of deoxyribonucleic acid (DNA). These molecules, neatly and elaborately coiled to form structures called chromosomes, are found in the nucleus of the fertilized egg from which the animal or plant grows, and exact copies are stored in the nuclei of all the cells that are formed by division of the fertilized egg.

The sequence of nucleotides that represents the instructions to form a particular protein is called a *gene*, and the code that translates the sequence of nucleotides into a sequence of amino-acids – the so-called *genetic code* – is remarkably simple. As only four different nucleotides are involved, but there are twenty different amino-acids, it is clear that the coding cannot be one-to-one: if each amino-acid were specified by a single nucleotide only four amino-acids could be specified unambiguously. Even if pairs of nucleotides were used to specify each amino-acid, it would be possible to specify only 16 (i.e. 4×4) different amino-acids. It is therefore not surprising that the code actually employed uses triplets of nucleotides (each triplet being known as a *codon*) to specify each amino-acid. That, of course, gives 64 (i.e. $4 \times 4 \times 4$) codons, which is more than is needed, but many amino-acids can be specified by any one of several codons, and three codons are used as an instruction to the protein synthesizing machinery to stop.

To trace evolutionary pathways by comparing the molecular anatomy of different species, you can look either at variations in the sequence of amino-acids in a given protein, or at variations in the sequence of nucleotides in the portion of DNA that controls the synthesis of that protein – conventionally referred to as 'the gene encoding that protein'. Whichever method is used, comparison of sequences has the important advantage over the comparison of larger structures that you know precisely how many codons must have been changed to get from one sequence to another – this number therefore provides a measure of the length of the evolutionary pathway between the two sequences. By looking at the amino-acid sequences in the same protein – or the nucleotide sequences in the gene coding for the same protein – in a series of organisms, it is possible to work out the most likely family tree connecting those organisms. Studies of this kind have strongly supported the picture that had already emerged from traditional comparative anatomy, and they have been particularly useful in sorting out relationships between very primitive organisms, where the traditional methods of comparative anatomy or the study of fossils are of little use. They have also been useful in determining the closeness of relationships between humans, chimpanzees, gorillas and orang-utans, where the results of comparative anatomy are controversial. It turns out that our DNA is closest to that of chimpanzees and pygmy chimpanzees, differing in only about 1.6 per cent of the nucleotides; it is slightly more different from the DNA of gorillas (about 2.3 per cent), and still more from the DNA of orangutans (about 3.6 per cent).[18] (It is, of course, this closeness that justifies the title of Jared Diamond's book, *The Rise and Fall of the Third Chimpanzee*[19]).

Evidence from embryology

The third large group of facts that Darwin presented as evidence for evolution – and the group that he himself regarded as the weightiest – concerned the stages by which individual animals grow from the fertilized egg to the adult form. Early in the nineteenth century Karl Ernst von Baer, in Königsberg, noticed a striking resemblance between the early embryos of fish, amphibia, reptiles, birds and mammals; indeed, having failed to label two little embryos that he had preserved, he was later quite unable to tell whether they were embryos of lizards, small birds or very young mammals.[20] The resemblance is not restricted to the external appearance of the embryos. Just as all the early vertebrate embryos have obvious gill pouches, although gills will develop only in the fish and amphibian, so the heart and arterial system in all early vertebrate embryos are arranged in a way that makes sense only for an animal that uses gills to oxygenate its blood. In particular, on each side of the embryo there are six arterial loops that, in the fish, will later take blood to and from the gills. In amphibia, reptiles, birds and mammals, as the embryo grows, the six-loop arrangement is modified to give the adult pattern characteristic of each class. And this strange sequence of events, in which the individual embryo passes through stages reminiscent of some of those through which its ancestors must have evolved, is not limited to vertebrates.

The close resemblance between embryonic forms of animals whose adult forms are quite different is particularly striking in some animals with larval forms. Just looking at the adults, it is difficult to believe that segmented worms such as earthworms and ragworms, and unsegmented shelled animals such as snails and clams, have a common ancestor. Yet their larval forms, which are quite unlike the adults, are almost identical.

Evidence from fossils

The last great group of facts brought forward by Darwin are those concerned with fossils. It is curious that to most people the existence of fossils would seem to be the most obvious and perhaps the strongest evidence for evolution. That is so far from being the way Darwin regarded them that he called the chapter in the

Origin dealing with fossils 'On the imperfection of the geological record'. Fossils were, of course, already well known by Darwin's time, proving that species different from any now existing once existed and had become extinct. In 1821 Cuvier had pointed out that, if species did gradually change, one ought to be able to find evidence for intermediate forms; and the lack of such evidence – for example the lack of forms intermediate between the fossil tapir-like three-toed *Palaeotherium* and the modern single-toed horse – was one of the reasons he did not believe that evolution had occurred. By the time of the *Origin*, however, Richard Owen had shown that intermediate forms did exist – in particular, the extinct *Hipparion*, with one main toe and two small extra toes in each foot, fitted nicely between the *Palaeotherium* and the modern horse with its single toe and two splint bones. Ironically, although Owen provided the classical example of evolution through a succession of fossil forms, he rejected both Lamarckian and Darwinian explanations of such evolution, preferring to believe that, at their creation, species were endowed with innate tendencies to change to a pre-ordained goal – a process he called 'ordained continuous becoming'. (It was perhaps this tendency to mysticism and a cultivated obscurity of exposition, rather than disagreement on substantive issues of comparative anatomy, that so irritated Huxley, who, hearing Owen referred to as 'the British Cuvier', remarked that he bore the same relation to the French article that British brandy did to French cognac.)

In the middle of the nineteenth century, modern methods of dating fossil-bearing rocks were of course not available. It was possible to get some idea of the age of strata in sedimentary rocks by extrapolating from current rates of deposition, but calculations based on such estimates seemed to show that, so far from evolving, many fossils stayed constant or nearly constant over very long periods of time. If that meant that evolution was very slow, had there been enough time? Estimates of the age of the earth based on rates of cooling suggested that there had not.

None of these objections was fatal. The geological record was bound to be imperfect – the chance that a dead animal would be fossilized was slight and, even if it were, the chance that the fossil would be discovered was also slight. If the formation of new species was more likely in small communities that had somehow got isolated, the transitional forms would not be expected to occur in the main centres of population. If, later, the new species reinvaded the area from which their ancestors had come, the fossil record would show the appearance of a new species without transitional forms.

It was not too difficult to think of excuses; but as the years went by excuses became less and less necessary. Two years after the publication of the *Origin*, the fossil that became known as *Archaeopteryx* was discovered in the Upper Jurassic limestone at a quarry in Solnhofen, Bavaria. It had the socketed teeth and the long articulated tail of a reptile, and it had claws on its wings like a pterodactyl, but it also had the feathers and the wishbone of a bird. It was the ideal transitional form* and Owen acquired it for the British Museum for what then seemed the exorbitant sum of £400.[21] Eventually, too, it was realized that the worryingly low estimate of the age of the earth based on rates of cooling could be disregarded, because radioactive decay inside the earth must have generated sufficient heat to make cooling of the earth much slower than had been assumed.

Even now, the geological record is very imperfect, and there are not many species whose history can be traced like that of the horse. But what the study of fossils has provided is a reliable time-scale for evolution. In the nineteenth century, this time-scale was only relative. The introduction of radioactive dating has made it possible to give absolute ages with confidence, because of the extraordinary constancy of the rates at which radioactive atoms disintegrate and their breakdown products accumulate. It is the reliability of radioactive dating that allows palaeontologists to state with such assurance that the first vertebrates appeared between 440 and 500 million years ago, that the dinosaurs flourished for nearly 150 million years and were already extinct 65 million years ago, that the first horse appeared about 54 million years ago, and so on.

DARWIN'S 'COLD SHUDDER'

We have now looked at the main lines of evidence for evolution. What about the difficulties?

One difficulty that caused Darwin great anxiety was the need to account for the evolution of organs such as the eye – 'organs of extreme perfection and complication', as he called them. 'The eye to this day gives me a cold shudder', he wrote to Asa Gray in 1860. For natural selection to produce such an organ, it was necessary to suppose that the 'perfect and complex eye' had evolved by a series of steps from something 'very imperfect and simple, each grade being useful to its possessor'. In the invertebrates, Darwin argued, such a series of steps did in fact exist. The argument, though impressive, was a little sketchy, but the great increase in our knowledge of invertebrate vision since 1859 shows not only that it was correct but also that eyes – organs forming useful images of the outside world – have evolved independently many times.[22] The light-sensitive pigments that make vision possible in our own eyes are closely related to light-sensitive pigments found in the eyes of other vertebrates and of invertebrates, and also in light-sensitive spots in the alga *Volvox*. In all these cases the pigment allows the organism to use light as a source of information; in certain bacteria, a high concentration of a similar pigment makes it possible for light to be used as a source

*Recent fossil evidence suggests that the archaeopteryx and modern birds both evolved from small predatory dinosaurs that lived on the ground – see Padian, K. & Chiappe, L. M. (1988) *Scientific American*, Feb., pp. 28–37.

of energy. (In spite of all this evidence, the evolution of the eye continues to be presented as a stumbling block by those who doubt the efficacy of natural selection. Elegant and convincing arguments supporting the Darwinian case have been given by Richard Dawkins.[23])

The notion that natural selection cannot account for the evolution of 'organs of extreme perfection' has been resurrected by Hoyle and Wickramasinghe.[24] They argue that the complex structure and subtle behaviour of the haemoglobin molecules that carry oxygen and carbon dioxide in our own blood could never have evolved without intelligent guidance. Their argument is invalid for the same reason that the argument about the eye was invalid – that it is possible to point to a series of stages, of increasing complexity, that lead to the human haemoglobin molecule, each stage being of value to the organisms possessing it.[25]

THE SURVIVAL OF FEATURES THAT THREATEN SURVIVAL

Another difficulty faced by Darwin was the need to explain the evolution of features that seemed to confer negligible benefit, or even to threaten the survival of the organism possessing them. The most puzzling example was altruistic behaviour. Surely, we feel, natural selection should eliminate genes that predispose to altruistic behaviour, yet such behaviour is well known in many species. Most dramatically, various insect species have developed groups of individuals who do not reproduce at all but whose foraging, fighting or nursing greatly augments the reproductive success of a much smaller number of fertile members.

In analysing this kind of situation it is helpful to adopt the viewpoint of Richard Dawkins,[26] who pointed out that the basic competition is between genes (which are almost immortal), and that individual organisms can be thought of simply as short-lived vehicles built by genes to ensure their own survival. An altruistic gene would indeed be eliminated by natural selection, but a 'selfish gene' – to use the metaphor Dawkins made famous – can help ensure its own survival by making altruistic individuals that promote the survival of other individuals that contain the same gene.

The explanation of most altruistic behaviour is probably *kin selection*. The notion that, by affecting the behaviour of the individual possessing it, a gene might affect not only the chances of survival of that individual but also of related individuals carrying the same gene had occurred to both Ronald Fisher and J. B. S. Haldane early in the 20th century.[27] Its full significance though, was not realized until 1964 when William Hamilton published material that had been so little appreciated by the London University examiners the previous year that it had failed to get him his PhD.[28] On average, a mammal will share 50 per cent of his (or her) genes with a brother or sister or child, 25 per cent with a nephew or niece or grandchild, 12.5 per cent with a first cousin, and so on. If all that matters is maximizing the chances of survival of the genes that you carry, then, other things being equal, it is worth taking up to an even chance of being killed to save the life of a relation who is likely to share 50 per cent of those genes. If gene sharing is only 25 per cent, it would be worth taking up to a one-in-four chance; and so on. The rule is to take what Robert Trivers has called 'a gene's eye view'.[29] Altruistic behaviour is particularly likely to evolve when the pattern of reproduction is such that the altruist and the beneficiary share more than 50 per cent of genes. In ants, for example, queens and workers share 75 per cent of genes.* In some aphids, asexual reproduction leads to the formation of soldiers and fertile forms (protected by the soldiers) that are genetically identical and differ only because they have been fed differently.

Altruism can also develop by reciprocity.** Vampire bats have had a consistently bad press, but they behave admirably to each other.[30] A bat cannot survive for more than two or three days without a meal of blood, yet failure to find a victim for this long is not uncommon. But starved bats do not die; on returning to their roosting places they are fed by successful bats, not necessarily related, who will in turn be fed when they miss out on a meal. Such behaviour can be favoured by natural selection because the cost to the altruist is less than the return benefit. A tendency to cheat by failing to reciprocate is presumably selected against, since a bat that fails to reciprocate too often may find that its fellow bats are less forthcoming when their help is needed:

> The man who has plenty of salt peanuts, and giveth his neighbor none,
> He shan't have any of my peanuts when his peanuts are gone.

On the other hand cheating is crucial to the development of a third form of altruistic behaviour. Here the beneficiary of the altruism induces altruistic behaviour by pretending to be the appropriate recipient. The young cuckoo in the nest of a hedge sparrow or wagtail is the most obvious example.

A likely explanation of the persistence of some inherited features that are disadvantageous is that the gene responsible for them is also responsible for a feature that is advantageous. The classical example is sickle-cell anaemia, a disease caused by an inherited abnormality of haemoglobin that makes the red blood cells more fragile, but which also makes them a less suitable environment for malarial parasites. If you inherit the sickle-cell gene from both parents, the increased fragility dominates and the severe anaemia outweighs any advantage you might gain from increased resistance to malaria. But if you inherit the gene from only one parent (and you avoid the oxygen lack associated with high

---

*This is because males develop from unfertilized eggs and so have half the number of chromosomes. Each worker or queen will have received all of her father's genes and half of her mother's. (See pp. 177–9 of Trivers' *Social Evolution*.)

**See also Chapter 24 of this book, p. 408.

mountains and unpressurized aircraft) it is unlikely to cause any trouble, and the increased resistance to malaria actually increases your chances of survival if you happen to live in a malarial district.

AS UNLIKE AS TWO PEAS

Of all the difficulties that Darwin faced, the most serious was the then current ignorance of the nature and causes of the variation between individual members of a species, and the belief that the characters of parents were simply blended in their children, so that any small favourable variation would get diluted away in successive generations.[31] We now know that this does not happen because, although in general the child receives half of its genes from each parent, the genes themselves do not fuse but retain their identity. The offspring will be a blend in the sense that it may show some characteristics of each parent; but as far as the genes are concerned – and it is the genes that carry almost all the instructions for making the next generation – they are not blended but merely shuffled. Even a gene whose effects are not apparent in an individual because of the presence of a dominant gene* from the other parent, is nevertheless present and will be passed on to future generations. Favourable variation in a gene will not be diluted, and may, by natural selection, gradually spread through a population.

In Darwin's day, knowledge of genes lay far in the future. But in a monastery garden in Brno in Moravia, evidence that inheritance works as if characters were transmitted in discrete packages was being produced by Gregor Mendel over a period of about ten years spanning the publication of the *Origin*. Mendel looked at the results of crossing and recrossing varieties of pea that possessed contrasting characters – tall and dwarf, smooth seeds or wrinkled seeds, red flowers or white flowers – and because these characters are determined by single genes he obtained very striking results which could be simply interpreted. Unfortunately, he published his results and interpretations, in 1865, in the Proceedings of the Brno Society for the Study of Natural Science,[32] and they remained unnoticed until 1900. Mendel died an abbot, but with his great scientific work ignored; Darwin spent a great deal of time speculating unprofitably about inheritance, and died without being aware that the most serious of all his problems had been largely solved. The irony of the situation is that Mendel possessed a copy of the *Origin*, and had annotated it; if only he had written a letter. And copies of the proceedings of the Brno Society were sent to both the Royal Society and the Linnean Society in London; if only Darwin had not incarcerated himself at Down and had read one of them.

*A dominant gene produces its characteristic effect in the individual organism even if only one copy is present, i.e. if it is inherited from only one parent. A recessive gene produces its characteristic effect only if two copies are present, i.e. only if it is inherited from both parents.

The shuffling of genes can promote evolution by bringing together genes, or combinations of genes, that confer particular advantages; but the effect of natural selection and shuffling alone will be limited. The most it can achieve is to create an organism with whatever combination of genes *present in the original population* is optimal for the current environment. If evolution is to continue beyond that point it is necessary to suppose that at least some of the genes can change (mutate), so that natural selection acts not only on the original gene pool but also on any mutated genes. If the mutation makes the new gene harmful it is likely to be eliminated by natural selection; if it is beneficial natural selection will tend to spread it through the population. How fast will it spread?

In the 1920s and early 1930s this question was tackled by J. B. S. Haldane and by R. A. Fisher in England and by Sewall Wright in the U.S.A. The answer (making certain assumptions)[33] is that even a very small selective advantage can be expected to promote the fairly rapid spread of a gene through a population. For example, if 1 per cent of a population carry a mutant gene that causes those carrying it to produce, on average, 1 per cent more offspring than those lacking it, then after about 560 generations (about 8 days for bacteria, 15 years for *Drosophila*, 280 years for rabbits, 14,000 years for humans), about half of the population will be carrying the mutant gene. After about 1580 generations the proportion will have risen to 90 per cent. The proportion then increases more slowly, eventually approaching 100 per cent. If the selective advantage conferred by the mutant gene is 10 per cent rather than 1 per cent the spread is ten times as fast. If the mutant gene is recessive, no benefit is conferred by its presence unless two copies are present, so the spread is initially very slow; but eventually it accelerates dramatically because, once a large fraction of the population is carrying a single copy of the gene, acquisition of a second copy will be advantageous.

The crucial question is whether mutation rates, and the rates of spread of beneficial genes, are likely to have been sufficient to account for the enormous variety of living things that appear to have evolved in a period of less than four billion years. And if four billion years seems a comfortingly long time, we must also ask whether we could ourselves have evolved from apes in the five to seven million years which both the geological record and more recent biochemical studies suggest is the period that was available for that change.[34]

A difficulty in answering questions of this kind is that even if we consider a well-defined change – say the increase in the size of brain between apes and ourselves – we do not generally know how many genes must have changed, in what way they must have changed, or at what rate those changes are likely to have occurred. It is important to realize, though, that very far-reaching and coordinated changes can be caused by the modification of very few genes. For example, experiments with fruit flies have shown that the development of an extra pair of wings, or of an extra pair of legs replacing the antennae – the kind of changes that you might think would require great reorganization involving very many genes – can each be caused by the mutation of a single gene. The explanation of these remarkable effects is that the genes concerned are regulatory genes – that is, they control the synthesis of proteins that control the activity of other genes. The need for this kind of control arises because, although the nucleus of every animal or plant cell contains an almost complete set of all the genes necessary for making the animal or plant – perhaps 100,000 genes in mammals – only some of these genes will be needed in any given cell at any one time. For example, the gene for making haemoglobin will be 'switched on' only in developing red blood cells; the gene for making insulin will be switched on only in certain cells in the pancreas. For the animal or plant to develop properly, and to function properly once it has developed, there must be a mechanism for switching genes on and off at appropriate times. And since this switching mechanism must itself consist of genes (and proteins coded by them) – because there is no other way of transmitting instructions of this kind from one generation to the next – it is not surprising that a hierarchy of regulatory genes has evolved, the genes at each level producing proteins that, alone or in conjunction with other regulatory proteins, control the activities of genes at lower levels. It seems likely that the rapid evolution of man in the last few million years has involved mutations of regulatory genes.

EVOLUTION AT THE MOLECULAR LEVEL

We have already seen that comparisons of amino-acid sequences or of nucleotide sequences can supplement more traditional comparative anatomy in tracing the evolution of individual species. But the interest of comparative molecular anatomy is not just that it extends the findings of comparative gross anatomy; it also provides information about the timing of significant evolutionary events in the past, and it reveals the kind of changes in genes that favour such events.

In looking at a protein – or, more precisely, at a single chain of amino-acids in a protein – we are looking at the product of a single gene. Suppose we compare one of the amino-acid chains in our own haemoglobin with the corresponding chain in the haemoglobin of, say, a shark. The geological record tells us that our ancestry and the shark's ancestry diverged about 460 million years ago. Mutations no doubt occurred in both branches so the differences between our haemoglobin and the shark's haemoglobin are the result of 2×460 million years of evolution. Of the mutations that occurred, some would have been harmful, some neutral and some beneficial; but because the harmful ones would have been eliminated the differences we now see must be the accumulated result of non-harmful mutations. By dividing the number of differences by the number of years of evolution we therefore get the average rate of non-harmful mutations over the period since our ancestry diverged from the shark's. If, instead of haemoglobin from a shark we use haemoglobin from a dog – whose ancestry diverged from our own only about 90 million years ago – we shall get estimates of the average rate of non-harmful mutations over the much shorter period since our ancestry diverged from the dog's. Comparisons of this kind, made on several different proteins from a range of different animals, have led to a surprising conclusion. It turns out that, *though rates of change have varied widely between different proteins, for any one protein the rate of change has been roughly constant over enormous periods of time.*[35] As it is difficult to believe that the selection of favourable variations could have occurred at such a uniform rate in animals of different kinds, and over periods of time that cover vastly different conditions, some other explanation is needed.

It has been supplied by Motoo Kimura.[36] He points out that because beneficial changes are likely to have been rare, and harmful changes would have been eliminated, the great majority of the differences in amino-acid sequence that exist now must have resulted from mutations that had little effect on the function of the protein. The roughly constant rate of evolution of a given protein could, then, simply reflect a roughly constant mutation rate, and a roughly constant rate of elimination of harmful mutations.

This interpretation is supported by a comparison of the rates of evolution of different proteins. The accuracy of the machinery responsible for protein synthesis is likely to be the same irrespective of which protein is being synthesized; so if Kimura is right, the differences in the rates of accumulation of changes in amino-acid sequence should reflect the extent to which changes are eliminated because they are harmful. A protein that can tolerate many changes in amino-acid sequence without its function being affected should therefore change composition much faster than a protein that can tolerate very few changes – because fewer of the changes in the less fussy protein will lead to the elimination of the organism in which they occur. (You can open a locked door with an irregularly shaped jemmy but not with an irregularly shaped key.) It is therefore significant that much the fastest changes are found in the fibrinopeptides – small proteins that are split off from the protein fibrinogen when blood clots, and whose function before they are split off is simply to prevent the fibrinogen molecules from coming together to form a clot. The fibrinopeptides show a 1 per cent change in amino-acid sequence in about 800,000 years.* In contrast, haemoglobin shows a 1 per cent change in about five million years; and some of the small proteins around which the DNA in the nucleus is wrapped evolve so slowly that 500 million years are required for a 1 per cent change in sequence. This is presumably because almost any change in the protein interferes with the functioning of the DNA that is wrapped round it.

*This is roughly equivalent to the observed rate of change of nucleotides in that portion of DNA that is known not to code for protein, and so is presumably not liable to elimination by natural selection as the result of making ineffective protein.

The steadiness, over long periods of time, of the rate of change of the amino-acid sequence of a given protein – or of the nucleotide sequence in the DNA encoding that protein – makes it possible to use changes in either sequence as a molecular clock. Such clocks show that the human ancestral line diverged from the ancestral line leading to chimpanzees – probably our closest living relatives – something like five to seven million years ago, and from the line leading to gorillas a little earlier – perhaps eight to eleven million years ago.[37] These estimates necessarily depend on the accuracy of the dating of the fossils used to calibrate the clocks.

Kimura's hypothesis does not, of course, deny the crucial role of favourable variations in evolution. It accepts that *significant* changes in evolution are largely the result of positive Darwinian selection that brings about adaptation of the organism to its environment. The point is simply that the great majority of the individual changes that can be detected in the amino-acid sequences of proteins (or the nucleotide sequences in genes) will not have had significant effects on the organism; they are the result of random drift.

The same surprising conclusion can be reached by another route. We saw earlier (p. 27) that many amino-acids are coded for by several codons (triplets of nucleotides), so that sometimes a substitution of one nucleotide within a codon does not change the amino-acid specified. A substitution of this kind has no effect on the protein produced by the gene, so it can neither be selected for nor selected against. If most nucleotide substitutions had been accumulated as a result of positive selection, such impotent substitutions should be rare. If Kimura is right they should be the commonest type of substitution. And they are.

If the great majority of changes that have accumulated in a given protein in a particular species are neither beneficial nor harmful, the rate at which they have accumulated will bear no particular relation to the rate at which the species itself has evolved. And this is what is found. In the last 90 million years frogs have changed very little while mammals have diversified enough to stock a zoo. Yet comparisons of individual proteins or DNAs suggest that molecular evolution has gone on at a comparable rate in the two groups.

Why then are there great fluctuations, both at different times and in different organisms, in the rate of *significant* evolution – the accumulation of changes that affect the organism's chances of survival? In the first place, the rare favourable variations are unpredictable both in their timing and in the extent of their effects – we have already seen that mutations in regulatory genes can have effects that are particularly far-reaching. Secondly, selective pressures will vary widely: a change in the environment may make conditions more adverse or may create opportunities; extinction of one group may leave ecological niches ready to be filled by members of other groups. Thirdly, it is wrong to think of the adaptive evolution even of a single protein as simply the result of the gradual accumulation of favourable 'point mutations', that is, changes in single nucleo-tides. All organisms possess machinery capable of cutting and splicing nucleo-tide chains, and this machinery makes possible two kinds of molecular event of great evolutionary significance: it allows individual genes to be duplicated within a single nucleotide chain, and it allows bits of genes to be shuffled, or spliced into other genes.

An advantage of gene duplication is that one copy of the gene can continue to make the normal gene product, while the other can evolve unrestricted by the need to fulfil its normal role. In this way old functions can be retained while new ones are added. Sometimes the products of the old and the new gene can cooperate. The efficacy of mammalian haemoglobin depends on the interaction of two almost identical kinds of subunit which presumably arose by slight modification following gene duplication.

An advantage of the shuffling and splicing of bits of genes is that a structure that has evolved for a particular function in one protein can become available for a similar function in another protein. Many enzymes, for example, need to bind – i.e. attach to themselves – calcium ions or molecules of ATP* or of nucleic acid, and different parts of these enzyme molecules – different 'domains' – are specialized for these functions. Comparisons of domains for the same function in different proteins often show resemblances which are too close to be explained merely on the basis of the shared function. Such resemblances are thought to have resulted from the transfer of bits of genes coding for the relevant domains.

Transfers of this kind, which seem to have occurred frequently in the course of evolution, involve only genes of a single species. On two occasions in evolu-tionary history, however, momentous consequences seem to have followed the transfer of genetic material between species. In all higher animals and plants, the production of ATP at the expense of energy from oxidative processes is brought about in organelles called mitochondria, and the mechanism involves two steps. In the first step, energy from the oxidations is used to charge an electrical battery in the mitochondrial membrane; in the second step, energy stored in this battery is used to synthesize ATP. Precisely the same mechanism is used in certain aerobic bacteria, and this led Lynn Margulis to suggest that mitochondria had evolved from aerobic bacteria living symbiotically in the cells of organisms that were otherwise unable to use energy from oxidations.[38] Striking support for this apparently wild hypothesis came from the discovery that mitochondria contain their own DNA, and protein-synthesizing machinery (closely resembling that found in bacteria) which makes some, though not all, of the mitochondrial pro-teins. What started as a symbiotic relationship between two kinds of free living organism has turned into complete mutual dependence. In an almost identical

*Adenosine triphosphate, the fuel for most energy-requiring reactions that occur in living cells.

fashion, free-living, photosynthetic bacteria are thought to have given rise to chloroplasts – the organelles in green plants that are responsible for generating ATP using the energy from sunlight. The major energy-yielding processes in both animals and plants therefore depend on machinery whose origin involved the wholesale transfer of genes from bacteria. In the struggle for existence there is great advantage in being able to use devices that have evolved in one context to serve in another. The next chapter shows how such adaptability has helped in our own evolution, and particularly in the evolution of our brains and our minds.

4   'The Descent of Man'

Is man an ape or an angel?
—Benjamin Disraeli

WALLACE'S DOUBTS

We have seen that Darwin deliberately said very little explicitly about the evolution of humans in the *Origin*, holding his fire until the publication of his *Descent of Man* twelve years later. There is no doubt, though, that he felt strongly that his arguments applied to humans. A few days after completing the correction of the final proofs, he wrote, 'I would give absolutely nothing for the theory of Natural Selection if it requires miraculous additions at any one stage of descent.'[1] So he was particularly upset when, ten years later, Wallace declared that he did not believe that natural selection alone could account for the evolution of the human brain or of human mental abilities.[2] Just as humans, for their own ends, had supplemented natural selection in producing the dray horse and the Guernsey milch cow, so, Wallace imagined, a 'Higher Intelligence . . . for nobler ends', had supplemented natural selection in producing the human brain and the human mind.

Wallace's argument was that natural selection could not lead to the modification of any organ or character beyond the point at which such modification was helpful in the struggle for existence. Early humans could, he felt, have had no need for a brain three times as large as that of any ape; a brain slightly larger than that of a gorilla would have been sufficient. Eventually this oversize brain would make possible the great mathematical, musical, aesthetic and moral abilities of modern humans; but though rudiments of these abilities might be found in primitive tribes, they were of no help in a way of life that demanded few faculties not possessed in an equal degree by animals. In the capture of game or fish, primitive men were, he thought, not more ingenious than the jaguar, who drops saliva into the water and seizes the fish as they come to eat it.

If natural selection was not the cause of the excessively big brain there must be some other cause; hence the invocation of the Higher Intelligence and the nobler ends – the assumption that primitive humans were given brains too big for their needs so that we who followed them could behave in ways that were intellectually, aesthetically and morally admirable.

Even if Wallace had been right in thinking that natural selection was incapable of explaining the size of the human brain, it was rash of him to assume that the actual cause of enlargement was related to the way in which the brain was to be used several million years later. Living organisms are nothing if not adaptive, and the essence of adaptation is to use for one purpose what originally served (and may still serve) for another. The point is nicely made, in a different context, by S. J. Gould and R. C. Lewontin.[3] The four spandrels of the arches that support the great dome of St Marks, in Venice, are each decorated with mosaics of one of the four evangelists; below each evangelist a man pours water from a pitcher, the four streams representing the four biblical rivers. The design is so splendid and appropriate that anyone unfamiliar with architectural construction might well assume that the purpose of the spandrels was to carry the design. In fact, of course, if a dome is to be supported on four arches, spandrels are inevitable.

But was Wallace right in his claim that the brains of human ancestors were larger than could have been useful to them? In the second chapter of his *Descent of Man*, Darwin points to the invention of weapons, tools, traps, rafts and canoes, to the use of fire, the adoption of social habits, and the use of language, as examples of 'the direct results of the development of . . . powers of observation, memory, curiosity, imagination and reason'. And he particularly emphasizes the likely dependence on brain power of even the smallest proficiency in language. In arguing in this way, Darwin tended to assume that our distinguishing features as human beings – a large brain, a fully upright stance, the making and use of tools and weapons, and the use of language – developed in concert; and in the absence of any knowledge of the intermediate forms that link us to the last common ancestor that we are presumed to share with modern apes this was a reasonable assumption. He was not surprised that no fossilized remains of these intermediate forms had been discovered, first because the chances of fossilization were small, and secondly because no one had yet looked in the right place. 'In each great region of the world', he argued, 'the living animals are closely related to the extinct species of the same region. It is therefore probable that Africa was formerly inhabited by extinct apes closely allied to the gorilla and chimpanzee; and as these two species are now man's nearest allies, it is somewhat more probable that our early progenitors lived on the African continent than elsewhere.'[4]

FILLING THE FOSSIL GAP[5]

> . . . *a wilderness of monkeys*
> —Shylock, in *The Merchant of Venice*

In 1924, little more than half a century after Darwin had suggested that humans originated in Africa, Raymond Dart, the young Professor of Anatomy at the

University of the Witwatersrand in Johannesburg, was presented by one of his students with a fossilized monkey skull.[6] The skull had been blasted out of a cliff at a limestone quarry at Taung, 80 miles north of Kimberley, and Dart was interested because this was an area where evidence had recently been obtained about a succession of stone ages in South Africa. Thinking that this lime deposit, like that at Broken Hill in Northern Rhodesia (now Zambia), might contain fossil remains of early man, he consulted his colleague R. B. Young, the Professor of Geology at Witwatersrand, who by a fortunate coincidence was called down to Taung at that time to investigate the lime deposits on an adjacent farm.

Young was able to inspect the site of discovery and to select further samples of fossil material from the same formation, including some rock fragments disclosing portions of bone, and two natural endocranial casts. Such casts are formed by the gradual deposition of lime within the cavity of a skull from lime-laden water trickling through it, and they show the size and shape of the cranial cavity, and can even give some idea of the convolutions on the surface of the brain that had once occupied the skull. The smaller of the two endocranial casts was clearly that of a monkey, but the larger one, both in size and in the pattern of convolutions, was more reminiscent of a chimpanzee or gorilla.

Dart noticed that the broken front of one of the casts fitted accurately into another of the rocks brought back by Young, a rock that showed the edge of a broken lower jaw projecting from it. He concluded that the face of the creature whose cranial features had been revealed by the cast must be embedded in the rock, and by a painstaking and protracted removal of rock fragments he was eventually able to expose that face, and to fit the cast into the back of it. He found he was then holding in his hand the skull of a child, but a very peculiar child (Figure 4.1). It had a cranium as long as that of an adult chimpanzee, but its orbits, teeth and lower jaw looked more human, and its face thrust forward less than that of an ape though more than that of a human child. It was clearly young, as all the milk teeth were in place and the first molars were just erupting; but the size of the cranial cavity was smaller than that of a human child at the same stage of tooth development. The hole at the base of the skull through which the spinal cord emerges from the brain was further forward than it is in apes – suggesting that the creature walked upright, not with its head poked forward like an ape.

Dart concluded that what he was looking at was a creature that belonged to an 'extinct race of apes *intermediate between living anthropoids and man*'; and he named his find *Australopithecus africanus* – the southern ape of Africa.* He promptly published a detailed description in the journal *Nature*,[7] in London, but his paper had a frosty reception from the distinguished London anthropologists – probably because the evolutionary story told by the Taung skull was the exact converse of the evolutionary story told by the (fraudulent) fragments of skull

---

Australo simply means south. The only connection with Australia is an etymological one.

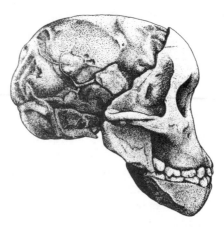

FIGURE 4.1 Raymond Dart's
'Taung child', *Australopithecus
africanus*. (From I. Tattersall.)

discovered at Piltdown, in Sussex, over a decade earlier. The Piltdown skull
showed an ape-like jawbone but a human-sized cranium – not surprisingly, since
the jawbone was that of an orang-utan, with parts removed and its teeth filed,
and the cranium was in fact human. The Taung skull suggested that, in the
evolution of humans, the jawbone and teeth became human-like before the great
expansion of the brain.

Following Dart's discovery, rather similar fossil skulls were found in limestone
caves at various other places in South Africa. Some were delicately built, closely
similar to the Taung skull. Other were more robust, with heavy outwardly projecting cheekbones, huge molars, and a midline crest that must have served to anchor
powerful muscles that closed the massive jaws; their discoverers assigned them to a
separate species *Australopithecus robustus*. Dating all these fossils was difficult,
because it depended largely on matching associated animal remains with well-
characterized collections of animal remains from sites of known date, but a rough
estimate put the age of the more delicately built fossils at two to three million years,
and the more robust ones at one to two million years.

From the late 1950s onwards, slightly different fossils, both of the delicate and
robust types, were found at a variety of sites along the East African Rift System –
a discontinuous series of rift valleys that runs northwards for 3000 kilometres
from southern Malawi, through Tanzania and Kenya to the Afar region of
Ethiopia near the Red Sea. This rift system is believed to have been caused by the
moving apart of two tectonic plates, the central African plate to the west, and the
Somali plate to the east.

The rift valley sites are all open sites, near the shores of lakes or rivers, or
on floodplains, and because of the tectonic origin of the rift system they also
tend to be in areas of past volcanic activity. The result of this combination of
circumstances is that layers of ash, which with time have hardened into 'tuff', tend
to be interleaved with layers of sand or silt. The tuff contains small quantities of

radioactive potassium and uranium, which can be used to determine the time at which the material forming the tuff solidified from the molten state.* For this reason, the dating of fossils in the rift valley sites is particularly reliable.

A skull of the robust form was discovered in 1959, by Mary Leakey, in the Olduvai Gorge in the rift valley east of the Serengeti Plain, Tanzania. It was even more heavily built than its South African counterparts – its massive lower jaw earned it the nickname 'nutcracker man' – and it seemed sufficiently different to be classified as a separate species.[8] Later, similar fossils were found in Kenya on both the eastern and western shores of Lake Turkana, and in Ethiopia near the Omo River, which flows into Lake Turkana at its northern tip. These robust ape-men seem to have appeared by about two-and-a-half million years ago, and to have been extinct by one million years ago.

The lightly built ape-men whose fossils are found in the eastern rift valley seem to have lived rather earlier, appearing about four million years ago and becoming extinct about a million-and-a-half years later. In 1974 the American anthropologist Donald Johanson and the French geologist Maurice Taieb led an expedition to the Afar region of Ethiopia. Here, near an ancient lake bed at Hadar, they found numerous bits of bone that had been washed out of the sloping side of a gully. Assembled, they proved to be a substantial part (about 40 per cent) of the skeleton of a single diminutive but adult individual, probably female. As no one had discovered such a complete early skeleton before, it was felt that she needed a personal name, and since the Beatles' *Lucy in the Sky with Diamonds* was being played in the camp, she was named Lucy. Lucy was only a little more than one metre tall, and she was apelike in having a small brain (relative to her body size), long arms, and short legs with curved foot bones. On the other hand, the shape of the pelvis and the angle between thigh and leg suggested strongly that she walked upright. The most recent radioactive dating shows that Lucy probably lived between three million and three-and-a-half million years ago.

In later expeditions to Hadar, Johanson and his colleagues found a large number of hominid fossils, including a collection of thirteen individuals possibly the victims of a flash flood. Many were much bigger than Lucy, and they showed a wide variation in size, and some variation in jaw shape – features that raise the possibility that more than one species was present. Johanson himself believes that the range of sizes simply reflects a large difference in the size of the two sexes, and he allocates all the specimens and Lucy to a single species, which he named *Australopithecus afarensis* (after the Afar region). Among the earliest fossils from Hadar are some which include the hand bones. They show features intermediate between those of apes and humans; in particular, the thumb is longer than in apes but not fully opposable to the fingers, and some of the finger

*Because the direction of the earth's magnetic field has changed in a known way, an estimate of the time at which solidification occurred can also be obtained from the direction of magnetization of magnetic elements in the tuff.

bones are curved. This curvature, and the existence on several of the hand bones of what appear to be sites of attachment of powerful muscles flexing the wrist and fingers, suggest that Lucy and her like, though they walked fully upright, were also good at climbing trees.

Whether or not all the Hadar fossils are correctly classified as a single species, they provide strong evidence that by about three-and-a-half million years ago hominids, whose brains (relative to body size) were not significantly bigger than those of modern great apes, were walking fully upright. This early evolution of a fully upright posture is confirmed by some remarkable footprints. About 3.7 million years ago, in Laetoli, in the Serengeti region of Tanzania, three individuals walked for over seventy feet along a path of soft volcanic ash. The ash hardened to form tuff, which later became covered, but in 1978 Paul Abell, a member of a team led by Mary Leakey, noticed what looked like a single heel print where the covering had been eroded.[9] Careful excavation during that season and again in 1979, revealed a large number of footprints with surprisingly human characteristics – a big toe about twice as large as the next toe, the absence of a gap between the big toe and the next toe, evidence of a stable longitudinal arch, and impressions of the heel and of the ball of the foot similar to those produced by modern humans walking in normal fashion on soft ground. At first sight the footprints appear to be those of an adult and a child walking closely side by side, with the adult adjusting the length of his or her stride to match that of the child. Closer examination shows a third set of footprints – those of a rather smaller adult – carefully placed within the footprints of the first adult. What the relationship between the individuals was, and why they should have proceeded in this peculiar fashion, we do not know, but the significant feature is that the footprints show that all three were walking upright. There is no firm indication of what sort of hominids left this astonishing and moving record of their walk; but the date of the footprints, and some similarity between fragments of fossil jawbones in Laetoli and jawbones found by Johanson at Hadar, make *Australopithecus afarensis* a possible candidate.

Until very recently there was no evidence that *Australopithecus afarensis* existed outside the area of the eastern rift valley, but in 1995 a fossilized lower jawbone, between three and three-and-a-half million years old, and with characteristics somewhat similar to those of *Australopithecus afarensis*, was discovered in Chad, about 2500 km further west.[10]

From the study of amino acid sequences in proteins, or of nucleotide sequences in DNA, we know that the split between the evolutionary line leading to us and the evolutionary line leading to chimpanzees, our closest living relatives, probably occurred between five and seven million years ago. The best evidence for the date of Lucy and her kind is three to three-and-a-half million years ago, so even if *Australopithecus afarensis* is on the direct line of descent – which is by no means certain – there is still a gap of at least one-and-a-half million years, and perhaps as much as four million years, before we get to our common ancestor.

In 1994, Tim White and his collaborators closed that gap by about a million years when they described seventeen hominid fossils at Aramis on the River Awash in the Afar region of Ethiopia, about 50 miles south of the site where Lucy had been discovered.[11] All but one of the fossils were sandwiched between two layers of volcanic ash that could be reliably dated, proving that the fossils were between 4.3 and 4.5 million years old. The presence of fossil wood and seeds, and of monkey remains, suggested that the hominids – which White named *Australopithecus ramidus* (*ramid* meaning root in the Afar language) – probably lived in a wooded landscape. The remains are different from those of *Australopithecus afarensis*, being, on the whole, more chimpanzee-like in character, though with small canine teeth. It is not known whether *Australopithecus ramidus* walked upright.

It is easy to put the different species of australopithecines in order of age, and tempting to suggest in biblical fashion that *Australopithecus ramidus* begat *Australopithecus afarensis* which begat *Australopithecus africanus* which begat – through intermediate forms – us. And since, superficially at least, the robust australopithecines seem more different from us than does the Taung child or Lucy, it is also tempting to suppose that the robust forms lie on one or more side branches. Such a family tree is not necessarily wrong, but there is no reason to suppose that it is right. We are looking at a few fossilized bits of what may well have been a highly branched tree, and we cannot assume that any particular bit that we happen to know about is on the direct path to the twig that is *Homo sapiens*. Various alternative schemes have been suggested, but none has general support. What is clear though is, first, that the notion that there are intermediate forms between us and the last ancestor we share with the apes is entirely plausible; and secondly, *that the adoption of a fully upright stance preceded any major enlargement of the brain, and occurred at least a million years before the date of the earliest known manufactured stone tools.*

This second conclusion was unexpected, and it shattered the long-held assumption, going back to Darwin, that our large brains, our fully upright stance and our use of tools, evolved in concert. But if a fully upright stance did not evolve in our australopithecine ancestors because it freed their hands to use the tools or weapons that their greatly enlarged brains enabled them to make, why did such a stance evolve? There is no generally accepted answer. Peter Rodman and Henry McHenry have pointed out that although bipedalism is less efficient than the quadripedalism seen in horses and dogs it is more efficient than the semi-upright stance and knuckle walking used by chimpanzees; its development would therefore have helped early hominids at a time when the environment was becoming less wooded and they needed to travel further in search of food.[12] Owen Lovejoy has suggested that the fully upright habit evolved because it allowed fathers to carry food to both mothers and their offspring who could remain in greater safety near a home base.[13] He believes that these changes had far-reaching effects on sexual patterns and on family and social life, though Robert Foley reminds us that a picture 'of nuclear happy families with the males off at work collecting food and the females staying at home to look after the young' is unlikely to be an accurate picture of the social lives of early hominids.[14] Other suggestions are that bipedalism enabled the female to carry her young more securely; that it led to less heating from the overhead tropical sun;[15] and, perhaps less plausibly, that it made it easier to see predators over tall grass.

THE ORIGIN OF *HOMO*

Having traced the australopithecines as far back to the common ancestor as we can, let us now look forward and consider the fossil evidence linking the australopithecines to ourselves.

In the early 1960s Louis Leakey identified a new kind of fossil hominid from the Olduvai Gorge. It had a thin cranium, smaller molars than the australopithecines, and a more human-looking face, but the most remarkable thing about it was the size of its brain – about 50 per cent more than those of the australopithecines, though still only about half the size of a modern human brain (average size about 1350 millilitres). Because of the large brain, and because the bed from which the remains had been excavated also contained very primitive stone tools, Leakey assumed that the new hominid was the maker of the tools, so his find appeared to provide the earliest sign of regular tool-making among our putative ancestors. (But see extra note on page 448.)

At the suggestion of Raymond Dart, Leakey named the new hominid *Homo habilis* – 'handy man' – (Figure 4.2a). At first, both its placing in the genus *Homo*, and the assumption that it made tools, proved highly controversial, but when similar and more complete fossils, also accompanied by primitive stone tools, were discovered at Koobi Fora, on the east side of Lake Turkana, Leakey's views prevailed.* Dating at both sites suggested that *Homo habilis* existed for rather less than a million years, becoming extinct by a million-and-a-half years ago.

For the next step we have to go back to the nineteenth century. In 1891–2, Eugène Dubois, a young physician in the Medical Corps of the Royal Dutch East Indies Army, aided by a gang of convicts, unearthed an unusual skullcap together with two teeth, and later and separately a thighbone, from the bank of the Solo River at Trinil in Java. He claimed that they represented a creature halfway between humans and the higher apes; but his claim had an unenthusiastic

---

*Variation in form and size between the different supposedly *Homo habilis* fossils led some anthropologists to suspect that two species were present – *Homo habilis* and a separate species *Homo rudolfensis*.

reception, and recognition of what we now know as 'Java man' had to wait until the 1930s. By then, several years of excavation in caves at Zhoukoudian (then known as Choukoutien) near Peking had revealed the fragmented remains of about 40 individuals. The skulls were particularly striking, with cranial capacities around 1000 millilitres – in one case as much as 1200 millilitres, i.e. well within the bottom half of the modern human range – and with huge brow ridges that met in the midline. At that time there was no way of dating these remains of what became known as 'Peking man', but estimates made more recently suggest that the oldest of them are a little less than half a million years old.[16]

In 1940, Franz Weidenreich, a German anthropologist based in Peking, classified both Java man and Peking man as a single species, *Homo erectus*. In the half century since then it has become clear that hominids whose characteristic features place them in that species or close to it had already existed for more than a million years by the time 'Peking man' was living in the caves at Zhoukoudian, and that such hominids were widespread and extraordinarily successful. The earliest fossil, thought to be about 1.8 million years old, and with a skull capacity of about 850 millilitres, was found on the eastern shore of Lake Turkana (Figure 4.2b), and the most complete specimen – almost the entire skeleton of a twelve year old boy about 5ft 6in tall – was discovered near the opposite shore. Other African examples have been found in Tanzania, the Transvaal, Ethiopia, Algeria and Morocco. (All these fossils were originally regarded as examples of *Homo erectus*, though recently there has been a tendency to regard the early African specimens, with their slightly smaller skulls, as a separate species, probably ancestral to *Homo erectus*.)* It has long been assumed that the spread of hominids from Africa into Asia, and southern Europe, could not have occurred much earlier than about a million years ago. Within the last few years, though, the finding of a lower jaw, estimated to be 1.6 million years old, in Dmanisi in Georgia,[17] and the redating to an even earlier period of known skull fragments in Java,[18] have suggested that the date at which hominids first spread beyond Africa was much earlier.[19]

WHAT'S THE GOOD OF A BIG BRAIN?

It is one thing to discover that between one and two million years ago our putative ancestors had relatively large brains, and another to know what they did with them. We have seen that at Olduvai, in Tanzania, and at Koobi Fora, near Lake Turkana, *Homo habilis* fossils were associated with primitive flaked stone tools. About 1.4 million years ago, larger and more sophisticated stone tools appear,

*A species referred to as *Homo ergaster*. Some anthropologists regard *Homo ergaster* as the ancestor of all subsequent human species, and *Homo erectus* as a local East Asian development that led nowhere. See Tattersall, I. (1997) *Scientific American*, 276, Apr. 46–53.

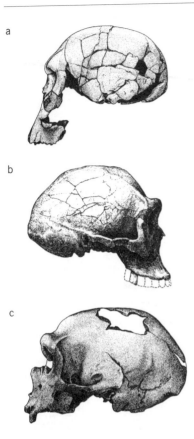

FIGURE 4.2 Fossil skulls of (a) a Kenyan *Homo habilis*, (b) an early Kenyan *Homo erectus*, and (c) the Neanderthal 'old man' found at La Chapelle-aux-Saints, France. (Drawings (a) and (c) are from I. Tattersall, (b) is from A. Walker and R. E. Leakey.)

such as teardrop-shaped 'hand axes' whose symmetry and conformity to a standard pattern show that their makers must have had a clear idea of the finished product in mind while the tool was being made. By then, judging by the fossil record, *Homo habilis* was already extinct, and *Homo erectus* (or its immediate precursor) seems to have been responsible for this advance in technology, an advance which set the pattern for stone tools in Africa and parts of Asia for the next million years, reaching Europe something like half a million years ago.*

By making it possible to cut through all but the toughest animal hides, primitive stone tools are thought to have added meat to the diet of our early ancestors. In any event, there is evidence from cut marks on fossilized animal bones at both Olduvai and Koobi Fora that simple stone tools were used to dismember animals. What is not clear is whether, initially, the meat that was eaten was scavenged from the kills of animal predators or was obtained by hunting. Clearly some scavenging went on; Pat Shipman has drawn attention to the

*Though not reaching parts of China, including Zhoukoudian.

frequency with which cut marks on ancient bones cross over marks left by carnivore teeth.[20]

It would be tidy to attribute the improvement in the design of stone tools to the enlargement of the brain between *Homo habilis* and *Homo erectus*, and to attribute the success of *Homo erectus* to the availability of better tools. Both attributions are almost certainly too simplistic. In the first place, there is a gap of something like four hundred thousand years between the appearance of the earliest African *Homo erectus* and the dramatic improvement in tool design. Secondly, it is at least as likely that the selective advantage conferred by the larger brain of *Homo erectus* lay less in an ability to design better tools than in an ability to develop the more complex social patterns (possibly involving the beginnings of language) that were required to take full advantage of even primitive stone tools.

Complex and flexible social interactions would be needed in organizing the hunting or scavenging, the carrying and storing and sharing of the food – and also in the preparation of the tools themselves, for at some sites of tool production the nearest sources of stones suitable for flaking must have been several kilometres away.[21] The role of social relations as a dominant force enlarging the brains of our hominid ancestors was suggested independently by the zoologist Richard Alexander, in Michigan, and the psychologist Nicholas Humphrey, in Cambridge.[22] 'Social primates', Humphrey wrote, 'are required by the very nature of the system they create and maintain to be calculating beings; they must be able to calculate the consequences of their own behaviour, to calculate the likely behaviour of others, to calculate the advantage of balance and loss – and all this in a context where the evidence on which their calculations are based is ephemeral, ambiguous and likely to change, not least as a consequence of their own actions.' In this situation, with members of the same species competing against one another, there is no obvious upper limit to the usefulness of mental abilities, such as Wallace imagined when he argued that a brain a third larger than that of a gorilla would be adequate for an early human. As Matt Ridley puts it: 'What Alexander and Humphrey described was essentially a Red Queen chess game. The faster mankind ran – the more intelligent he became – the more he stayed in the same place, because the people over whom he sought psychological dominion were his own relatives, the descendants of the more intelligent people from previous generations.'[23]

And Ridley goes on to suggest that one of the changes that could conceivably be associated with the enlargement of the hominid brain, that would certainly have helped us to excel as the calculating beings that Humphrey describes, and which would therefore have been favoured by natural selection, is the acquisition of our remarkable ability to be aware that other individuals have minds – to realize that they have desires, beliefs, fears and knowledge, and often to know what these are. Just how important this ability is in our own lives is made clear by the seriousness of *autism*, a condition in which (despite the normal use of language) some of the skills needed for such 'mind-awareness' and 'mind-reading' have been shown to be impaired, and in which the patient has great difficulty in forming social relationships.[24] It is unlikely that we have a complete monopoly of this 'mind-reading' ability, and there is a considerable literature discussing the extent to which apes, and particularly chimpanzees and pygmy chimpanzees, are able to appreciate the existence of mental states in others,[25] but it is clear that the chimpanzee's grasp of the idea of others' minds is not comparable with our own, and there is no way of knowing at which stage in the evolution from our common ancestor our skill was acquired.

Where there are social relationships that involve reciprocal benefits, there is the possibility of cheating, and therefore the need to avoid being cheated (or being discovered cheating). In his book *Social Evolution*, Robert Trivers points out that, in human reciprocal altruism, the *gross cheat* – the beneficiary who fails to reciprocate at all – will be readily detected and barred from future benefits. But the *subtle cheat* – the recipient of benefits who routinely attempts to give less than he received, or than the partner would give if the situation were reversed – presents that partner with a problem.[26] For he still benefits, even if less than he would do if the relationship were equitable. Trivers argues that, because human altruism continues for long periods and may involve thousands of transactions, keeping track of transactions is difficult, and even where possible often leads to difficult decisions. On the principle that half a loaf is better than no bread it may be better to continue an inequitable relationship – or perhaps cheat a little oneself. The 'right' decision, in any individual case, obviously depends on the extent to which the relationship is exclusive, on the availability of other actual or potential partners, and on an assessment of the likely response of the partner to any change in behaviour – an assessment that is both facilitated and complicated by our ability to read others' minds. It is the need to cope with problems of this kind, Trivers believes, that originally led to the development, through natural selection, of complex psychological systems capable of friendship, a sense of fairness, gratitude, sympathy and guilt. This sounds plausible, and we shall return to the question of morality at the end of the book.

But, less flatteringly, others have argued from the many reports of deliberate deception by chimpanzees and pygmy chimpanzees, that the need to cope with such problems during hominid evolution probably also led to the development of what they have called Machiavellian intelligence.[27] In any event, Leda Cosmides and John Tooby, of the University of California at Santa Barbara, have reported experiments showing that humans are particularly good at detecting cheating on social contracts.[28] They gave students conditional rules – of the form: if P then Q – then told them stories and asked whether the characters in the stories had violated the rules. In some of the stories violation of the rules could be interpreted as cheating on a social contract, in others it could not; and it turned out that detection of the violations was far more frequent when the cheating interpretation was available. In contrast, whether the stories dealt with situations that were familiar or unfamiliar made little difference to the detection rate. Of course, the subjects in Tooby and Cosmides's experiments were sophisticated members of modern *Homo sapiens*, and there is no way of knowing to what extent members of *Homo erectus* could be Machiavellian in their thoughts.

THE LAST MILLION YEARS

By one million years ago all the australopithecines were extinct and so was *Homo habilis*. The only hominid species known to exist was *Homo erectus*, which, wherever it originated, seems by then to have been present in Africa, Asia and probably southern Europe. In Zhoukoudian there is evidence that, half a million years ago, *Homo erectus* lived in caves for long periods and, judging by the large accumulations of ash, may have used fire.

The history of *Homo erectus* in Europe is almost unknown. At sites near Nice, in the south of France, and at Isernia, in central Italy, there are collections of stone tools, thought to be nearly a million years old, which are attributed to *Homo erectus*; but (apart from the ancient fossil in Georgia mentioned above) there are no fossils. And when further fossils do appear – half a million years ago at the earliest – they have characteristics intermediate between those of *Homo erectus* and of *Homo sapiens*, with larger braincases, smaller browridges, flatter faces, and a generally more human look. How these fossils should be interpreted is not at all clear. They differ markedly among themselves; they come from widely separated sites in Europe,[29] and fossils with similar characteristics have been found at numerous sites in Africa and at two sites in China.[30] Because all these fossils have characteristics intermediate between those of *Homo erectus* and of *Homo sapiens*, they have collectively been called '*archaic Homo sapiens*'. There is, though, no justification for putting them in the same species as ourselves, and the current tendency is to refer to them as *Homo heidelbergensis* – Mauer, near Heidelberg, being one of the places where they have been found. (The name does not imply that the species originated in Europe.)

The Neanderthals

About 130,000 years ago, the picture in Europe changed dramatically with the appearance of the Neanderthals. The fossilized bones in the Neander Valley, near Düsseldorf, that gave Neanderthals their name were discovered three years before Darwin published *The Origin of Species*. They were discovered by Johann Fuhlrott, a schoolteacher in the neighbouring town of Elberfeld, who was excavating a cave in the valley. He consulted Hermann Schaafhausen, the Professor of

Anatomy at Bonn, and together, they presented the finds to a meeting of the Lower Rhine Medical and Natural History Society. In their presentation, they drew attention particularly to the shape of the skull (Figure 4.2c), which, with its sloping forehead and formidable eyebrow ridges, was, they said, unlike anything seen in even the most barbarous races. They noted that the left ulna had a poorly healed fracture, but also – and this is significant in view of what happened later – that there was no sign of degeneration due to rickets. They concluded that the remains were those of a representative of an early and primitive race, but later this view was hotly disputed. Professor Mayer, a colleague of Professor Schaafhausen at Bonn, argued that the fossilized bones were from the skeleton of a Mongolian cossack, probably a relic of Tchernitcheff's army that had crossed Germany to attack France in 1814. The rather curved thigh bones indicated a life-time in the saddle; the fractured ulna was the result of childhood rickets; and the eyebrow ridges were caused by the vigorous action of the muscles of the forehead. A view that the peculiarities of the skull were the result of disease was, surpris-ingly, supported by Virchov, the founder of modern pathology; but the notion that Neanderthal man could be dismissed as 'nothing but a rickety, bow-legged, frowning cossack'* became untenable when further skeletons whose skulls showed identical peculiarities were found, first at Spy in southern Belgium, and then further and further afield in Europe and western Asia – from the Iberian peninsula in the west to Iraq and Uzbekistan in the east, and at sites as far north as southern England and northern Germany.

The low sloping brow, the massive eyebrow ridges, the large nose, the strong yet receding lower jaw, and the very robust limb bones made the Neanderthals seem to their discoverers both brutish and stupid; in fact, although the cranium is low it is also very long, and the size of the cranial cavity is, on average, slightly larger than our own – a fact that was discussed by Darwin in his *Descent of Man*. Whether the Neanderthals had any language is not known and has been the subject of some controversy – see Chapter 18. What we do know is that they lived as hunter-gatherers, used advanced stone tools, hunted with spears, and knew how to make fires. Their front teeth, which were very large, are peculiarly worn down and scratched, as if they used them to grip objects which were then cut with a stone tool. They are the first hominids thought to have buried their dead, and, from the number of skeletons with serious but healed fractures, it is clear that they must also have cared for their injured.

The Neanderthals are thought to have evolved (probably via *Homo heidelbergensis*) from a part of the *Homo erectus* population that entered south-ern Europe between two and one million years ago. Neanderthal features begin to appear in the fossil record about 230,000 years ago and were fully developed by about 130,000 years ago. For nearly another 100,000 years the Neanderthals

*The quotation is from a mocking account of Mayer's views by T. H. Huxley.

flourished, then they disappeared completely from the fossil record, leaving modern *Homo sapiens* the sole surviving hominid (but cf endnote 31).

Neither the reasons for their success nor the reasons for their extinction are fully understood. It is conventional to relate their stocky build (giving them a low surface-to-volume ratio) and large noses (better able to warm and humidify inspired air) to the cold dry conditions of ice-age Europe, but it would be wrong to envisage Neanderthals as always living in near-glacial conditions. The pattern of alternation between glacial and interglacial conditions over the last quarter of a million years is now very well understood, and though it is likely that the characteristic Neanderthal features appeared as the result of selection during the mainly glacial period between 180,000 and 130,000 years ago, the succeeding 100,000 years during which the Neanderthals mainly flourished began with 15,000 years of rather warm conditions, with the northern deciduous forests at their maximal extent, hippopotamuses, elephants and rhinoceroses in southern England, and the waters round Britain probably rather warmer than they are today. This warm period was followed by mainly temperate or cool conditions until about 75,000 years ago, conditions which probably favoured hominids by thinning the forests to provide more open landscapes with grazing animals which could be hunted or scavenged. The climate then continued to get cooler, becoming fully glacial about 30,000 years ago, by which time the Neanderthals were all extinct.

The extinction of the Neanderthals presents two major puzzles. Why, after more than a hundred millennia, should they become extinct, leaving only modern *Homo sapiens* to carry on the hominid line? And what was the ancestry of the survivors – sometimes known as Cro-Magnons after the village, in the Perigord region of south-west France, where, during excavations for a railway line in 1868, the fossilized remains of modern *Homo sapiens* were first discovered?*

The first question can be rephrased by asking: What was the nature of the competition between the previously very successful Neanderthals and the thinner skulled, less robust *Homo sapiens*? Was it simply a competition for limited resources, in which the less robust moderns proved to be more able? If so, in what way were they more able? Did they have more cunning or better memories? Could it be that the Neanderthals lacked language, and were therefore less able to co-operate in hunting or scavenging? Did the two kinds of hominid fight one another? If they did, how was it that the weaker and less robust species survived? Was that a matter of language and better strategy too? Were Neanderthals more susceptible to infectious diseases? Did Neanderthals and moderns ever interbreed? To none of these questions are there reliable answers. It is worth remembering, though, that the ultimate success of the moderns does not imply

---

*The remains consisted of the skeletons of an old man, a young man, and a woman whose skull bore signs of a severe wound.

that there was any dramatic or lurid genocide. Over the many millennia during which Neanderthals and moderns overlapped, a very marginal advantage could have led to the elimination of the disadvantaged species.

The origin of modern Homo sapiens: 'Candelabrum' or 'Noah's ark'

The second puzzle, the ancestry of modern *Homo sapiens*, has been the subject of vigorous controversy for the past two decades – a controversy all the more vigorous because the two main rival hypotheses have different implications for the date of origin of racial differences. Until the late 1970s it was generally believed that, leaving aside known migrations, modern humans in the different broad regions of the Old World had evolved from the separate populations of *Homo erectus* (or the successors of *Homo erectus*) that had become established in those regions (Figure 4.3). Thus modern Europeans would have evolved from Neanderthals (or possibly from some form intermediate between *Homo erectus* and the Neanderthals), modern Chinese and Indonesians from *Homo erectus* in Asia, modern Africans from *Homo erectus* in Africa. Supporters of this *regional continuity hypothesis* – also known as the multiregional-evolution hypothesis,[32] or sometimes as the candelabrum hypothesis, because of a fancied resemblance of the family tree to a candelabrum – point out that there are consistent differences between *Homo erectus* fossils in China, Java and Africa, and they claim that

FIGURE 4.3 The evolution of the genus *Homo*: a version of the 'candelabrum' hypothesis.

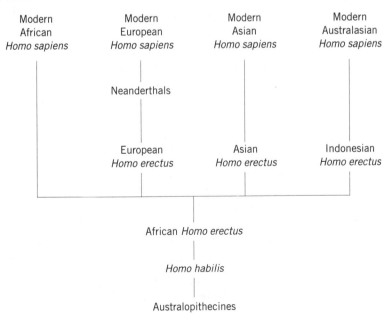

parallel differences are found in the corresponding modern populations. The difficulty is in knowing how much weight to give to resemblances between regional *Homo erectus* fossils and modern humans in the same regions.[33] A strong argument *against* the candelabrum hypothesis is that, since the different races of modern humans resemble one another much more than they resemble *Homo erectus*, the hypothesis implies that, for something like half a million years, evolution has occurred along parallel paths in many different and distant parts of the world. That is plausible only if there has been a continued interchange of genes between the distant populations, to an extent that seems unlikely.

The alternative hypothesis (Figure 4.4) is that *Homo sapiens* evolved from *Homo erectus* (? via *Homo heidelbergensis*) only in Africa, and then spread into Asia, Europe, and eventually the rest of the world, completely replacing the descendants of local *Homo erectus*, including, in Europe and eastern Asia, the Neanderthals.

This hypothesis, known as the *single recent origin hypothesis* or, less prosaically, as the 'out of Africa hypothesis'* – or sometimes as the 'Noah's ark hypothesis', because it supposes that an earlier population has been entirely replaced by descendants of a small part of that population – was proposed partly on fossil evidence but more convincingly on the basis of studies of mitochondrial DNA. We saw at the end of the last chapter that nearly all animal cells contain organelles called mitochondria which are thought to have had their evolutionary origin as symbiotic bacteria, and which still retain a small amount of DNA and the ability to synthesize some of their own proteins. The amount of DNA contained in each of our mitochondria is minute compared with the amount in the nucleus of each of our cells, but for studying evolutionary pathways mitochondrial DNA has an important advantage.

Unlike nuclear DNA, mitochondrial DNA is inherited solely through the female line. When a sperm and an egg fuse, the nucleus from the sperm enters the egg, so the sperm and the egg contribute equally to the *nuclear* DNA of the off-spring; but the tail of the sperm, which contains at its base the few mitochondria the sperm possesses, does not enter the egg, so the sperm contributes nothing to the *mitochondrial* DNA of the offspring. All the mitochondrial DNA in the fertilized egg therefore come from the mother; there is no shuffling of genes each generation. It is this that makes the investigation of mitochondrial DNA such a powerful tool.

Because there is no shuffling, children of a common mother will contain identical mitochondrial DNA, for they will have had very little time in which to accumulate mutations. Grandchildren of a common maternal grandmother will have had twice as much time; great-grandchildren of a common maternal

---

*Sometimes as the 'Out of Africa 2' hypothesis, to distinguish the migration of *Homo sapiens* from the hypothetical earlier migration of *Homo erectus* from Africa to Asia and Europe.

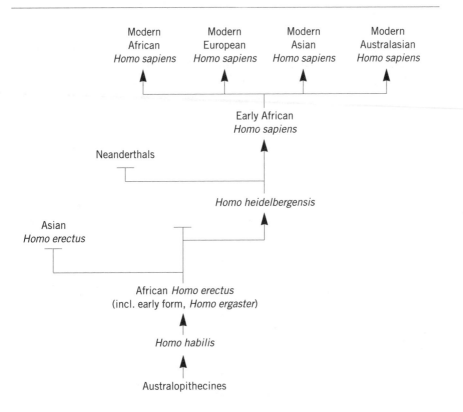

FIGURE 4.4 The evolution of the genus *Homo*: a version of the 'Out of Africa' hypothesis.

great-grandmother three times as much, and so on down the generations. Assuming that mutations accumulate at an average rate that is fairly constant, the number of nucleotides that differ when the mitochondrial DNA of two individuals is compared will therefore be proportional to the time that has elapsed since their last common ancestor *in the strictly female line.*

In the late 1970s and the 1980s Wesley Brown, Allan Wilson and their colleagues[34] looked at human mitochondrial DNA from a large number of people of various races and geographical regions, and they compared the divergence of the human mitochondrial DNA with the divergence between human and chimpanzee mitochondrial DNA. It turned out that the divergence in humans was less than 1/25 of the divergence between humans and chimpanzees. Assuming that the rate of accumulation of mutations is the same in human and in chimpanzee mitochondrial DNA, this implies that the time that has elapsed since present-day humans last had a common ancestor in the female line is less than 1/25 of the time that has elapsed since the line leading to humans separated from the line leading to chimpanzees. But, as we saw in the last chapter, work on amino-acid sequences in proteins has shown that that separation must have occurred something like five to seven million years ago. It seemed to follow that the most recent common ancestor in the female line of all the humans studied must have lived as recently as 200,000–280,000 years ago. (Later work suggested the lower limit could be as little as 140,000 years ago.) This was a startling conclusion. Not only was it incompatible with the candelabrum hypothesis but it also implied that all racial differences had arisen relatively recently.

What about these racial differences? Using the similarity of the nucleotide sequences in mitochondrial DNA as a measure of relatedness, Wilson and his colleagues attempted to construct the simplest family tree that would account for the 398 mutations that they detected in 133 types of human mitochondrial DNA. They noted that there was greater divergence among mitochondrial DNA samples from sub-Saharan Africans than among samples from anywhere else, and they explained this by supposing that the human population of the world outside Africa was derived from a fraction of the African population that had migrated into Asia. They therefore put the root of the tree in Africa.

That there was respectable scientific evidence that we all, irrespective of race, share a common great great great . . . great grandmother who lived in Africa, something like 200,000 years ago, was too exciting for the popular press to ignore. Inevitably, she was called Eve, but she differed from the biblical Eve in important ways. In the first place, unlike the biblical Eve, she was not the first woman or the only woman; she was merely our most recent common ancestor in the strictly female line – the strictly female lines from all her female contemporaries having died out. Secondly, although all our mitochondrial DNA (somewhat modified by accumulated mutations) derives from her, this is not true of our nuclear DNA; and it is our nuclear DNA that accounts for over 99.9 per cent of our genes. And thirdly, because Eve is defined as *the most recent* common ancestor in the strictly female line, her identity will change with time as female lines die out.

A problem with the mitochondrial evidence is that there is no sure way of calculating the 'most parsimonious' tree capable of fitting the data. To an extent that was appreciated only later, the computer programs originally used for this purpose gave ambiguous answers, and in particular could not definitely identify the root of the tree. Many equally plausible trees have since been published, and in some of these the root is in Asia rather than Africa. Another difficulty is caused by uncertainties about the mutation rate. In comparing human and chimpanzee mitochondrial DNA, Wilson and his colleagues assumed that the mutation rate was the same in both. But because the pattern of nucleotide substitution differs in the two species, it is conceivable that the rates of substitution might differ too.

In the last few years most of these problems have been at least partly solved but, in any case, they don't affect the important conclusions drawn from some very recent studies by Svante Pääbo and his colleagues, at Münich. These provide convincing evidence that, contrary to the candelabrum hypothesis, modern

European *Homo sapiens* are not descended from Neanderthals.[35] The Münich group succeeded in extracting a minute amount of a segment of the mitochondrial DNA from a few grams of the upper arm bone of the original Neanderthal skeleton, and compared the sequence of nucleotides in it with the sequences in the corresponding segment of mitochondrial DNA in modern humans and in chimpanzees. The results suggest that the line leading to Neanderthals and the line leading to modern humans must have separated something like 600,000 years ago. As the earliest fossils with Neanderthal-like features do not appear in the fossil record till about 230,000 years ago, Neanderthals could not have been our ancestors. What is more, the modern sequences closest to the Neanderthal sequence came from Africa rather than Europe – the opposite of what would be expected if modern Europeans were descended from Neanderthals.

There is also fossil evidence that favours the out-of-Africa hypothesis. This is of two kinds. First, it is only in Africa that there is a clear sequence of transitional forms linking *Homo erectus* and modern *Homo sapiens*.[36] Unfortunately, there are still doubts about the interpretation and dates of some of the significant African fossils and the old anthropological maxim 'absence of evidence is not evidence of absence' is a reminder that caution is needed in assessing the significance of the dearth of transitional forms in other parts of the world.[37] The second line of fossil evidence is that the oldest early-modern forms are found in Africa and the Middle East, with those from the Border Cave and the Klasies River Mouth in South Africa, those from Omo-Kibish in Ethiopia, and those from caves in northern Israel, all estimated to be around 100,000 years old.[38] The oldest remains of modern *Homo sapiens* found in Europe or in the rest of Asia are less than half this age.

The Israeli fossils are particularly intriguing. Remains of what look like Neanderthals have been found in caves at sites on Mount Carmel, and at Amud, near the Sea of Galilee. Remains of what look like early modern *Homo sapiens* have been found in caves at a separate site on Mount Carmel, and at Qafzeh, south of Nazareth.[39] Until the 1980s dating of these fossils was uncertain since it depended mainly on radio-carbon dating, which is not reliable beyond about 40,000 years, but this situation changed dramatically when two new dating methods were introduced, thermo-luminescence and electron spin resonance. Both depend on the remarkable fact that when crystalline substances such as flint or pottery or tooth enamel are exposed to very weak radiation – from trace amounts of radioactive substances in the sample or its vicinity, or from cosmic rays – electrons become trapped permanently at faults in the crystalline lattice. The longer the flint or potsherd or tooth has been exposed to radiation, the more electrons will have been trapped. By measuring the number trapped, and estimating the rate at which they have been accumulating, you can calculate how long ago the process started.

When these techniques were used to date the Israeli fossils, most of the

Neanderthal fossils appeared to be roughly 50,000–70,000 years old, much as expected, though those from one of the Mount Carmel sites seemed to be about 100,000–120,000 years old (with less secure dating). But the startling finding was that the early modern fossils from Mount Carmel and from near the Sea of Galilee turned out to be at least 80,000 and possibly as much as 120,000 years old, that is, much older than most of the Neanderthals.[40] As the period from 120,000 years ago to 80,000 years ago is known to have been a warm phase, and to have been followed by cooling, it looks as though early modern humans, perhaps moving north out of Africa, occupied the Levant during the warm phase, and then disappeared during the succeeding cold phase, being replaced by Neanderthals, perhaps moving east from Europe.

THE 'EAST SIDE STORY'[41]

In the penultimate chapter of Jane Austen's *Pride and Prejudice*, Elizabeth asks Darcy to account for his having ever fallen in love with her. 'How could you begin?' she says, 'I can comprehend your going on charmingly, when once you had made a beginning; but what could set you off in the first place?' I want to complete this chapter by trying to answer a similar question about hominid evolution.

We have seen how a rough sketch of our own evolutionary tree can be made, starting with the hypothetical last ancestor shared with the chimpanzees about six million years ago, progressing through various australopithecines to *Homo habilis* by two million years ago, to *Homo erectus* a little later, and then, starting about half a million years ago, through intermediate forms to modern *Homo sapiens* about 100,000 years ago. The question I want to consider is: What set our ancestors off in the first place? What caused the split between the hominid line and the line that leads to chimpanzees?

A plausible and attractive answer to this question has been proposed by Yves Coppens.[42] He points out that, although something like 2000 hominid fossils and more than 200,000 fossils of other vertebrates have now been found in Africa east of the East African Rift System, this region contains no signs of chimpanzees or gorillas or their ancestors for the whole of the period in which the australopithecines were evolving. As the molecular evidence makes it abundantly clear that chimpanzees are our closest relatives, and that the split between our two lines was not more than something like five to seven million years ago, this absence of ape ancestors needs explaining. Either the earliest hominids must have evolved from the common ancestor somewhere else and then migrated to where their fossils are found, or they evolved in East Africa but, for some reason, the line leading from the common ancestor to the modern apes did not progress in this region.

Coppens prefers the second explanation. He suggests that the sinking of the rift valley resulting from tectonic activity about eight million years ago, and the simultaneous formation of the line of peaks along its western rim, would have split the population of common ancestors into two, at the same time as it altered climatic conditions – the west remaining humid, the east becoming progressively more desiccated to give, first, open woodland and later savannah – i.e. grassy plains with scattered trees in a tropical or subtropical environment. In these circumstances, the western descendants of the common ancestors would continue evolving to suit a humid forested environment, giving rise eventually to chimpanzees and gorillas, with cusped molars suited to a diet of soft fruit. The eastern descendants would evolve to suit a drier and more open environment, giving rise to the australopithecines, with molars adapted for grinding grain, nuts and hard fruits, and with a more upright habit better suited to foraging in sparsely wooded countryside. Eventually, and for reasons that are still debated, the australopithecines would produce descendants with larger brains who were capable of escaping from the confines of Africa, first, more than a million years ago, as *Homo erectus*, and much later – assuming that the 'out-of-Africa' story is right – as *Homo sapiens*. With our large brains and more powerful minds, we represent the latest chapter in the East Side story; but whether the frightening combination of capabilities and limitations displayed by those brains and those minds will make the latest chapter the final chapter, or will allow humans to survive and evolve further, only time will tell.

5　The Origin of Life

It is mere rubbish, thinking at present of the origin of life; one might as well think of the origin of matter.
—Charles Darwin, 1863[1]

It is a measure of Darwin's intellectual self discipline that he was prepared to spend a large fraction of his life thinking about the origin of species, where he had a good theory and plenty of facts, but he was not prepared to speculate about the origin of life, where he had neither. Such reluctance was justified at that time. But I want to show how recent discoveries have changed the position so radically that, though it is of course not possible to give a historical account of the way in which life started, it is possible to put forward a series of more or less plausible steps that could have accounted for that start – and reasonable to suggest that a series of steps something like these probably did. If you are not a biologist you may find this claim surprising; but it is no less surprising that so profound an advance in knowledge has occurred with so little awareness by the general public. One result of this advance is that the origin of life, though it still presents major difficulties, is now less puzzling than the origin of consciousness within living organisms.[2]

WHAT IS LIFE?

Living and partly living
—T. S. Eliot

Before we explore the series of steps thought to have led to the origin of living things, we need to be clear what life is. There are, of course, many features that are common to most living things, but here we are concerned only with features that are universal. Of these, the most striking is that, as a class (though not always individually) living things reproduce themselves, 'each after its kind'. Furthermore, this reproduction is autonomous: the organism itself provides all the information required for reproduction – instructions without which the off-spring cannot be made – and in its enzymes and associated smaller molecules it also provides all the catalysts necessary for bringing about the chemical changes that are involved. (This provision of catalysts is missing in the reproduction of viruses, which provide detailed instructions for making new viruses, but which rely on the catalytic machinery in the cells of the host on whom they are parasitic for carrying out those instructions.)

To these two general properties, we can add a third. So far as we can tell, the behaviour of all organisms living now – their entire behaviour, not just their reproductive behaviour – tends to be of a kind that will promote their survival and reproduction. To the outsider watching the behaviour, the organism appears to have a purpose. This feature is, of course, just what we should expect in organisms whose behaviour is largely the product of very long periods of natural selection.

OUR COMMON ORIGIN

The universal distribution of these rather general features of living organisms does not imply that all living organisms share a common origin. But when we look at the detailed structure and internal workings of living organisms, and see that there are many other features that are found universally, a common origin does seem likely.

All known forms of life store all their genetic information – the instructions for making the offspring's proteins (and therefore, indirectly, of almost every-thing else in the offspring) – as long chains of nucleotides in DNA. During protein synthesis, the synthesizing machinery moves along copies of the nucleotide chain interpreting each group of three nucleotides (each codon) as an instruction to add a particular amino-acid to the growing protein. But though the geometric and chemical characteristics of nucleotides make them ideally – and perhaps uniquely – suitable for storing genetic information, neither these characteristics nor any characteristics of the individual amino-acids explain why particular codons should correspond to particular amino-acids. The code defin-ing the correspondence seems to be almost entirely arbitrary. Yet despite this arbitrariness, in all living organisms in which the matter has been examined, the code used is virtually the same; with few and slight exceptions, a given codon always corresponds to the same amino-acid.

Further overwhelming evidence for a common origin comes from compar-isons of the detailed sequences of amino-acids in the enzymes of all organisms in which those sequences have been looked at. Of course, enzymes catalysing similar reactions in different organisms will be expected often to have resem-blances in structure where the structure is related to the function, but the resemblances between the amino-acid sequences in such proteins are far too extensive to be explained in this way.

That all known forms of life have a common origin does not imply that life arose only once; it is possible that living things of independent origin once existed but died out, through competition or for other reasons. But the common origin of all known forms of life does simplify matters when we come to the second and third parts of the story – the origin of the building blocks and the origin of the self-reproducing machinery.

EXORCIZING THE 'VITAL FORCE'

Until 1828 – the year that Darwin came up to Cambridge as an undergraduate – it was believed that living things not only behaved differently from non-living things but were also composed of substances that could not be made from those found in non-living things. Some 'vital force' was assumed to be necessary for their synthesis. But in that year, in Berlin, Friedrich Wöhler succeeded in synthesizing urea (a substance found in blood and urine) from ammonium cyanate, thus breaching the supposed barrier between the 'organic' and the 'inorganic'. The breach was less convincing than it might have been, because the ammonium cyanate was itself derived indirectly from animal material, but the synthesis of urea was soon followed by the synthesis of other organic substances from starting materials whose inorganic provenance was above suspicion.

The notion that chemical processes in living organisms might resemble those that occur in the inanimate world was much older. In 1780, nine years before the French Revolution, the chemist Lavoisier and the mathematician Laplace, suspecting (correctly) that respiration – the intake of oxygen and the expulsion of carbon dioxide – was the result of combustion going on within the body, measured the amount of heat and the amount of carbon dioxide produced by a guinea pig in ten hours. They found that the amount of heat was roughly the same as that produced by burning just enough carbon to produce the same quantity of carbon dioxide as the guinea pig had produced.

The next major step in exorcizing the notion of vital force, was the discovery of enzymes. Pasteur had proved that various kinds of fermentation, including the fermentation of sugar to alcohol, were the result of the action of yeasts or other micro-organisms; if these were excluded no fermentation occurred. Yet forty years after Pasteur's demonstration, Eduard Buchner showed that when yeast was ground up with sand and filtered, a clear juice could be obtained which contained no yeast cells, or visible fragments, or other microorganisms, but which was, nevertheless, capable of converting sugar to alcohol. It was clear that whatever in the yeast was responsible for fermentation had been extracted. Later, it turned out that the active material that had been extracted was not a single substance but a collection of protein catalysts (and other substances with much smaller molecules) whose combined effect was to catalyse a series of reactions that together convert sugar to alcohol and carbon dioxide. Protein catalysts are now known to exist in all living cells, but because their existence was postulated first in yeasts, they are called 'enzymes', from the Greek words meaning 'in yeast'.

The last flicker of the idea of vital force was extinguished in the early 1940s, when biochemists in the USA – in particular Fritz Lipmann and Herman Kalckar – discovered the way in which living organisms trap the energy released by the metabolism of foodstuffs and make it available later for various energy-requiring processes. The biochemical machinery involved was unfamiliar, but the thermodynamic laws that described its behaviour were no different from those that described events in the non-living world. Here was no mystery.

THE ORIGIN OF THE BUILDING BLOCKS

The discoveries that organic substances can be synthesized outside living organisms, that sequences of reactions that occur in living organisms can also occur in the test tube, and that living organisms follow normal thermodynamic laws, are important because they clear the ground of a series of false beliefs. The next step is to ask: What is the historical origin of the relatively small molecules that are the building blocks from which living organisms are constructed – the amino-acids, nucleotides, lipids, organic phosphates and so on? A complete answer to this question would state, for each of the building-block molecules, the starting materials from which it was first synthesized, the reactions involved in that synthesis, and the source of energy that enabled the synthesis to occur – for making complex molecules from simpler ones generally needs energy. We are so far from being able to give a complete answer, that it is not possible to make firm statements of this kind even for a single molecule. In the last four decades, though, it has become possible to make plausible suggestions. Increasing knowledge of the conditions during the earliest period of the earth's history, and laboratory experiments in which such conditions have been simulated, have allowed those who work in this field to suggest possible pathways of *prebiotic synthesis* – the synthesis of molecules characteristic of living things before there were any living things.

It is now generally agreed that the earth is about four and a half billion years old, but the time at which life began is uncertain.[3] The earliest undisputed fossil remains – a collection of filamentous microbial fossils near the town of Marble Bar in Western Australia, were found sandwiched between rock strata that can be dated to about 3.5 billion years ago.[4] Life in anything like its existing forms is unlikely to have started earlier than about 3.8 billion years ago, because it is thought that before that time heat generated by the impact of large asteroids would have destroyed any emerging life.[5] There is therefore a period of about 300 million years during which life as we know it probably started, and it is the conditions during that period that we need to consider.

Unfortunately, the composition of the earth's early atmosphere is very uncertain.[6] Our present atmosphere contains 21 per cent oxygen, but that high concentration is almost certainly the result of photosynthesis, first by blue-green algae and later by green plants. It is now thought that the earth was formed, not directly by condensation from a mass of hot gas, but by the accretion of chunks of solid material,[7] and that the atmosphere, at the time we are interested in, was mainly the result of the accumulation of gases escaping from inside the accreted mass. The trouble is that the probable nature of those gases depends crucially on the time at which the 'outgassing' occurred.

The accepted view is that if it occurred while there was still plenty of metallic iron near the surface, the strongly reducing effect of the iron would have ensured that in the atmosphere the elements carbon, nitrogen and sulphur existed mainly as methane (CH_4), ammonia (NH_3) and hydrogen sulphide (H_2S).* If it occurred after most of the iron had sunk to the earth's core, these three elements are more likely to have existed in the atmosphere as carbon dioxide, free nitrogen and sulphur dioxide. There is also a possibility that methane may have been formed by the action of heat and pressure on hydrocarbons in the chunks of the solid material whose accretion formed the earth.[8] The general view now is that at the relevant time conditions at the earth's surface were probably only 'mildly reducing', and that the atmosphere is likely to have been rich in carbon dioxide, nitrogen and water vapour, with smaller amounts of carbon monoxide, hydrogen, hydrogen sulphide and other gases, and very uncertain amounts of methane and ammonia.

*   *   *

We are on surer ground in considering the sources of energy that were available for the synthesis of biologically relevant molecules. 'Your serpent of Egypt', says Lepidus in Shakespeare's *Antony and Cleopatra*, 'is bred now of your mud by the operation of your sun: so is your crocodile.' Accuracy in matters of natural history is not to be expected of a Roman consul,** but it is true that the sun provides much more energy than all the other sources put together. However, the importance of each source of energy depends not only on its size but also on the efficiency with which it is used. Ultraviolet light accounts for only a small fraction of the energy in sunlight but it is more likely to have contributed to prebiotic synthesis than visible light because the energy in each quantum of light is much greater. Lightning, shock waves from the impact of meteorites[9] and from thunder, and heat from volcanoes could all have contributed.

*In effect the iron removes oxygen from water, leaving the hydrogen atoms to combine with other atoms.

**The idea that life arose from the action of the sun on something moist can, in fact, be traced back to Anaximander in the fifth century BC.

If life started not on the surface of the earth or in shallow water, but, as has recently been suggested, in the deep submarine hydrothermal vents that are associated with volcanic activity at the growing edges of tectonic plates,[10] energy could have been obtained as heat from hot magma. Theoretically, energy could also have been obtained from the combination of hydrogen and sulphur to form hydrogen sulphide, or the combination of hydrogen and carbon dioxide to form methane and water, but the extraction of useful energy from these reactions requires machinery that can both catalyse the reactions and ensure that the energy released is not simply dissipated as heat. Such machinery exists in bacteria living now in hydrothermal vents and hot sulphurous springs, but these are sophisticated organisms with DNA, enzymes and lipid membranes. It seems unlikely that the energy-yielding reactions used by these bacteria could have been used for making the building blocks that were needed before life could start.

In 1953, S. L. Miller, a graduate student at Chicago who had ignored advice not to pursue what was regarded as an over-ambitious project, published the results of an experiment in which he attempted to simulate the probable conditions on the earth at the time life started.[11] A large flask was filled with a mixture of methane, hydrogen and ammonia. Boiling water in a small connecting flask added water vapour to the mixture and circulated the gases through the apparatus; a cooled tube leaving the large flask condensed some of the water vapour to liquid water. At the end of a week during which electric sparks were passed continuously through the gases, the condensed water was analysed. Astonishingly, it contained a score of simple compounds accounting for about 15 per cent of the carbon that had originally been present in the methane. These compounds included half a dozen amino-acids, several other acids of biological interest, and urea. Later experiments showed that the amino-acids were not formed directly by the electric discharge; the discharge formed hydrogen cyanide and simple aldehydes (including formaldehyde), and these reacted together in aqueous solution to form the amino-acids. (It may seem bizarre that hydrogen cyanide, which Agatha Christie has taught us to associate with sudden death, should be cast for a crucial role in the origin of life, but it is the simplest molecule containing carbon, nitrogen and hydrogen, and it is also extremely reactive. It is deadly to higher animals and plants because it combines avidly with the haem pigments that are crucial to respiration in most modern living cells. At the time we are considering, an atmosphere rich in oxygen, and the evolution of haem pigments, were both far in the future.)

The mixture of gases used in the early experiments of Miller is not much like the mixture now thought to have formed the early atmosphere of the earth. In later experiments he and others therefore tried other mixtures. Replacing most of the ammonia with nitrogen gave smaller yields but more diverse products. In one experiment of this kind, twelve of the twenty amino-acids found in proteins were produced, and if a little hydrogen sulphide was added to the mixture the sulphur-containing amino-acid methionine was also produced. Substituting carbon dioxide and carbon monoxide for methane has also been tried. With both gases, and with nitrogen but no ammonia, one or two amino-acids were obtained in rather small amounts, but the yields were improved tenfold by the addition of ammonia.

Experiments along the lines of Miller's but supplying energy in other forms have also been tried. Heat gave a good yield of a variety of amino-acids, in proportions that were very sensitive to the temperature.[12] Ultraviolet light was much less effective, apparently because it is less good at synthesizing hydrogen cyanide.[13] Experiments using shock waves (which produce large but very transient increases in both pressure and temperature) and bombardment with electrons, also had some success.[14] As a result of these various approaches, prebiotic syntheses have now been suggested for all but three of the twenty amino-acids found in proteins.

So far we have considered only the synthesis of amino-acids, the building blocks of proteins. What about the building blocks of the nucleic acids, the nucleotides? Present-day nucleic acids (see Figure 5.1 on p. 73) fall into two classes: DNA (deoxyribonucleic acids), whose role is to store genetic information, and RNA (ribonucleic acids), which serve as the templates used by the machinery that synthesizes proteins. Both DNA and RNA consist of long strings of nucleotides, each nucleotide consisting of a phosphate group attached to a sugar – deoxyribose in DNA, ribose in RNA – which is attached to a nitrogenous base. Phosphates in small quantities are distributed widely in rocks, so here we need consider only the source of the sugars and the nitrogenous bases.

We have already seen that formaldehyde was one of the intermediate products in Miller's experiments. It has been known for over a century that, in alkaline conditions and in the presence of a suitable catalyst, formaldehyde undergoes a series of complex reactions yielding a variety of different sugars. And calcium carbonate, among the most plentiful of all minerals, is an effective catalyst for these reactions. Admittedly, the fraction of ribose in the mixture of sugars produced is rather small, but alternative pathways for the synthesis of ribose have been suggested.[15] It is also possible that the first nucleic acids contained a linking molecule other than ribose in their nucleotides.[16]

That leaves us with the nitrogenous bases. DNA always contains four kinds of nitrogenous base: adenine, guanine, cytosine and thymine. RNA contains the first three, but thymine is replaced by the closely related compound, uracil. We therefore need prebiotic syntheses for five compounds: adenine, guanine, cytosine, thymine and uracil.

Adenine and guanine are both members of a class of compounds called purines, of which much the most familiar member is uric acid – the white material in bird droppings and the 'chalky' excrescences on gouty knuckles. The core of all purines is a pair of linked rings each consisting of a mixture of carbon and nitrogen atoms, and this structure is rather easily made by reactions between hydrogen cyanide molecules in concentrated solution, or even in dilute solution under the influence of ultraviolet light.[17] Other plausible synthetic routes start with formaldehyde and hydrogen cyanide, or ammonia and hydrogen cyanide, all substances likely to have been available prebiotically.[18]

The other three nitrogenous bases – cytosine, thymine and uracil – belong to a class known as pyrimidines. They are simpler than purines, with a core that consists of a single ring of mixed carbon and nitrogen atoms, and several prebiotic syntheses have been suggested, using methane, nitrogen and cyanides as starting materials.[19]

Suggesting a plausible way in which nitrogenous bases, ribose and phosphate might have joined together to form nucleotides presents more of a problem. Purines and ribose react together slowly when heated dry in the presence of certain inorganic salts, but no likely prebiotic method for linking pyrimidines and ribose has yet been described. Once the base and sugar are linked the addition of phosphate occurs readily in hot, dry conditions, so it may be that the first nucleotides were indeed formed 'out of your mud by the action of your sun'.

HEAVEN-SENT EVIDENCE

It seems then that there are plausible hypotheses which can account for the origins of most (though not all) of the building block molecules necessary for forming the two great classes of complex molecules characteristic of life as we know it – the proteins (responsible for much of the fabric, and for the extraordinarily specific catalysis that makes living possible), and the nucleic acids (responsible for storing and using the genetic information).

Plausible hypotheses, though, are not facts. It is one thing to know that various building-block molecules could conceivably have been synthesized in the absence of living organisms, and another to know that they have been. To ask for direct evidence that syntheses of this kind have indeed occurred might seem unrealistic; but it arrived, quite unexpectedly, on 28 September 1969, in the form of a shower of rocks that fell out of the sky onto an area of about five acres near the town of Murchison in south-eastern Australia. The Murchison meteorite weighed about 100 kilograms and fragmented as it passed through the atmosphere. Analysis of some of the pieces revealed an astonishing array of organic molecules.[20] Of the twenty amino-acids that are found in all proteins, eight were detected in the meteorite. In addition there were eleven less common biological amino-acids, and fifty-five other amino acids not occurring naturally on earth. Besides amino-acids, there were traces of urea, of several short-chain fatty acids (though no fatty acids with long chains such as are found in natural fats), of oxalic and lactic acids (and many related acids) and of adenine, guanine and uracil, i.e. three of the five nitrogenous bases found in nucleic acids. The amino-acids even included gamma-aminobutyric acid (GABA for short), which, as we shall see, has an important role in the human brain (p. 132).

It was, of course, necessary to exclude the possibility that the presence in the meteorite of biological amino-acids was the result of contamination from living organisms after the meteorite had reached the earth's atmosphere. Fortunately, this could easily be done. Many amino-acids are asymmetric molecules and, like all asymmetric molecules, exist in two forms, each the mirror image of the other. In solution, the two forms (known as *dextro* and *laevo*) can be distinguished from each other by their opposite effects on a beam of polarized light. With one exception (which lacks asymmetry) all amino-acids in proteins are of the *laevo* form, whereas the same amino-acids made in the laboratory (without the use of enzymes) always consist of mixtures of the two forms. Samples of the amino-acids from the meteorite invariably included both forms, showing that they are very unlikely to have come from terrestrial organisms that had contaminated the fragments.

LIFE FROM OUTER SPACE?

If molecules of biological importance have been brought to the earth by objects from outer space, can life itself have arrived on the earth in this fashion? This was first proposed by the great Swedish physical chemist Svante Arrhenius, who made major contributions to our understanding of electrolysis, and whose equation is still used to predict the effect of temperature changes on the rate of chemical reactions. In 1907, long before the discovery of biologically interesting substances in meteorites, Arrhenius suggested that life is constantly emitted from all inhabited worlds in the form of spores, which travel through space. Most of these spores, he thought, are ultimately destroyed in blazing stars, but a few come to rest on other worlds which have reached the stage of being habitable.[21]

Theories of this kind have seemed unattractive to most scientists, partly because they have lacked convincing evidence, and partly because, rather than solving the problem of the origin of life, they make it more difficult: the origin has to be explained in circumstances even more obscure than those that existed in the early history of the earth. The recent and much publicized claim by David McKay, of the Johnson Space Center in Houston, and his colleagues, that a meteorite found in the Antarctic twelve years ago, and believed to have arrived from Mars about 13,000 years ago, contains fossil evidence suggesting that a primitive form of life once existed on Mars now seems very doubtful.[22] There is evidence that amino-acids and other organic material found in the meteorite are of terrestrial origin.

ENTER SELF-REPLICATING MOLECULES

Natural selection could not begin to operate until there were self-reproducing entities in competition. The step from building-blocks to the formation of such entities is therefore a crucial step in the origin of life. It is also the most difficult to explain.

Because all living organisms seem to have had a common origin, an obvious approach to the problem is to ask how self-reproduction is achieved in those living now. The answer to this question lands us in an immediate and serious difficulty. In all known living organisms reproduction depends on the ability of very long molecules of nucleic acids to make almost exact replicas of themselves, and on the ability of the daughter nucleic acids to initiate and control the synthesis of proteins. The replication of the nucleic acids can be thought of as the preparation of a copy of the instructions necessary for making a new organism, and the synthesis of proteins as the first step in the carrying out of those instructions. The difficulty is that the replication of the nucleic acids is a complicated process that involves a large number of enzymes, and those enzymes, being proteins, can only have been made with the help of instructions provided by a nucleic acid – the classic chicken and egg problem.

There seem to be three possible solutions. The first, and much the most attractive, is that the earliest nucleic acid molecules were formed without the help of proteins and replicated without the help of proteins.[23] The second is that early proteins replicated themselves (or each other),[24] and that the use of nucleic acids to store information and ultimately to control protein synthesis arose only later. The third possibility is that the first self-replicating system was not related either to nucleic acids or to proteins,[25] and that the nucleic acid/protein system arose later.

*   *   *

When Watson and Crick, in 1953, published their famous paper proposing that DNA was a double helix, what made their work so exciting was not just that they had discovered the structure of the molecule that was known to store genetic information, but that that structure immediately suggested a possible mechanism for copying the information. The structure that Watson and Crick proposed resembled a twisted ladder in which the uprights consisted of alternate sugar and phosphate groups and the rungs each consisted of paired purine and pyrimidine bases – *either* adenine and thymine *or* guanine and cytosine – see Figure 5.1. Because the bases in each rung are held together only by very weak bonds, the two halves of the ladder (i.e. the two complementary strands of the double helix) can be separated. Because of the specificity of the base-pairing, each half contains all the information that was present in the whole; each half can therefore act as a template. By pulling the two halves of the ladder apart at one point (i.e. by partially separating the two strands of the DNA), allowing each dangling base in succession to pair with the complementary base of an appropriate nucleotide, and then joining that nucleotide to its predecessor in the growing strand, it should be possible to replicate the whole DNA molecule.

This turns out to be essentially what living organisms do, with the help of appropriate enzymes. If both strands of the original DNA are used as templates, the end result is the production of two daughter molecules of double-stranded DNA (see Figure 5.2). In each of these molecules one strand will be from the parent molecule and the other a newly formed strand. This process occurs before every cell division, and makes it possible for both daughter cells to have a complete copy of the organism's genetic information. Alternatively, with different enzymes, and using nucleotides containing ribose instead of deoxyribose, one strand of the DNA can be used to form a complementary strand of RNA, which is then detached. The end result is the restoration of the original DNA and the production of a single strand of RNA complementary to one of the strands of the DNA. It is such single strands of RNA, known as *messenger* RNA, that are used as templates by the machinery that synthesizes proteins. (You can think of the double stranded DNA as the precious hardback library copy of the genetic code, whereas the single stranded messenger RNA is a paper-back working copy – containing the same information – designed for use in the factory that is actually making proteins.)

Joining nucleotides together by linking their sugar and phosphate groups requires energy, and in living organisms this is supplied by the use of 'activated' nucleotides, i.e. nucleotides with a chain of three phosphate groups where ordinary nucleotides have one. When these activated nucleotides are being linked, the splitting off of the last two phosphate groups provides the energy* for joining the remaining phosphate group to the sugar of the preceding nucleotide.

The question is whether this kind of replication – or something like it – could have occurred first in the absence of enzymes. In 1983 T. Inoue and Leslie Orgel reported the results of an astonishing experiment.[26] They took strands of RNA made up wholly from nucleotide molecules containing cytosine as the base and incubated them with a mixture of 'activated' nucleotides** containing each of the four bases normally found in RNA. *Although no enzymes were present*, new strands of RNA were formed alongside the original strands – each old and new together forming a double helix – and in the new strands all of the nucleotides

---

*This energy must, of course, come from somewhere. It comes from the ATP that was consumed in adding the two extra phosphate groups to the nucleotide to activate it.

**In these experiments the nucleotides were 'activated' by linking a methylated imidazole group to the phosphate group of the nucleotide. Although it is not likely that 2-methyl imidazole was abundant on the primitive earth, the conclusion that template-directed synthesis is possible without enzymes is unaffected.

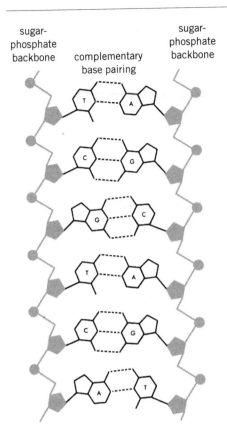

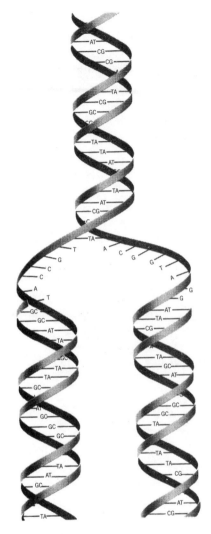

FIGURE 5.1 The structure of DNA.
A portion of the double helix is drawn
uncoiled to show the sugar-phosphate
backbone of each chain (sugar shown as
pentagons, phosphate as circles) and the
way that pairing between the bases links
the two chains. 'A' and 'G' label the
(purine) bases adenine and guanine;
'C' and 'T' label the (pyrimidine) bases
thymine and cytosine. Adenine always
pairs with thymine, and guanine always
pairs with cytosine. The sequence of bases
in each chain therefore contains all the
information present in its partner. (From
J. D. Watson, J. Tooze and D. T. Kurtz.)

FIGURE 5.2 Replication of DNA.
Each strand is used as a template to
form a complementary strand. The
result is two new molecules of DNA
each containing one of the original
strands and a complementary new
strand. (From J. D. Watson, J. Tooze
and D. T. Kurtz.)

contained guanine – the base complementary to cytosine. In other words, starting with a half-ladder and an assortment of loose half-rungs (each with a bit of
upright attached), they obtained a whole ladder with normal rungs. When the
original (template) strands contained a small proportion of other bases in
addition to cytosine, the new strands contained a small proportion of the
complementary bases in addition to guanine. No base was incorporated into the
new strands unless the complementary base was included in the template. In the
right conditions, new strands containing more than 30 nucleotides could be
synthesized.

What Inoue and Orgel observed was not quite self-replication. The new strand
of RNA was not identical to the original strand but complementary to it –
guanine interchanging with cytosine and adenine with uracil. If the process had
started again with the new strand it should have formed an exact copy of the
original strand, but this did not happen – partly because the original strand and
the newly formed strand failed to separate.

But even if it was not self-replication, what Inoue and Orgel observed was
remarkable enough. Without any enzymes, a strand of RNA acted as a template
for the formation of a new strand of RNA that contained all the information
stored in the sequence of bases in the original strand.[27] If this could happen, it
was not too fanciful to suppose that, in the right conditions, self-replication of
an RNA-like molecule could occur without the help of enzymic catalysis.

RIBOZYMES

The next step in the story was almost as surprising as the observations of Inoue
and Orgel. It turned out that looking for self-replication in the absence of the
kind of catalytic activity we associate with enzymes was not necessary. For in the
early 1980s, Thomas Cech and his colleagues at the University of Colerado, at
Boulder, and Sidney Altman and his colleagues at Yale, described experiments
showing that, at least for a limited range of reactions, and contrary to all expec-
tation, molecules of RNA can themselves act as powerful and selective catalysts.[28]
The long held assumption that such catalytic activity is a monopoly of proteins
is not quite true.

Proteins are very effective catalysts because they can have very complicated
shapes and surfaces, and because the side chains of the twenty different amino-
acids provide a great variety of reactive groups. The structure of RNAs offers
much less scope, with a much smaller choice of reactive groups, but pairing
between complementary bases situated in different parts of a single RNA chain
can ensure that the chain gets folded into complicated and reproducible shapes
– see Figure 5.3 on p. 77. In any event, it is now clear that some naturally occurring RNAs can cut other molecules of RNA at selected points, others can join the

ends of two pieces of RNA together, yet others can catalyse the elongation of an RNA chain. Some 'self-splicing' RNAs can even cut a section out of themselves and join the two remaining bits together. RNAs with catalytic activity are often referred to as *ribozymes*, to distinguish them from conventional enzymes.

The crucial question is whether any ribozyme can catalyse its own replication; and here there is a snag. To act as an enzyme, an RNA molecule must be folded into a well-defined, and probably complicated, three-dimensional structure. To act as a template, the RNA must have all its bases free to pair with the bases of individual nucleotides. To do one job it needs to be tied in knots; to do the other it must not be. One possible solution to this difficulty exploits the ability of RNA molecules to exist in both folded and unfolded forms: it is conceivable that, in conditions in which both forms existed, one form could act as template and the other as catalyst. Another solution would be to have a ribozyme made up of several subunits which could associate to form a structure with catalytic activity, but which, when not associated, could serve as templates. Experiments by Jack Szostak and his colleagues, in Harvard, have shown that such behaviour is feasible.[29]

SO WAS THERE AN 'RNA WORLD'?[30]

The discovery that RNA molecules could have significant and rather specific catalytic activity offered an attractive solution to the chicken and egg problem. Early living things would manage with ribozymes. Later, as methods evolved for linking amino-acids in a particular order, enzymes would prove superior and would replace ribozymes. To be in a position to say snap to this hypothesis we need answers to three questions: Where does DNA fit in? Could ribozymes really have been up to the job? And how did ribozymes first manage to control protein synthesis?

DNA has no catalytic activity – not surprisingly, since being double stranded it lacks unpaired bases and can't tie itself into fancy shapes – but it is much more stable than RNA. And because each strand contains the same information, damage to one strand can be repaired using the information in the other strand. It is therefore reasonable to assume that it evolved as a more secure way of storing precious information. The precious library copy is printed on acid-free paper that won't rot.

Whether ribozymes could have been up to the job, is a question that has recently been given much attention. Known naturally occurring ribozymes certainly are not, but we should not expect them to be. Proteins are so much more effective at selective catalysis, that current naturally occurring ribozymes must be regarded as living fossils. What we need to know is not what those ribozymes can do but what ribozymes as a class can do. This question is beginning to be answered by recently developed techniques, of fiendish ingenuity and power, for the *in vitro* selection and 'evolution' of synthetic RNA molecules.[31] And the answer seems to be that the versatility of the class of catalytically active RNA molecules is far greater than that of current naturally occurring ribozymes and may well have been adequate.[32]

How ribozymes came to control protein synthesis is a more difficult question. In organisms living now, the synthesis of each protein involves a battery of about fifty enzymes and many different kinds of RNA, but the essence of the operation is rather simple. The order of amino-acids in the protein is determined by the order of the codons in the messenger RNA. So all that needs to be done is to line up the amino-acids alongside the RNA strand, making sure that each codon has the right amino-acid adjacent to it, and then to go down the line linking each amino-acid to its immediate neighbours. The difficult part of the operation is getting the right amino-acids in the right places. In cells living now this is done by having many little RNA molecules each shaped like a four-leafed clover (Figure 5.3). The tip of one leaf has an affinity for a particular amino-acid; the tip of the opposite leaf bears a codon that is complementary to the codon on the messenger RNA that specifies that amino-acid. The result is that the four-leafed clovers – they are called *transfer RNAs* – attach themselves to specific amino-acids with one of their leaves and to the appropriate codons with the opposite leaves, and hey presto the job is done.

A mechanism of this kind could only have evolved in a stepwise fashion. The initial machinery may have involved fewer bases, it probably specified only broad classes of amino-acid, and it must have worked without using the catalytic activity of protein. But however simple the initial machinery was, once it started working, the advantage of possessing amino-acid chains *with specified rather than haphazard sequences* would have ensured that those entities in which the amino-acid sequence was determined even loosely by the RNA stood a better chance of reproducing themselves. Ultimately, as the result of continued natural section, RNA would gain complete control of protein synthesis.

HOLDING THINGS TOGETHER

The ability to initiate and control the synthesis of catalytic proteins would give a self-replicating molecule an advantage over other self-replicating molecules in the competition for raw materials and for sources of energy. When particular building blocks became scarce, with suitable catalysts they might be made from more plentiful precursors. If such reactions required energy they might, with suitable catalysts, be coupled to other reactions that produced energy. Energy from such reactions would also be available for the synthesis of nucleic acids and of the proteins themselves. There is one problem, though, that would be aggravated by a dependence on catalytic proteins, and that is the risk of dispersal.

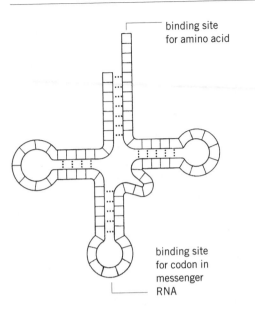

binding site
for amino acid

binding site
for codon in
messenger
RNA

FIGURE 5.3 The 'clover-leaf' structure of a molecule of transfer-RNA. Pairing between complementary bases in different parts of the chain gives the molecule a highly reproducible shape. The structure is, in reality, not planar but highly twisted. (From L. Stryer.)

The need to prevent dispersal of the molecules necessary for life must have existed at all stages in the evolution of life. That is why those who have considered these problems tend to speak of isolated pools filled with prebiotic soup, or evaporation to form a scum, or adsorption on the surfaces of clays. An attractive feature of the notion that life started in submarine hydrothermal vents is that, in such vents, sea water extracts minerals from fractured rock covering the hot magma and then proceeds upwards in a vast network of narrow channels so that any reactions can proceed without the products being widely dispersed.[33] Even Darwin, on a rare occasion when he forgot that thinking of the origin of life was 'mere rubbish', spoke of conceiving 'in some warm little pond, with all sorts of ammonia and phosphoric salts, light, heat, electricity &c., present, that a protein compound was chemically formed ready to undergo still more complex changes . . .'.[34] But the dependence of self-replicating molecules on associated proteins adds another dimension to the problem; for if the proteins are to be of use to the self-replicating molecule that controls their formation they must remain in the vicinity, and so, initially at least, must be the products of the reactions they catalyse. Clearly some kind of compartmentation is necessary.

At first sight, this presents little difficulty. All cells living now are enclosed by membranes whose basic structure is a double layer of closely packed molecules of fatty material. The individual molecules have one or more long, hydrophobic (i.e. water-repellent) hydrocarbon tails, and a hydrophilic (i.e. water-loving) head. It is characteristic of molecules of this kind that, in water, they associate spontaneously to form double-layered sheets (bilayers), in which the molecules are so arranged that the tails are in contact only with each other, and the heads are in contact with the water at the two surfaces of the bilayer. The capacity for self-assembly of such molecules is astonishing, and very easily demonstrated. If a blob of lecithin – the most abundant fatty material in cell membranes – is placed in a warm dilute salt solution and stirred gently, over the course of a few hours it will grow out into long sausage-like tubes made up of many concentric bilayers of lecithin separated by very thin layers of water. If the sausage-like tubes are agitated they break up into onion-like structures, and if the agitation is violent enough the onion-like structures break up to form many small hollow spheres, the wall of each consisting of a single lecithin bilayer – the basic structure of the cell membrane. Given that some kind of molecule similar to lecithin could have been formed prebiotically, or with the help of ribozymes or very primitive proteins, the formation of basic cell membranes does not present a major problem.

The snag is that basic cell membranes alone would have been useless. Surrounding a self-replicating molecule with a lipid bilayer would prevent dispersal of most of the products of any reaction in which it was involved; but it would also prevent access to essential raw materials. Lipid bilayers are permeable to water, to dissolved gases, and to lipid-soluble substances, but to nothing else. Real cell membranes, though their basic structure is a lipid bilayer, contain a host of pumps, channels, gates and pores which regulate transport across the membrane, and without which the cell could not survive; and all these pumps, channels, gates and pores are proteins. Again we seem to have a chicken and egg problem. Without a cell membrane, the self-replicating molecule would not be able to take advantage of the proteins whose synthesis it controlled; without proteins to facilitate the passage of necessary substances across the cell membrane, the membrane would kill the cell.

An ingenious solution to this problem has been proposed by Günter Blobel.[35] He suggests that what first developed was a kind of 'inside out' cell, a hollow sphere consisting of a single lipid bilayer, with the replicating machinery, the machinery for making proteins, and some metabolically useful proteins, all attached to the *outside*. Such an arrangement would serve to keep those three sets of macromolecules together, so that the unit on which natural selection acted was not the self-replicating molecule but the association of molecules. If part of the cell membrane invaginated, some of the useful machinery would be facing inwards with access only to an enclosed volume separated from the outside environment by two lipid bilayers. The evolution of proteins that enabled essential raw materials to pass through the lipid bilayers would then allow the inward-facing machinery to act. The final result would be a cell with a double membrane and with the cell machinery in its conventional place – inside the cell. The advantage of invagination is that useful metabolic products of the enzymes that were originally on the surface of the sphere could accumulate in the enclosed space, and undergo further reactions there, catalysed either by enzymes attached to the membrane or by soluble enzymes in the enclosed space. Subsequent loss of the outer membrane would give an arrangement similar to that in a conventional cell, except that the genetic machinery would not be segregated in a nucleus.

CONCLUSION

We have now considered the various kinds of evidence that are thought to throw some light on the origin of life. It is painfully clear that our knowledge of that origin is much less certain and much more fragmentary than our knowledge of the evolution of living forms that was to follow. This is inevitable. The earliest forms of life would have left no fossils, and the records represented by the nucleotide sequences in the simplest present day organisms serve for sorting out the relations between those organisms but not for going beyond that. That is why this chapter is full of statements of what could have happened rather than what is known to have happened, and with extrapolations from brief laboratory experiments under well-defined conditions to hypothetical events over vast periods under very ill-defined conditions.

But the aim of this chapter was not to give a doubtful historical account of what actually happened, but to answer a question that is both simpler and more profound. Could life have arisen as the result of a series of plausible events of the kind that we know do occur – events that are compatible with the known laws of physics – or do we need to invoke some other agency, some vital force or higher intelligence? The answer seems to be clear. However obscure the origin of life was in Darwin's day, or even in the 1920s when A. I. Oparin[36] in Russia and J. B. S. Haldane[37] in England first speculated seriously about the earliest stages, the work of the last seventy years has produced a remarkable clarification. There is very good evidence that life on earth started between three-and-a-half and four billion years ago, that all known forms of terrestrial life have a common origin, that the substances found in living organisms can, in principle at least, be synthesized from materials not derived from living organisms, and that the behaviour of such substances can be described by the same thermodynamic laws that describe the behaviour of substances not found in living organisms.

The experiments of Miller and those that followed him showed that many of the crucial building-block molecules could have been formed in the absence of living organisms, and the analysis of meteorites proved that such abiotic synthesis had indeed occurred within the solar system. The discovery of the structures of the nucleic acids DNA and RNA led to the elucidation of their role in storing, replicating and making use of genetic information, though the role of proteins in the replicating machinery created a chicken and egg problem. No demonstration of self-replication of a nucleic acid in the absence of proteins has yet been achieved, but the experiments of Orgel and his colleagues, and the dramatic discovery by Cech and Altman and their colleagues that some RNAs have catalytic activity, strongly suggest that it is achievable. With self-replicating information-storing molecules controlling the synthesis of powerfully catalytic proteins, with self-assembling lipids able to form membranes that could serve as barriers to diffusion, and with potential sources of energy both in light and in energy-yielding reactions between substances in the environment, the emergence of life could be left to natural selection. There is no need to postulate any other agency.

Let us, then, turn to the nerves and nervous systems without which minds could never have evolved.

II NERVES AND NERVOUS SYSTEMS

"*Mr Bennett, . . . you have no compassion on my poor nerves.*"
"*You mistake me my dear. I have a high respect for your nerves. They are my old friends. I have heard you mention them with consideration these twenty years at least.*"
—Jane Austen, *Pride and Prejudice*

Not all animals have nervous systems. If you prod an amoeba with a needle, it will move off in the opposite direction by forming a bulge at the side remote from the needle, and then withdrawing into it. It is clear that the amoeba is *irritable* – in the sense that it responds actively to an adequate stimulus – and that the response is appropriate. But as the same kind of response is obtained whichever part of the amoeba is prodded, and any part of the amoeba can form a bulge, it is also clear that the response does not depend on specialization within the amoeba into visible organs for receiving information – *receptors* – and for producing a response – *effectors*. And although the bulge is formed at the side of the cell remote from the prodding needle, there is no visible pathway to account for the conduction of the influence of the prod across the cell. There must be molecular mechanisms responsible for the sensitivity, the conduction, and the response, but there is no visible nervous system.

Unlike the amoeba, some unicellular organisms do have obvious receptor and effector organs – light-sensitive pigmented spots, contractile threads, and beating 'hairs', for example – but, as in the amoeba, the machinery that enables the information received by one part of the cell to influence the behaviour of other parts is molecular in scale and is not visible. In multicellular animals, too, the individual cells usually possess elaborate molecular machinery that enables them to detect and respond to changes in their environment; but, in addition, specialization has led to the development of separate tissues or organs with different functions. In particular, there are tissues or organs for receiving information (such as the retina in the eye), for carrying information about the body (the nerves), and for producing responses (the muscles and glands).

In the present context, though, the most significant difference between an amoeba and highly developed multicellular organisms such as ourselves is not a matter of structure but of behaviour. Our behaviour is enormously more complicated; and it is more complicated not in a random way but in a way that makes us better able to survive in a changing outside world. To achieve these qualities, an organism needs something in addition to receptors, nerves and effectors. It needs some kind of storing and sorting device to store the instructions for many patterns of behaviour, to set in motion the particular pattern that is appropriate, to modify that pattern continuously as the situation changes and, ideally, to learn – in other words to modify its responses to events appropriately in the light of past experience. This device is, of course, the central nervous system – that is, in vertebrates, the brain and spinal cord. By receiving information from the receptors via *sensory* nerves, and transmitting appropriate instructions to the effectors via *motor* nerves, the central nervous system tends to ensure that the organism's behaviour is suitable for the circumstances in which it finds itself. To an intelligent observer, such behaviour will appear 'purposive', though the purposive character of an organism's actions does not, of course, imply that the organism itself is conscious.

In the course of this book we shall be much concerned with brains, but brains consist mainly of nerve cells, and our understanding of the way brains work has been transformed by the elucidation of the way nerves work. I therefore want to begin by discussing nerves.

6   The Nature of Nerves

What nerves do, and how they do it, are questions that have puzzled people for nearly two thousand years. In the course of the 20th century, the problem was largely solved, and I shall describe the solution in the next chapter. First, I want to show how the slow development of ideas over a very long period made that solution possible; for, as has so often happened in the history of physiology, questions that were asked very early had to await the gradual development of new fields of science before they could be answered. By giving the historical background, I also hope to help readers who know little science to appreciate the extraordinary intellectual achievement that lies behind our present understanding.

ANIMAL SPIRITS

From the second century AD, right up to the second half of the 17th century, the standard answer to the question: 'How do nerves work?' was that nerves, both motor and sensory, produce their effects through a flow of *animal spirits*. What animal spirits were was not clear, but they were thought to be generated at the base of the brain from *vital spirits* brought in the blood from the heart. And the vital spirits were supposed to be generated in the heart by the action of a component of inhaled air on *natural spirits* that reached the heart in blood from the liver.

Why should a theory couched in terms so obscure, and unsupported by evidence, have been believed by intelligent and knowledgeable men for one and a half millennia? Part of the explanation lies in this very obscurity. Since, initially at any rate, animal spirits were thought of as weightless, intangible and invisible, they could be detected only by the effects their flow was assumed to produce. To say that nerves produced their effects by a flow of animal spirits was to say little more than that the effects were the result of the passage along the nerve of something otherwise undetectable; and for the whole of the period we are considering that was true. But this alone could hardly account for the persistence of a theory which, even if it could not be disproved, explained so little. The main reason for the theory's persistence lay in the acceptance, by both the Christian and the Muslim worlds, of the teachings of the man who gave it its final form. That man was Galen.

Galen[1] was born in 129 AD, in Pergamum, Asia Minor, to a distinguished architect father and a mother who, her son tells us, was so bad-tempered that she sometimes bit her servants, and screamed and quarrelled with her husband worse than Xanthippe with Socrates.[2] After twelve years of medical study, including periods in Smyrna, Corinth and Alexandria, he was appointed surgeon to the gladiators at Pergamum, but he also spent much time investigating the anatomy and physiology of animals. Then at the age of 32 he went to Rome, where eventually he was to be physician to three emperors. As well as practising medicine, he studied philosophy, lectured in the public theatre, performed experiments on animals before large audiences, and wrote incessantly. In his medical practice, he combined a respect for the teaching of Hippocrates with a detailed knowledge of anatomy, often based on his own dissection – knowledge that he was sometimes able to exploit to dramatic effect. One of his famous cures – of a Persian philosopher – depended on his realization that the recalcitrant numbness in certain fingers of the Persian's hand was not the result of anything wrong with the fingers but of an injury to nerve roots at the base of the neck.

Galen's animal studies were equally impressive, and often, like his diagnosis of the Persian, have an extraordinarily modern feel. He had a surprisingly accurate knowledge of the detailed anatomy of the eye (though he was quite wrong in his ideas about the functions of the lens and the retina). He studied, in great detail, the effects of damage to the spinal cord at different levels. He distinguished between motor and sensory nerves,[3] showing, for example, that of the three nerves to the tongue, one was concerned with movement and the other two mainly with touch and taste. While still surgeon to the gladiators in Pergamum, he had settled a controversy about whether the larynx was responsible for producing the voice, by cutting the nerve to a pig's larynx and observing that the pig could no longer squeal. He showed that, contrary to the then current belief, arteries contained blood rather than air;* and, most dramatically, his observations on the heart and vascular system so nearly led him to discover the circulation of the blood, that William Harvey, the actual discoverer fifteen centuries later, was puzzled that Galen had not got there first.

In all of these examples, Galen was basing his ideas on his own experiments or on his own observations. His notions about animal, vital and natural spirits could not, of course, be based on either. In fact, they represented an extension of the ideas of Erasistratus, an Alexandrian physician who had flourished about

---

*This rather surprising belief arose from observations on dead animals. Erasistratus had noticed blood spurting from cut arteries in wounded animals, but had supposed that this was because the escape of 'pneuma' from the damaged artery caused a vacuum which sucked blood from the veins.

four and half centuries earlier, ideas that can themselves be traced back to the sixth century BC. Unfortunately, neither Galen's contemporaries nor, until the 17th century, later physicians, distinguished between those of his views that were based on solid evidence and those that were not.

Paradoxically, it is an aspect of Galen's writings far removed from experimental evidence that was probably responsible for the persistence of his ideas. By the second century AD it was difficult for educated people to accept the ancient Greek myths as literally true. Galen was an enthusiastic participant in the rites associated with Aesculapius, the god of medicine, but in some of his writings he refers frequently to an all-wise all-just God or Creator (*demiourgos* in Greek).[4] This tendency to monotheism has sometimes been attributed to the influence of Judaism,[5] which at that time had numerous sympathizers in Rome, but Galen's writings contain few references to Judaism or Christianity,[6] and he is explicitly critical of the Mosaic view of God's omnipotence, believing 'that some things are naturally impossible, and that God does not attempt these at all but chooses from among the possible what is best to do'.[7] Galen did, however, take a more generous view of God's wisdom and justice. In a long, strange book *On the Usefulness of [the] Parts [of the Body]*, he attempts to explain a vast array of anatomical facts, from the length of eyelashes to the position and construction of the penis, by arguing that the observed arrangements are the best possible, and are, therefore, those to be expected from an all-wise all-just Creator. The arguments are reminiscent of Aristotle's dictum 'In Nature there is nothing superfluous', and of Plato's argument that the earth is a globe because *like* is fairer than *unlike*, and only a globe is alike everywhere. But though both Aristotle and Plato also talked about a God, as well as gods, Aristotle's God – the 'prime mover, himself unmoved' – seems unlikely to have thought about the length of eyelashes. And though Plato's God was very much concerned with 'the Good', Plato himself, far from seeing the body as an example of the perfection of the divine plan, saw it as a source of endless trouble through its need for food, its lusts and its diseases.[8] In contrast, Galen's *On the Usefulness of Parts* reads as a massively sustained 'argument from design',* and he refers to it as a 'sacred discourse'. The result of this emphasis on a single creative God was that not only this book but all Galen's medical writings were later found acceptable by those who did take over monotheism from the Jews, first the Christians and then the Muslims.

With the emergence of Christianity as the dominant religion in the west, and later of Islam as the dominant religion further south and east, Galen's views on medical topics became almost canonical over a very large area, and his works were translated into Aramaic, Arabic, Latin and Hebrew. But the tradition of experiment and of dissection had vanished. Throughout the medieval period,

---

*An argument for God's existence, that relies on the apparent presence of design or purpose in the universe.

learning, in nearly all fields, was a matter of the study of texts; and the man who had said that anyone who reads books on anatomy but omits inspecting with his own eyes is like a seaman who navigates out of a book,[9] was destined to have his own books on anatomy treated for many centuries as infallible.

It was not until the early 14th century that the tradition of dissection was revived; and it was not until the Renaissance that ideas about the workings of our bodies escaped from the Galenic straitjacket, and then only gradually. In 1490, Leonardo da Vinci had to defend himself against those who had criticized his anatomical drawings for being contrary to the accepted authorities. He boasted to Cardinal Louis of Aragon that he had made 'anatomies of more than thirty bodies', and there is no doubt that his drawings would have weakened the hold of Galen had they been published. He intended to publish them in a textbook of anatomy and physiology that he planned to write with a young collaborator at Padua, but the collaborator died, the book was never written, and the drawings did not become generally known for another two centuries. Meanwhile, in 1543, Andreas Vesalius, a Belgian who had studied at Louvain and Paris and was then working in Padua, published his *De Humani Corporis Fabrica* (*Of the Structure of the Human Body*). This magnificent book, exquisitely illustrated with engravings by a pupil of Titian, was the first modern textbook of anatomy and one of the most influential medical books ever written. Vesalius showed that Galen could be wrong in matters of anatomy; and though Vesalius too made mistakes he succeeded in establishing that, where there was disagreement, the court of appeal was the dissecting room rather than the library. On questions of function he was much more reticent, being content on the whole to repeat Galen's views. This reticence was deliberate. 'And so the learned anatomist trained in the dissection of dead bodies, and tainted with no heresy, will readily understand how little I should be consulting my own interests were I to lecture on the results to be obtained by the vivisection of the brain, which otherwise I would most willingly have done, and indeed at great length.'[10] Despite Vesalius's caution, and despite the book's ultimate gigantic success, the immediate and local effect of its publication was disastrous. It led to a controversy, with his medical colleagues at Padua, so bitter that he burnt his remaining manuscripts, resigned his post, and retired, not yet thirty, to a life of affluence and intellectual frustration as physician to Charles V.[11]

THE MECHANISTIC APPROACH

The year that saw the publication of Vesalius's *Fabrica* also saw the publication of Copernicus's *De Revolutionibus Orbium Coelestium* (*On the Revolutions of the Celestial Spheres*). Copernicus had died shortly after receiving his first printed copy, and at the time of his death he was still having difficulties with his theory.

In the last years of the 16th century, and early in the seventeenth, these difficulties were removed, first by Galileo's discovery of the laws of motion and inertia, and secondly by Kepler's discovery of the laws of planetary motion – particularly that orbits were elliptical rather than circular. When, in 1610, Galileo, using his own much improved version of the new Dutch invention of the telescope, discovered four satellites orbiting Jupiter, the heliocentric theory appeared to be illustrated in miniature. The more plausible the theory became, though, the more it seemed to threaten traditional theology. Galileo was warned by the Inquisition in 1615; in 1616 his views were pronounced heretical; and in 1633, at the age of 70, he was forced to recant. But coercion of Galileo could not prevent the collapse of the traditional view of the universe; nor could it conceal the lessons that ancient and venerable beliefs could be wrong and that the combination of observation, hypothesis and experiment was powerful.

The same lessons, in the field of physiology, were taught slightly later, but almost as dramatically, by the experiments of William Harvey* showing that blood circulated round the body driven by the pumping action of the heart. But though Harvey made notes for a treatise on animal locomotion, and he expressed some doubt about the flow along the nerves of 'motive and sensory spirits', the treatise never got written, and the notes do not reveal significant advances in understanding nerves.

In continental Europe, though, the success of mathematical and mechanistic approaches to the problem of understanding the movements of the heavenly bodies, and to more mundane problems such as the flight of projectiles, led a number of scholars to believe that similar approaches might be fruitful in the areas of physiology and medicine. Those who took this view became known as *iatrophysicists* (from *iatros*, Greek for physician). With time, the notions of the iatrophysical school became increasingly bizarre, but the views of two early iatrophysicists, Descartes and Borelli, are worth considering in some detail, because of the novelty of their approach and because the ideas of Descartes provide the first example of a phenomenon that has occurred repeatedly in the history of investigations of the nervous system – the enthusiastic adoption of some complex but understandable system: hydraulic machinery, telephone exchanges, digital computers – as a model for the working of the brain.

René Descartes is now remembered for his mathematics, and as the founder of modern philosophy; but he was also concerned with the workings of the body, and in particular of the nervous system. In 1629, when he was 33 and living in Holland, he told his friend Father Mersenne, in Paris, that he was writing a work in which 'I have resolved to explain all the phenomena of nature ...'.[12] By 1633 the work was nearly complete, but while he was engaged in the final revision he

*As a young man, Harvey had visited Padua in the first years of the 17th century, and he might have heard Galileo lecture.

heard that Galileo had been condemned and every copy of his book burnt. Descartes immediately gave up all idea of publication, explaining to Mersenne, 'I would not for all the world want a discourse to issue from me that contained the least word of which the Church would disapprove, and so I would prefer to suppress it than to have it appear in mangled form.' Four years later, Descartes published (anonymously) a brief summary of part of his suppressed book, including interesting observations on optics, in his *Discourse on the Method, and Essays*. Further topics are discussed in his last work, *The Passions of the Soul*, but publication of a substantial fraction of the book had to wait until after his death. This fraction appeared as two separate treatises, and it is the second *L'Homme de René Descartes*, usually referred to as the *Treatise on Man* that concerns us here.

In this treatise, Descartes gives a very readable account of his views on physiology. The striking thing more than three-and-a-half centuries later is that, though he was often wrong in his facts, sometimes grotesquely so,* his writings reveal a general aim that was to become the aim of modern physiology – to assume that the body is a machine and to attempt to explain its workings solely on the basis of laws that apply to non-living matter. Each animal, he believed, was nothing but a machine – a machine created originally by God – but in man this 'earthly machine' was governed by a rational soul. For Descartes, the soul was not something that conferred life on the body; the soul leaves when the body dies, but this departure was an effect of death not its cause. The difference between the body of a living man and of a dead man, he said, was the difference between a going watch and a broken watch.[13] Animals were alive but had no souls. What the soul conferred was the ability to have thoughts. And because 'we have only one simple thought about a given object at any one time',[14] it seemed appropriate that the soul should be symmetrically placed. Almost all parts of the brain appeared to be paired, so Descartes concluded that the 'principal seat of the soul' – the place 'where the soul directly exercises its functions' – was 'a certain very small gland situated in the middle of the brain's substance'. This pineal gland, to give it its modern name, is a small structure shaped like a pine cone that sits in the midline just below the front of the *corpus callosum* – the great bundle of nerve fibres that connects the two halves of the brain.

Although Descartes admired Harvey and accepted his views about the circulation of blood, he did not share Harvey's dismissive attitude to spirits. Descartes talked about 'animal spirits', which he described as 'a very fine wind, or rather a very pure and lively flame',[15] but he thought of them as a real fluid which behaved according to the physical laws that describe the behaviour of other fluids. He believed that it was the distension of muscles by animal spirits

---

*As when, having said that Harvey deserved the highest possible praise for discovering the circulation of the blood, he rejects the notion that the heart acts as a pump, and argues that the visible changes in the heart at each beat are caused by the heating and expansion of the blood it contains. (See Part 2 of *Description of the Human Body*.)

conveyed to them through hollow nerves that caused muscles to shorten. He assumed (wrongly) that the nerves contained valves, and he compared the nerves' actions with what could be observed in the fashionable grottos and fountains in the royal gardens, where 'the mere force with which the water is driven as it emerges from its source is sufficient to move various machines, and even to make them play certain instruments or utter certain words depending on the various arrangement of the pipes through which the water is conducted'. And he continues:

> Indeed, one may compare the nerves of the machine I am describing with the pipes in the works of these fountains, its muscles and tendons with the various devices and springs which serve to set them in motion, its animal spirits with the water which drives them, the heart with the source of the water, and the cavities of the brain with the storage tanks ... External objects, which ... stimulate its sense organs ... are like visitors who enter the grottos of these fountains and unwittingly cause the movements which take place before their eyes. For they cannot enter without stepping on certain tiles which are so arranged that if, for example, they approach a Diana who is bathing they will cause her to hide in the reeds ... And finally, when a *rational soul* is present in this machine it will have its principal seat in the brain, and reside there like the fountain-keeper who must be stationed at the tanks to which the fountain's pipes return if he wants to produce, or prevent, or change their movements in some way.

The unavoidable tiles that caused the Diana to hide were presumably connected to levers that actuated valves, and Descartes postulated an analogous mechanism to account for the working of sensory nerves. He imagined that the supposedly hollow nerves contained tiny fibres 'so arranged in each part of the machine that serves as the organ of some sense that they can easily be moved by the objects of that sense'. The effect of the movement was, he claimed, the simultaneous opening of pores in the internal surface of the brain – 'just as when you pull one end of a string you can cause a bell hanging at the other end to ring at the same time'. The opening of the pores was supposed to allow animal spirits to flow into appropriate nerves and so to activate appropriate muscles. To illustrate this mechanism, Descartes showed a picture of a kneeling, chubby boy who moves his foot away from a fire as the result of a chain of automatic events starting with the agitation of the skin of the boy's foot by the rapidly moving particles of the fire.[16]

The chubby boy, being provided with a 'rational soul', would not only have removed his foot automatically; he would also have felt the heat of the fire. A consequence of Descartes' denial of such a soul to animals was that he believed them incapable of thought or feeling, of pain or pleasure, of fear or anger: '... there is no preconceived opinion to which we are all more accustomed from our earliest years', he wrote, 'than the belief that dumb animals think'.[17] And although he accepted that animals could see, he believed that they were unconscious of what they were seeing:

> my view is that animals do not see as we do when we are aware that we see, but only as we do when our mind is elsewhere. In such a case the images of external objects are depicted on our retinas, and perhaps the impressions they make in the optic nerves cause our limbs to make various movements, although we are quite unaware of them. In such a case we too move like automatons . . .[18]

Descartes' surprising belief that animals do not experience fear is made dramatically clear in a passage in his reply to objections made by the Jansenist priest Antoine Arnauld:

> when people take a fall, and stick out their hands so as to protect their head, it is not reason that instructs them to do this; it is simply that the sight of the impending fall reaches the brain and sends the animal spirits into the nerves in the manner necessary to produce this movement even without any mental volition, just as it would be produced in a machine. And since our own experience reliably informs us that this is so, why should we be so amazed that the 'light reflected from the body of a wolf onto the eyes of a sheep' should equally be capable of arousing the movements of flight in the sheep.[19]

To almost anyone nowadays, such views seem so incredible that it is worth considering how it was that Descartes could hold them. He accepted that because animals have sense organs like ours there is a *prima facie* case that they have sensations like ours; but though he admitted that this argument was obvious and universally believed, there were, he thought, less obvious but stronger arguments suggesting that it was wrong. If consciousness required a rational soul, and such souls were immortal, the assumption that some animals were conscious implied that they had immortal souls. Quite apart from the impiety of such a belief, it was, he thought, unreasonable to suppose that some animals had souls while others did not, so one was driven to the conclusion that oysters and sponges had souls – which was absurd.[20] The alternative was to suppose that animals moved automatically, like machines; and although much animal behaviour was of a kind that seemed very far from being automatic, Descartes pointed out that such behaviour could not be regarded as being beyond the capacity of an automaton created by God.[21] Furthermore, 'real speech', which he regarded as 'the only certain sign of thought hidden in a body',[22] was never found in animals, even those such as parrots which were capable of producing sequences of words.[23] He accepted that there was no proof that animals were not conscious,[24] but he held that the balance of probabilities suggested strongly that they were not. Nowadays we take the opposite view, partly because Darwin and his successors have made the argument from similarity of sense organs (and parallel arguments from similarity of behaviour, of anatomy and of physiology) overwhelmingly strong, and partly because we are no longer convinced by arguments based on the untestable properties of hypothetical souls.

*   *   *

Giovanni Alfonso Borelli, like Descartes, was impressed by Galileo's use of mathematics and physics to explain natural phenomena. A mathematician by profession, he had become interested in medical matters in connection with the spread of a pestilence in Sicily when he was a professor in Messina. Later, in Pisa, his friendship with Marcello Malpighi, the discoverer of blood capillaries, encouraged an interest in anatomy and physiology.

Borelli was particularly interested in animal movement. Believing (wrongly) that the apparent swelling of a muscle when it shortens reflects an increase in volume, he asked what it was that passes along the nerve to the muscle and causes the (supposed) increase in volume. He rejected any kind of incorporeal spirit, arguing that the 'mass of the muscle, possessing as it does three dimensions, cannot be inflated and increased in bulk by any wholly incorporeal influence having like an indivisible point no magnitude'.[25] He also rejected the then current notion that the muscles were inflated by the entry of 'extremely attenuated corporeal animal spirits like air', arguing that: 'When the muscles of a living animal are divided lengthwise while the animal is submerged under water, and . . . is struggling violently . . . one would expect that innumerable bubbles of gas would burst forth from the wound and ascend through the water, whereas nothing of the kind takes place.' Two possibilities, he said, remained: '. . . some corporeal [non-gaseous] substance must be transmitted along the nerves to the muscles or else some commotion must be communicated along some substance in the nerves, in such a way that a very powerful inflation can be brought about in the twinkling of an eye.' And because the nerves themselves showed no sign of inflation or hardening, he preferred the second alternative.

This conclusion left Borelli with two questions: What is the nature of the 'commotion' that passes along the nerve? And what takes place in the muscle that can account for the (supposed) increase in volume? He was not in a position to answer either question, but he gave interesting analogies. He thought of the nerves as 'canals filled with . . . spongy material like elder-pith . . . moistened with the spirituous juice of the brain . . . and saturated to turgescence'. If one of the extremities of the nerve were struck or pinched, 'the . . . concussion or undulation ought to be communicated right to the other end'. In motor nerves the concussion would be initiated centrally 'by that gentle motion of the spirits by which the commands of the will are in the brain carried out'; and the arrival of the concussion at the far end of the nerve would lead to the discharge into the muscle of a few droplets of the spirituous juice. These droplets, he supposed, reacted with something in the muscle, causing a 'sudden fermentation and ebullition' so 'filling up the porosities of the muscle' and inflating it. In sensory nerves, the concussion would be initiated peripherally in the skin or sense organ and would be conveyed to the brain. Here 'the faculty of the sensitive soul according to the region of the brain thus percussed, according to the vehemence of the blow and the fashion and mode of the motion, is able to form a judgement concerning the object causing the movement'. Borelli was, understandably, vague about what went on in the brain; but he had no doubt that the events in the nerves were purely physical.

Arguments about whether the nervous fluid inflated the muscle directly, as Descartes and others believed, or merely led to some 'fermentation and ebullition' that inflated the muscle, as Borelli believed, had to be abandoned when it became clear that muscles do not, in fact, increase in volume when they shorten. This was shown independently by Jan Swammerdam[26] in Holland, and by Jonathan Goddard[27] in England. Swammerdam's experiment is illustrated in Figure 6.1. The calf muscle of a frog and its attached nerve were enclosed in a glass vessel. A wire passed through the cork and was looped round the nerve in such a way that pulling on the wire stimulated the nerve and made the muscle twitch. A fine glass tube came out of the top of the vessel, and a drop of water was trapped in the middle of this tube. If the muscle swelled when it twitched, the twitch should cause the drop of fluid to rise in the tube. In fact, Swammerdam noted, during a twitch the drop didn't move.

THE ROLE OF ELECTRICITY

If nerves did not act as hydraulic pipes how did they act? Borelli's analogies were colourful but they were only analogies. The rationalism of the 17th century was more successful at exorcizing incorporeal spirits than at finding something real to put in their place. With hindsight, we can see that the difficulty was that so little was understood about electricity.

It had been known at least since the time of Thales of Miletus (about 600 BC) that amber rubbed with wool attracted light objects such as feathers or dry leaves. The ancients were also familiar with the numbing shocks that could be given by the Mediterranean fish known as the torpedo or stingray;[28] in one of the Socratic dialogues, Meno, having been so confused by Socrates' arguments that he does not know how to reply, accuses Socrates of being 'like the stingray in the sea, which benumbs whatever it touches'. But until the 18th century there was no reason to connect the attraction of feathers by rubbed amber with the shocks of a torpedo, or to connect either with the normal workings of nerves.

In 1600, William Gilbert, physician to Queen Elizabeth, published an account of the first systematic investigation of the phenomena of static electricity.[29] Gilbert showed that amber was just one of a large class of substances (including glass, sulphur and sealing wax) which, when suitably rubbed, attracted light

loops with loops of even thinner brass wire, the experiment again failed. It seemed that what was critical was not the thickness of the loops but the nature of the material forming them.

These experiments led to the division of substances into two classes: 'conductors' (including metals and slightly damp hemp thread) and 'non-conductors' – or, as we should now call them, insulators – (including dry silk). Gray also showed that a soap bubble, a live chicken, a small boy, and he himself, could all, if suitably suspended, be 'electrified' (i.e. made capable of attracting light objects) by being attached by a suitable conductor to a glass tube that was then rubbed.

Some of Gray's experiments were repeated and extended by M. Du Fay, the Director of the Jardin du Roi, in Paris. In one of these experiments Du Fay had himself suspended on silk cords and found that 'as soon as he was electrified, if another person approached him and brought his hand within an inch, or thereabouts, of his face, legs, hands, or cloaths, there immediately issued from his body one or more pricking shoots, attended with a crackling noise . . . in the dark those snappings were so many sparks of fire'.[31] These sparks, according to the Abbé Nollet, who collaborated with Du Fay, were the first to be drawn in this way from the human body, and such demonstrations became a fashionable party-trick – see Figure 6.2.

More important than the eliciting of sparks was Du Fay's startling discovery that electricity could be of two kinds. A piece of gold leaf that has been charged with electricity by being touched with a rubbed glass rod is subsequently repelled by the rod. Similarly, a piece of gold leaf that has been touched by a rubbed stick of sealing wax is repelled by the sealing wax. But what Du Fay found, to his great surprise, was that a piece of gold leaf that had been touched by a rubbed glass rod was subsequently *attracted* by a rubbed stick of sealing wax; and a piece of gold leaf that had been touched by rubbed sealing wax was subsequently *attracted* by a rubbed glass rod. He was forced to conclude that the electricity found on rubbed glass, which he called *vitreous*, was different from the electricity found on rubbed sealing wax, which he called *resinous*;[32] objects bearing the same kind of electricity repelled each other whereas objects bearing opposite kinds attracted each other.

The relation between these two kinds of electricity was greatly clarified by some experiments of Benjamin Franklin, done five years before his more famous and more dangerous kite-flying.[33] He had noticed that it was impossible for a man to electrify himself by rubbing a glass tube that he was himself holding, even if he stood on an insulating sheet of wax; and he suspected that this was because the glass tube transferred to the man no more electricity than it received from him in the act of rubbing. To test this idea he had two people stand on a sheet of wax; one of them rubbed a glass tube and from time to time transferred the electric charge[34] on the glass to the other by touching the glass to the other's knuckle. Franklin made four crucial observations:[35] (1) after the rubbing was complete, *both* people on the wax appeared to be electrified to someone standing on the floor beyond the wax – i.e. a spark could be obtained by approaching either with

FIGURE 6.2 'Electrification' as an 18th century party trick. The friction machine transfers electric charge to the suspended boy, who transfers it to the girl on the insulating platform, whose outstretched hand attracts fragments of paper from the little table. (Drawn from a photograph – reproduced in an article by Mary Brazier – of an illustration in F. H. Winckler's *Essai sur la Nature. Les effets et les causes avec description de deux nouvelles machines à Électricité,* Jorry, Paris, 1748.)

a knuckle; (2) if the people on the wax were touching each other during the rubbing of the glass tube, neither subsequently appeared to be electrified; (3) if the people on the wax touched each other after the electrification was complete, a spark passed between them which was bigger than the spark that could otherwise have been obtained between either and a person on the floor; and (4) after this strong spark, neither of the people on the wax appeared to be electrified to the person on the floor.

To explain these observations, Franklin postulated that rubbing *did not create electric charge but merely shifted it between the rubber and the object rubbed,* so that one ended up with an excess and the other with a corresponding deficit. The former he called positively electrified and the latter negatively electrified.* At the

*Franklin had no way of knowing whether, when glass was rubbed by the hand, charge passed from the hand to the glass or vice versa, but he assumed that it passed from the hand to the glass, in other words that the glass was positively charged. This was an unlucky choice because we now know that, in fact, the rubbing leads to a flow of electrons – almost weightless particles bearing a uniform charge – from the glass to the hand. To fit in with the nomenclature established by Franklin, the charge on an electron is therefore defined as negative.

start of the experiment there would be nothing to make charge flow between the three persons, but after the rubbing and the transfer of charge, one of the persons on the wax would have an excess and the other a deficit. These could not be obliterated by a flow of charge to or from the floor because of the insulating wax, but they could be obliterated by a flow to or from the knuckle of the person on the floor beyond the wax. (Any excess or deficit created in him would be rectified by flow of charge between him and the floor.) This explains why both persons on the wax appeared to be electrified. If the persons on the wax were touching during the rubbing of the glass, a flow of charge between them would prevent the excess and deficit from being built up; this explains the second observation. If the two persons on the wax touched each other after electrification was complete, charge would have flowed from the one with excess to the one with the deficit, thus accounting both for the spark (third observation) and for the subsequent lack of any evidence of electrification (fourth observation).

On Franklin's theory, then, there was only one kind of electricity but, by acquiring an excess or a deficit of it a body could be positively or negatively charged. The difference between Du Fay's vitreous and resinous electricity could be explained by supposing that, in Du Fay's experiments, when the glass was rubbed it became positively charged, whereas when sealing wax was rubbed it became negatively charged. (In each case, the object doing the rubbing would have acquired the opposite charge.) There was, though, a snag in the notion that there was only one kind of electricity. It was awkward to suppose that an excess of electricity had the same effect as a deficiency of electricity; yet two positively charged bodies repel one another just as do two negatively charged bodies. It was only in the second half of the 18th century that Du Fay's and Franklin's ideas were reconciled by supposing that uncharged bodies contain equal quantities of positive and negative electricity, which neutralize each other's properties, and that electrification by friction causes a transfer of charge leading to an unequal distribution of the positive and negative charge between the two bodies being rubbed together. More than a century later, the discovery that atoms consist of a positively charged nucleus surrounded by orbiting negatively charged electrons provided a physical basis for this view. It is, of course, the transfer of electrons between the rubber and the object being rubbed that causes the electrification of both.

Apart from the shocks delivered by the electric fish, all the electrical phenomena mentioned so far fall in the category often referred to as 'static electricity', because, unlike the electrical phenomena with which we are familiar domestically, they do not involve continuous flows of electric charge.

The discovery of the more familiar 'current electricity' arose from the work of Galvani and Volta in the last part of the 18th century. In January 1781, Luigi Galvani, Professor of Anatomy at Bologna, discovered by chance that sparks from a nearby 'electrical machine' – i.e. a machine for generating static electricity by friction – caused violent contractions of the leg of a frog whose sciatic nerve was being touched at that moment by an assistant holding a scalpel.[36] It was known that electric shocks applied directly to the frog would cause contractions, but these 'contractions at a distance' were puzzling. Five years later Galvani decided to see whether similar effects could be produced by atmospheric electricity. On a spring evening with 'black and white clouds to the south', he took his frogs and apparatus to the terrace of the Palazzo Zamboni in Bologna. A wire with one end touching the nerve was fixed to the wall of the house; another wire was arranged with one end touching the frog's muscles and the other dipping into the water in a well. When lightning flashed the muscles contracted; and even in the absence of lightning there were some contractions when thunderclouds were present. Later that summer, Galvani wanted to see whether muscles could be induced to contract by changes in atmospheric electricity during calm weather. To this end, he hung decapitated frogs by brass hooks (which passed through the spinal cord) on an iron garden railing. Occasional twitches occurred, particularly if he pressed the hooks against the railing. He first supposed that pressing the hooks against the railing somehow released accumulated atmospheric electricity; but this interpretation had to be given up when he found that he could get similar results indoors:

> when I brought the animal into a closed room, placed it on an iron plate, and began to press the hook which was fastened in the spinal cord against the plate, behold!, the same contractions and movements occurred as before. I immediately repeated the experiment in different places with different metals and at different hours of the day. The results were the same except that the contractions varied with the metals used; that is, they were more violent with some and weaker with others. Then it occurred to me to experiment with other substances that were either non-conductors or very poor conductors of electricity, like glass, gum, resin, stones, and dry wood. Nothing of the kind happened and no muscular contractions or movement were evident. These results surprised us greatly and led us to suspect that electricity was inherent in the animal itself.[37]

This hypothesis was not absurd, because by then it was generally suspected that the shocks of torpedoes and other fish were electrical. But Galvani was unable to account for a further observation: muscular contractions seemed to occur only when the conducting pathway was made up of two dissimilar metals, such as iron and brass.

This curious requirement was eventually explained by Alessandro Volta, then Professor of Physics in Pavia, to whom Galvani had sent an account of his experiments. Volta showed that if a disc of copper and a disc of zinc, each on insulated handles, were brought into contact and then separated, both were very slightly charged, and in opposite directions. He realised that the electrical force involved

– what we should now call the *electrical potential difference* or *voltage** – was extremely small, but by building a pile of paired copper (or silver) and zinc discs, with the metals always in the same order and with the discs of each pair separated by a conducting medium (discs of cardboard soaked in salt-water) he built what was in fact the first electric battery – see Figure 6.3. His 'pile' – the word is still used for an electric battery in France – not only produced a substantial voltage between the top copper plate and the bottom zinc plate, but also, in a way that he did not understand, was able to produce a continual flow of electrical charge, i.e. an electric current. Volta accepted that the cause of the contraction in Galvani's experiments was electrical, but he believed that the source of the electricity was not the tissues of the frog but the junction between the two metals – that the electricity was not *animal electricity* but *bimetallic electricity*. The bitter and protracted controversy that ensued was to lead ultimately, not only to modern ideas of current electricity and all the practical applications that have followed but also to modern ideas of the way in which nerves transmit messages.

It turned out that both Volta and Galvani were partly right. Volta was right in saying that the source of the electricity that had led to contractions in Galvani's experiments was the junction of the two metals; but Galvani was right in thinking that the tissues of the frog could themselves produce electricity, and that such electricity could excite muscles.

Although Volta realized that the continual production of electric current by his pile – the perpetual motion of the electric fluid, as he called it – required a source of energy, he failed to identify that source as chemical changes occurring in the pile. Within months of the publication of an account of the pile in the *Philosophical Transactions of the Royal Society*,[38] two Englishmen, Anthony Carlisle and William Nicholson, had constructed a pile, using zinc discs and (silver) half-crowns, and noticed that the production of electricity was accompanied by the corrosion of the zinc and the appearance of bubbles of hydrogen on the silver.[39] The overall chemical reaction that provides the energy for generating the electric current is, in fact, the combination of metallic zinc with water to form zinc hydroxide and gaseous hydrogen. But why should that reaction generate an electric current? And why should the bubbles of hydrogen appear on the silver? The answer to those questions[40] came only later, after the discovery by Faraday that in aqueous solution many substances exist to a greater or lesser extent not as neutral molecules but as mixtures of positively and

---

*For the reader unfamiliar with electrical terms it may be helpful to point out that electrical potential (or voltage) is equivalent to pressure in a hydraulic system. Just as, in a hydraulic system, the difference in pressure between two points is a measure of the work done in transferring a unit quantity of water from one point to the other, so in an electrical system the difference in electrical potential between two points is the work done in transferring a unit quantity of charge from one point to the other.

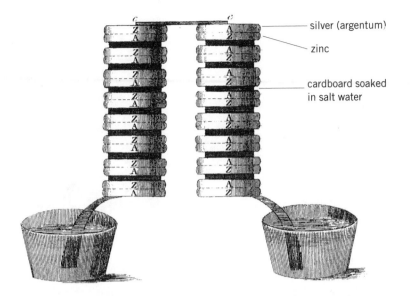

silver (argentum)

zinc

cardboard soaked
in salt water

FIGURE 6.3 The first electric battery. (From a letter written by Volta to the President
of the Royal Society, and published in 1800.)

negatively charged particles – particles which he called ions. More than a century
later, the behaviour of ions was to provide the key to the way nerves work.

DOUBTS ABOUT 'ANIMAL ELECTRICITY'

The evidence for *bimetallic electricity* provided by Volta's pile was so impressive
that, for the first few decades of the 19th century, Galvani's notion that animal
tissues such as nerve and muscle could produce *animal electricity* received little
attention. There was even some doubt about whether the much more dramatic
discharges of electric fish were really electrical. You might think that anyone who
had had a shock from a torpedo would not be disposed to argue the point, but
until the end of the 18th century electric phenomena were almost always those
of static electricity; and the characteristic feature of static electricity, with its very
high voltages and minute currents, is the ability to produce sparks. The shocks
from torpedoes felt like electric shocks, and they could be transmitted by conductors and stopped by insulators, but they appeared not to cause sparks. Henry
Cavendish, who was the first person to distinguish clearly between the quantity
of electric charge and the electrical 'force' (what we should now call the voltage
or electrical potential), had argued, in 1776, that the absence of a spark did not
imply that the shock from a torpedo was not electrical.[41] But despite Cavendish's
argument, and despite an unpublished demonstration by John Walsh that the

FIGURE 6.4 Matteucci's
demonstration that the
shock from a *Torpedo*
(stingray) can cause sparks.
(From M. Brazier.)

South American electric eel *Gymnotus electricus* could cause sparks to jump across a very small gap,[42] the electrical nature of shocks from fish was regarded as unproven until well into the next century. Then in the 1830s another Italian, Carlo Matteucci (who was later to play an active part in the Risorgimento), sandwiched a torpedo between two brass plates each fitted with a side arm bearing a small brass ball; watching in a darkened room, he twisted one of the plates to bring the brass balls close together and observed sparks passing between them (see Figure 6.4).[43]

Proof that nerves and muscles produced electricity required much more sensitive methods for detecting electric currents. Until the end of the first quarter of the 19th century, the most sensitive detector was a frog nerve with the muscle it supplied still attached. If the nerve was laid on top of another muscle, which was then made to contract (by pinching its own nerve), the overlying nerve was excited and the muscle that it supplied also contracted.[44] But though such observations supported Galvani's view that an excited muscle produced electricity, they did not prove it, for the muscle on which the nerve lay might have stimulated the nerve in contact with it in some other way.

Then, in 1820, Hans Christian Oersted discovered that a wire carrying an electric current deflected a compass needle, and within weeks of the publication of that discovery, a device for detecting electric currents by their magnetic effects had been invented. By 1827 such devices – called galvanometers in honour of Galvani – had been made sensitive enough to detect a flow of current along a moist thread connecting two vessels of salt solution, one in contact with the feet of a frog and the other in contact with its body.

THE DISCOVERY OF THE NERVE IMPULSE

In 1842 Matteucci, who had recently been appointed Professor of Physics at Pisa, found that when one wire from a galvanometer was pushed into a wound in a muscle, and the other wire touched the surface of the muscle, a small current passed through the galvanometer.[45] This observation of Matteucci's was extended by Emil Du Bois-Reymond,[46] a young man of part Swiss part Huguenot extraction working in Johannes Müller's famous physiological laboratory in Berlin. Du Bois-Reymond found that not only whole muscles but also nerves and small bundles of muscle fibres showed a current when a galvanometer was connected between the exterior and the cut surface. He thought that this was because the cut surface allowed access to the interior of the individual fibres in the muscle or nerve, and he concluded from the direction of the current that the interior of the muscle or nerve fibre was normally at a lower electrical potential than the exterior – see Figure 6.5. This turned out to be correct, and the potential difference across the membrane of the (unexcited) individual muscle or nerve fibre later became known as the *resting potential*. But Du Bois-Reymond noticed something much more striking. He noticed that when a nerve was excited – whether by giving it an electric shock, or by pinching it, or by putting strychnine on it – *the difference in electrical potential between the cut surface and the external surface was greatly but transiently diminished*. This transient diminution occurred first at the point at which the nerve was excited and passed (in both directions) along the nerve. Since the resting nerve membrane with a difference in electrical potential between its inner and outer surfaces may be said to be 'polarized', excitation of the nerve was said to cause a transient 'depolarization' to pass along it. This wave of depolarization was later called the *action potential* or *nerve impulse*.

Du Bois-Reymond had no doubts about the significance of his discovery: 'If I do not greatly deceive myself, I have succeeded in realising ... the hundred years dream of physicists and physiologists, to wit the identity of the nervous principle with electricity.'[47] Du Bois-Reymond's claim did not, of course, imply that conduction along a nerve was like conduction of an electric current along a telegraph wire.[48] The crucial question was: What is the relation between the electrical events occurring in an excited nerve and the transmission of a message along the nerve?

If the action potential (wave of depolarization) that passes along an excited nerve is identical with, or closely related to, the message being transmitted, the speed of the action potential and the speed of the message should be the same. Until 1850, neither speed was known, and Johannes Müller believed that the speed with which messages passed along a nerve was too great to be measured. But in 1850, another of Müller's pupils, Hermann von Helmholtz, succeeded in making just that measurement.[49] Three years after the publication of the famous essay that established his claim to be one of the discoverers of the law of conservation of energy, Helmholtz described a series of experiments in which he measured the time interval between the giving of an electric shock to the sciatic

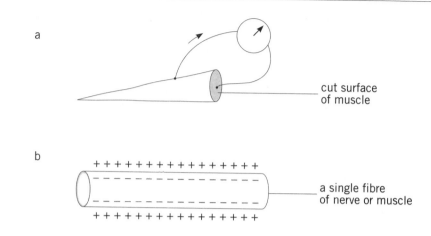

a

cut surface
of muscle

b

+ + + + + + + + + + + + + + + +

– – – – – – – – – – – – – – – –

– – – – – – – – – – – – – – – –

+ + + + + + + + + + + + + + + +

a single fibre
of nerve or muscle

FIGURE 6.5 (a) Du Bois-Reymond's discovery that if a galvanometer is connected between the outer surface and the cut surface of a muscle or nerve a current flows from the outer surface to the cut surface. (b) His (correct) interpretation of this discovery – the idea that in each individual resting nerve fibre or muscle fibre there is an electrical potential difference across the membrane, the outside being positive relative to the inside.

nerve of a frog and the resulting twitch of the gastrocnemius muscle – the main muscle in the calf. By stimulating the nerve at different distances from its junction with the muscle, and seeing how this distance affected the time interval, he was able to calculate what extra time was needed for conduction of the message along the extra length of nerve.* Hence he was able to calculate the speed at which the message is transmitted along the nerve. ('Message', in this context, means, of course, whatever it is that passes along the nerve to initiate contraction in the muscle.) His experiments showed that the speed was about 27 metres per second – less if the nerve was cooled. These values are characteristic of large motor nerves in the frog. Large mammalian nerve fibres, at body temperature, conduct at about 100 metres per second (220 miles an hour).

Helmholtz was able to measure *the speed of the message* because the timing of the stimulating shock and the timing of the onset of the muscle twitch were both easily determined. To measure *the speed of the action potential* it was necessary to determine accurately the time at which it arrived at a particular point along the nerve, and this could not be done directly because of the high inertia, and therefore slow response, of the galvanometers then available. In 1868, Julius Bernstein,

*He had to use this procedure (rather than dividing the length of nerve between the point of stimulation and the muscle by the total time interval) because he did not know how big a delay there was between the arrival of the message at the muscle and the contraction of the muscle.

a former pupil of Du Bois-Reymond and of Helmholtz, got round this difficulty with an ingenious mechanical device, and showed that the speed of the action potential was not significantly different from Helmholtz's estimate of the speed of the message.[50]

If the speed of the action potential was the same as the speed of the message, the obvious question was: Is it an incidental accompaniment of whatever it is that carries the message or is it itself the message? Is it like the smoke that accompanies a railway engine across the countryside, or is it like the heat generated by the combustion of each part of an ignited fuse of gunpowder? And if the action potential is itself the message, how is its existence explained and why does it travel along the nerve? Answers to all these questions came in the next century, and will come in the next chapter.

7 The Nerve Impulse

THE 'ALL-OR-NOTHING LAW'

Are her fond responses
All-or-none reactions?
—W. H. Auden, *Heavy Date*

Surprising facts about the nature of the messages carried by nerves emerged from some rather simple experiments that were done by Francis Gotch in Oxford and Keith Lucas in Cambridge before the First World War. They worked mainly with the isolated sciatic nerve of the frog, or with a preparation consisting of the sciatic nerve together with the main calf muscle that it controls. If the isolated nerve was stimulated with a brief electric shock, an action potential passed from the point of stimulation in each direction along the nerve. A second shock caused a second action potential unless it occurred within a few thousandths of a second of the first shock, when it was ineffective. It seemed that for a brief period following activity the nerve became inexcitable or 'refractory', and the existence of this *refractory period* has an important consequence. It implies that, if the action potential is indeed the message, the message transmitted along the nerve must consist of a number of discrete 'impulses' which cannot recur at more than a certain frequency. As E. D. Adrian put it, writing in 1927, 'the nervous message may be likened to a stream of bullets from a machine gun, it cannot be likened to a continuous stream of water from a hose'.[1]

And it turned out that there is an even more stringent limitation on the character of the message. It had long been known that by varying the strength of the electric shocks given to the frog's sciatic nerve it was possible to vary the strength of the twitch produced in the calf muscle – a gradual increase in the strength of shock leading to a gradual increase in the size of the twitch. But because the sciatic nerve contains a few thousand individual nerve fibres, there was no way of knowing whether this was because the stronger shocks caused a greater response in each individual nerve fibre or simply stimulated a greater number of fibres – or, perhaps, both. Keith Lucas solved this problem by using a very small muscle controlled by a nerve containing only eight or nine nerve fibres.[2] He found that as he increased the strength of shock, the size of the muscle twitch increased, but instead of increasing smoothly *it increased in a stepwise fashion; and there were never more steps than there were individual fibres in the nerve.* The obvious explanation was that each nerve fibre was responding in an 'all-or-nothing' fashion – either there was a full response or no response, an impulse or no impulse. The message was like a stream of bullets not only in its discontinuity but also in that the size of each unit did not vary with the size of the stimulus that initiated it.

Lucas and Gotch used only electric shocks as stimuli, and worked only with motor nerves, but later work showed that both the all-or-nothing character and the refractory period (the brief period of inexcitability following activity) were general features of nerve conduction. Nerves are not like telephone wires, which carry a continuously varying current reflecting the continuously varying pitch and intensity of the speaker's voice; they are not even like traditional telegraph wires, which carry short and long bursts of current corresponding to the dots and dashes of Morse code. So far as the nature of the message is concerned, they are more like what telegraph wires would be if they could transmit only dots. However complex the information that is to be sent along a nerve, it can be sent only as streams of impulses whose size cannot be controlled at the point of stimulation. Information relevant to the most subtle sensations or to the most complex actions can be transmitted to and from the brain in no other way. The answer to Auden's rhetorical question about 'fond responses' is that of course the responses are not all-or-nothing, but the messages in both sensory and motor nerves that make those responses possible are made up of discrete all-or-nothing units.

The all-or-nothing character of conduction along a nerve fibre, and the limited frequency at which the all-or-nothing impulses can be conducted, together impose a severe constraint on the amount of information that can be carried by a single fibre – a constraint that has far-reaching effects on the organization of the nervous system. There is, though, an important compensation. Transmission by all-or-nothing impulses has the advantage over transmission by a continuously varying current that Morse code telegraphy has over ordinary telephony. If conditions somewhere along the transmission path are unfavourable, Morse code can be used when speech cannot. The reason is that so long as dots and dashes can be detected, and distinguished from one another, it does not matter how much they are distorted; in contrast a relatively slight amount of distortion is sufficient to make speech incomprehensible. Transmission by all-or-nothing impulses is much more reliable than transmission by a continuously varying current because, in effect, even the most delicate nuances are conveyed as collections of the simplest possible 'black and white' statements.

Although the main significance of the all-or-nothing law lies in the way it limits the information-carrying capacity of nerves and increases their reliability, the fact that the size of the impulse is independent of the size of the initiating stimulus also provided an interesting clue to the way in which the impulse is propagated along the nerve. Disturbances can, in general, be propagated in two ways depending on whether the necessary energy is supplied at the beginning or along the route. When a stone is dropped into a pond and ripples spread across the surface, the energy is supplied by the stone, and the ripples get smaller and smaller as they spread. A bigger stone, or a stone dropped from a greater height, causes bigger ripples. In contrast when combustion spreads from a match along a fuse of gunpowder, each particle is ignited by heat from the combustion of particles upstream and produces heat to ignite particles downstream; so the behaviour of the fuse at some distance from the start is independent of the size of the match. It is characteristic of propagated disturbances in which the energy is supplied along the path that they tend to show all-or-nothing behaviour. The all-or-nothing character of nerve conduction therefore suggested that the energy for conducting the message is somehow supplied by each portion of nerve as the message passes along it. I want to follow up that clue because the elucidation of events responsible for nerve conduction led to an understanding not only of the way nerves work but also of the various ways in which nerve cells interact with one another to form devices that control, decide, remember and learn.

BERNSTEIN'S 'MEMBRANE THEORY'

In 1902, Julius Bernstein – the man who, thirty-four years earlier, had succeeded in measuring the speed at which an action potential travelled along a nerve – put forward an ingenious hypothesis to explain why there is an electrical potential difference across the nerve membrane, why there can be sharp but transient falls in the magnitude of this potential difference (in other words, action potentials), and why these action potentials travel along the nerve.[3] Although his hypothesis turned out to be partly wrong, the basic idea was right, and understanding it makes it much easier to understand the correct theory. But first I must say something about ions.

When a metal wire carries an electric current the movement of electric charge along the wire takes the form of a stream of electrons – almost weightless particles carrying a negative charge. When an electric current passes through an aqueous solution, the movement of charge occurs in a quite different way. Michael Faraday, in the 1830s, showed that it is the result of the movement, through the solution, of electrically charged particles – atoms or groups of atoms – that are formed by the splitting of some of the molecules in the solution in such a way that part of the molecule has a surplus of one or more electrons, and is therefore negatively charged, and the other part has a corresponding deficit, and is therefore positively charged. At the suggestion of William Whewell, the formidable and omniscient master of Trinity College Cambridge, Faraday called these charged particles 'ions' from the Greek word for wanderer. He assumed that the ions were formed by the action of the electricity on the molecules in the solution, but later it became clear that, without any application of an electric current, many substances exist in solution partly or wholly as ions. Common salt (NaCl), for example, in aqueous solution, consists wholly of equal numbers of positively charged sodium ions (Na^+) and negatively charged chloride ions (Cl^-). Potassium chloride behaves similarly, giving equal numbers of potassium ions (K^+) and chloride ions. Water itself consists almost entirely of uncharged molecules of H_2O, but a tiny fraction of the water molecules present will have split into hydrogen ions (H^+) and hydroxyl ions (OH^-).

Let us return to nerves. For the present purpose we can consider individual nerve fibres as long, very thin-walled tubes filled with a salt solution containing mainly potassium chloride with a little sodium chloride and bathed in a salt solution containing mainly sodium chloride with a little potassium chloride. Suppose the wall of the tube – the nerve membrane – were impermeable to all ions; the solutions on either side of the membrane are uncharged so there would be no electrical potential difference across the membrane. Now suppose that the membrane allows K^+ ions, but only K^+ ions, to pass through readily; the K^+ ions on both sides of the membrane move about randomly, but because they are in high concentration inside the fibre and low concentration outside, far more K^+ ions will leave the fibre than will enter – just on statistical grounds. But K^+ ions carry a positive charge, so a net loss of them from the fibre will leave the inside of the fibre negatively charged relative to the outside. In other words there will soon be an electrical potential difference, or voltage, across the membrane. As this potential difference builds up, it will become more difficult for K^+ ions to move outwards through the membrane and increasingly easy for them to move inwards. Eventually (and in this case eventually means almost instantaneously), a point will be reached when the electrical potential difference just compensates for the concentration difference, and the movements of K^+ ions in and out become equal. There will be a state of equilibrium. (Note that the actual quantity of ions that need to move across the membrane to build up a potential just suffic-ient to compensate for the difference in concentration is extremely small – not sufficient to cause measurable changes in the concentrations in the solutions bathing the two surfaces of the membrane. That is why the equilibrium is established almost instantaneously.)

Bernstein suggested that the electrical potential difference across the mem-branes of resting nerves (nerves not conducting impulses) could be explained in this way.

What about the action potential? Bernstein pointed out that if, with the equi-librium established, the nerve membrane were suddenly to change so that its per-meability to Na^+ ions increased to match its permeability to K^+ ions, there would be a sudden net influx of Na^+ ions, because their inward movement would be encouraged both by the high concentration of Na^+ ions outside and by the potential difference across the membrane (the outside being positive to the inside). Since Na^+ ions carry a positive charge, their sudden net influx would reduce that potential difference rapidly to zero. (You can see intuitively that the potential difference across the membrane must approach zero, because if the permeabilities were equal the membrane would not distinguish between Na^+ ions and K^+ ions, and there would be no effective asymmetry across it.) Bernstein therefore suggested that the action potential was caused by a sudden transient loss of selectivity, the permeability to Na^+ ions increasing to equal the permeability to K^+ ions.

Finally, why should the action potential travel along the nerve fibre? To explain this, Bernstein adopted a hypothesis – the *local circuit hypothesis* – proposed a few years earlier by Ludimar Hermann in Königsberg.[4] Hermann had pointed out that near the junction between the bit of the nerve carrying the action potential and the adjacent resting nerve, local electric currents would flow – see curved arrows in Figure 7.1. These currents would tend to reduce the electrical potential difference across the membrane of the resting nerve adjacent to the active region, and *he supposed that this reduction in potential difference excited that portion of the nerve*. In this way each bit of nerve would become active in turn, excited by local currents from the preceding bit, and producing currents capable of exciting the subsequent bit. All that Bernstein had to do to fit this hypothesis into his theory was to postulate that the reduction in the potential difference across the membrane caused a large but transient increase in the permeability of the membrane to sodium ions.

At the time Bernstein put forward his 'membrane theory' it was a plausible and ingenious speculation, with only a little evidence to support it.[5] And that remained the position until just before the Second World War.

PROOF OF THE LOCAL CIRCUIT HYPOTHESIS

Knowing whether the local circuit hypothesis was right was important because, if right, it would not only explain how action potentials travel along the nerve fibre; it would also finally dispose of the venerable question: Is the action potential itself the message or is it like the smoke above the railway engine? In the summer of 1938, Alan Hodgkin, then a young Cambridge physiologist working in the United States, was arguing with an American colleague. Since part of the local current flows in the fluid bathing the nerve fibre, altering the electrical resistance of that fluid should, if the local circuit hypothesis is right, alter the flow of electric current along the outside of the nerve fibre, and hence the velocity at which action potentials travel along the fibre. The American said that he would take the hypothesis seriously only if Hodgkin could demonstrate this effect.[6]

Hodgkin did demonstrate the effect by showing that single nerve fibres from crab or squid conducted action potentials more slowly if they were suspended in

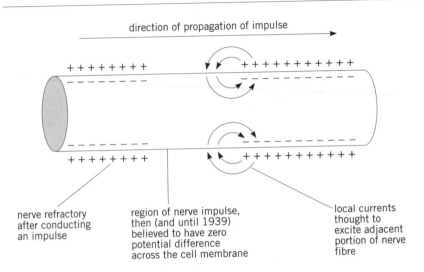

direction of propagation of impulse

+ + + + + + +   + + + + + + + + + + +
– – – – – – –   – – – – – – – – – –

– – – – – – –   – – – – – – – – – –
+ + + + + + +   + + + + + + + + + + +

nerve refractory after conducting an impulse

region of nerve impulse, then (and until 1939) believed to have zero potential difference across the cell membrane

local currents thought to excite adjacent portion of nerve fibre

FIGURE 7.1 Hermann's local circuit hypothesis, proposed in 1899.

oil or air so that there was only a thin layer of salt solution to conduct current outside the fibre. But the result was not conclusive because the slowing might have resulted from the deleterious effect of the oil or slight deterioration of the fibre when it was suspended in air.

The matter was finally clinched by an experiment of extraordinary simplicity and decisiveness.[7] Instead of trying to *increase* the electrical resistance to the flow of current along the outside of the fibre by reducing the volume of the bathing solution to a thin film, Hodgkin decided to *decrease* it by providing an additional metallic pathway for the electric current. A giant nerve fibre from a squid was arranged so that the part between the point at which it was stimulated and the point at which the action potential was to be recorded rested on a series of parallel platinum strips, in a moist atmosphere. The strips could be connected together electrically by raising a small trough of mercury – see Figure 7.2. Hodgkin found that raising the trough of mercury increased the speed at which an action potential was conducted along the nerve fibre; what is more, the effect occurred within the fraction of a second that it took to move the mercury trough. As nothing in direct contact with the nerve fibre had changed, and the only agent that could have travelled through a metallic short circuit in the time available is an electric current, the experiment provided unequivocal and elegant proof of the local circuit hypothesis.

THE HODGKIN–HUXLEY THEORY[8]

In England the following year, Hodgkin, working with Andrew Huxley, showed

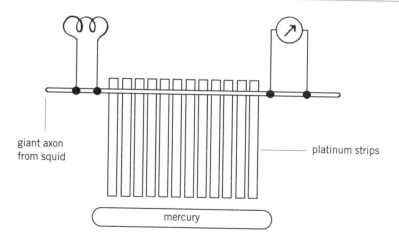

FIGURE 7.2 Hodgkin's proof of the local circuit hypothesis, using a squid's giant nerve fibre resting on a series of platinum strips. In the Figure, the coil represents a device for stimulating the nerve with a small electric shock, and the meter a device for recording the action potential. Raising the mercury to connect the strips electrically increased the speed of conduction. (It is necessary to have a series of strips because the return circuit is through the fluid in the nerve, and the resistance of the whole length of nerve would be too great.) (From A. L. Hodgkin.)

that Bernstein's theory needed an important modification. If, as Bernstein suggested, the action potential was caused by the transient loss of selectivity by the membrane, so that it became equally permeable to sodium and potassium ions, the potential difference across the membrane should drop to zero. This was not easy to test because with most nerve or muscle preparations there was, at that time, no way of making electrical connection to the interior of the fibre without damaging the membrane and so causing a partial short-circuit. Hodgkin and Huxley therefore took advantage of the size of the giant nerve fibres in the squid. Because these fibres are nearly a millimetre in diameter, they were able to poke a fine glass tube containing salt solution along the inside of a fibre, so that the (open) end of the tube was a long way from the cut end of the fibre; in this way they could measure the difference in electrical potential between the solution in the glass tube and the solution bathing the outside of the fibre. This gave them a measure of the potential difference across the membrane when the nerve was inactive. They then stimulated the fibre with an electric shock, and were astonished to find that instead of dropping transiently to zero, the potential difference across the fibre membrane transiently *reversed* in sign, so that momentarily the inside of the fibre was at a higher electrical potential than the outside.[9] The explanation of this reversal is that during the action potential the membrane does not lose its selective permeability as Bernstein had thought, but becomes, transiently, *much more permeable to sodium ions than to potassium ions.* The potential therefore approaches the value it would have if the solutions on either side of the nerve fibre membrane were separated by a membrane permeable only to sodium ions.

Three weeks after Hodgkin and Huxley's first successful impalement of a squid giant nerve fibre, Hitler invaded Poland; and it was not until six years later that they were able to resume their experiments. During the war, K. C. Cole, working in the United States, showed that during the passage of an action potential there is a drop in the electrical resistance of the nerve fibre membrane, as predicted by the membrane theory, but unequivocal evidence for that theory came only after the war. It came from a wealth of experiments of several kinds and by many people, but particularly from a series of now classical experiments reported by Hodgkin and Huxley between 1945 and 1952. This is not the place for a review of the evidence, much of which is highly technical, but the upshot of it can be described in a few paragraphs.

The electrical potential difference across the membrane of the resting nerve does indeed reflect the much greater permeability of the membrane to potassium ions than to any other ions, though the situation is not quite an equilibrium.

The action potential is more complicated than Bernstein envisaged. It involves two sets of channels, one selective for sodium ions and the other for potassium ions. Both kinds of channel are normally closed, but both can be opened by a reduction in the potential difference across the membrane. There is, though, an important difference. The sodium channels open immediately but stay open only for a very short time; they then close spontaneously. The potassium channels open with a slight delay, but remain open as long as the potential difference across the membrane is below its resting level. The significance of these curious differences will appear in a moment.

An action potential is initiated by a small (though not too small) reduction in the potential difference across the membrane. As soon as this reduction occurs it will open sodium channels, leading to a large influx of Na^+ ions, causing a further fall in the potential difference and hence the opening of more sodium channels . . . and so on. Eventually (actually in less than a millisecond) so many sodium channels will be open that the sodium permeability will be much higher than the potassium permeability and the potential difference across the membrane will have fallen so far that it reverses in sign – the inside of the nerve fibre will be positive relative to outside. By this time, though, the sodium channels are closing spontaneously, and the potassium channels are beginning to open. The result is that K^+ ions leave through the potassium channels, restoring the potential difference across the membrane to its resting level; as this level is approached the potassium channels close. We are now back where we started except that the nerve has lost a very small fraction of its K^+ ions and slightly increased its content of Na^+ ions. For a single action potential the amounts are almost negligible, but after a burst of action potentials the amounts are measurable.[10]

If the initial reduction in the potential difference across the membrane is too small, the spontaneous closure of sodium channels will be able to keep up with the opening of further channels, and the sodium permeability will never be sufficient to cause an action potential. That is why there is a threshold for excitation. Provided that threshold is exceeded, the sequence of reactions will proceed all the way – hence the all-or-nothing behaviour.

FASTER FIBRES

Because the passage of the impulse along the nerve fibre depends on the excitation of each bit of nerve by local currents from the preceding bit of nerve, anything that increases these currents will increase the speed of transmission. We have already seen how connecting together the platinum strips in Hodgkin's experiment increased the speed of conduction of the impulse along the giant nerve fibre of the squid. That increase in speed was the result of reducing the resistance to the flow of electric current along the outside of the nerve fibre. But the circuit along which the local currents flow also includes the fluid inside the fibre. The greater the diameter of the fibre, the less will be the resistance to the flow of current along the inside of the fibre and therefore, other things being equal, the faster the fibre should conduct.[11] And it turns out that fatter fibres do indeed conduct faster.

Speed achieved in this way, though, has a cost; it is bought at the expense of bulk. Where the message to be transmitted is very urgent, but also very simple, the increase in bulk is not a problem. When a squid finds itself in a threatening situation, impulses pass down the giant nerve fibres to the muscles, making the mantle contract, and squirting a jet of water forwards so that the squid is propelled sharply backwards. These giant fibres, which conduct impulses at about 25 metres per second, are nearly a millimetre in diameter, but this does not matter because a small number of fibres can conduct the very simple message which is all that is necessary to make the different parts of the mantle contract. However, there are many situations in the life of an animal – the catching of prey, the avoidance of predators – where it would be a great advantage to be able to transmit a great deal of information quickly. In such situations, the very limited capacity of each nerve fibre (imposed by the all-or-nothing law) means that to transmit a great deal of information a large number of nerve fibres are needed; and if, to be fast, the fibres needed to be large their bulk would become unmanageable. When we watch television, all the information we see on the screen has been transmitted down a single wire from the aerial, but to get the information from each eye to our brain requires an optic nerve containing about a million nerve fibres. If, like the largest giant fibres of the squid, each nerve fibre were one millimetre across, our optic nerves would be about a metre in diameter. So it is not surprising that many animals – including all vertebrates – have nerves that employ a more compact way of increasing the speed of conduction.

In Figure 7.3a, which shows the flow of current between the active and inactive regions of a nerve fibre conducting an impulse, the current density is greatest at the junction of the active and inactive regions, and the further we go from that region the smaller the current flow is – simply because the longer the path taken by the current the greater the total resistance. The result of this pattern of current flow is that each bit of nerve membrane is excited in turn; and the time required for this to happen makes the conduction of the impulse along the nerve rather slow. This difficulty is avoided in many vertebrate nerve fibres, by the presence of a fatty sheath, surrounding the fibre, which is interrupted at regular intervals of 1–3 millimetres – see Figure 7.3b. The consequence of this arrangement is that it is only at the points at which the sheath is interrupted – the *nodes of Ranvier* – that current can flow into the nerve and excite the nerve membrane.[12] The relatively slow changes in membrane permeability that give rise to the action potential therefore take place only at the nodes, and transmission between the nodes is simply by the conduction of an electric current through a salt solution, which is almost instantaneous. The result is that transmission is very much faster, and this increase in speed is obtained without a large increase in the overall diameter of the fibre. Because the fatty material of the sheath is called myelin, fibres of this kind are said to be myelinated.

In our own bodies, myelinated nerve fibres carry impulses from our sense organs to the brain, and from the brain and spinal cord to our voluntary muscles, at speeds of 10–100 metres per second. In contrast, small unmyelinated fibres involved in the control of our internal organs conduct at speeds of only 1–2 metres per second. It is the patchy loss of myelin from nerves in the brain and

FIGURE 7.3 Conduction along (a) an unmyelinated and (b) a myelinated nerve fibre. In the myelinated fibre, conduction *between* the nodes is purely by flows of electric current in the solutions inside and outside the fibre, and is therefore extremely fast. (Based on a Figure from A. L. Hodgkin.)

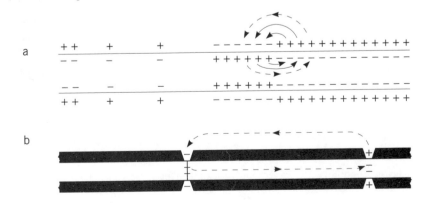

spinal cord that causes the progressive disabilities suffered by people with multiple sclerosis.

THE MOLECULAR MACHINERY

When Hodgkin and Huxley completed their work on nerve conduction in the middle 1950s nothing was known of the molecular machinery responsible for the changes in permeability to sodium and potassium ions. It was not even known whether there were separate kinds of channel for sodium ions and for potassium ions, or a single kind of channel which changed selectivity during the course of the action potential. The idea of a single kind of channel seemed less attractive once it became clear that it was possible to block the changes in permeability to the two species of ion independently. The Japanese puffer fish is a great delicacy, but it is served only in specially licensed restaurants because its ovaries, liver, skin and intestines contain a very powerful paralytic poison called tetrodotoxin. This toxin, and related toxins in scorpion venom and in certain red marine algae, block the sodium channels but have no effect on the potassium channels. Other toxins block the potassium channels but have no effect on the sodium channels. All the channels are protein molecules,* and because the various toxins bind very tightly to the channels that they block, it has been possible to use them to extract the channel proteins from the membranes of nerve fibres and then to purify the extracted proteins.

More detailed information has come from looking at the genes that contain the genetic information used to synthesize these channel proteins. In 1984, Shosaku Numa and his colleagues in Kyoto succeeded in isolating the gene that controls the synthesis of the main part of the sodium channel; by determining the sequence of nucleotides in this gene they were able to deduce the amino-acid sequence in the protein forming the channel.[13] It turned out that the protein consists of a single chain of amino-acids with four similar regions, which are believed to surround a water-filled space – the actual channel along which the sodium ions pass. These regions are also rich in amino-acids that carry positive charges, which makes them sensitive to the electrical potential difference (or voltage) across the membrane. Changes in this potential difference cause small movements of the charged regions, opening or closing the channel to the passage of sodium ions, though to account for the spontaneous closure of sodium channels (p. 113) the story has to be more complicated. Work on potassium channels suggests that the basic mechanism is similar.

During the 1980s and 1990s it became clear that, as well as the classical

---

*Somewhat confusingly, the word channel is used both for the passage along which the ions pass and for the protein molecule containing the passage.

sodium and potassium channels whose existence was deduced by Hodgkin and Huxley, the membranes of excitable cells may contain a variety of other channels permeable to particular ions. Some of these channels resemble the classical sodium and potassium channels in that the opening and closing of the channel is controlled by the potential difference across the membrane. In the rather expressive jargon of those who work in this field, they are said to be 'voltage-gated'. In other channels, the opening and closing is controlled by the binding or release of particular substances released by other cells; they are said to be 'transmitter-gated'. A large part of the modern pharmaceutical industry is concerned with the development of drugs that directly or indirectly affect the behaviour of such channels.

HOW DID THE MECHANISM EVOLVE?

At first sight the action potential mechanism, dependent on the existence of sodium and potassium gradients across the cell membrane, and on voltage-gated sodium and potassium channels with rather special properties, appears to present an evolutionary problem. But, like the eye or other 'organs of extreme perfection and complication', its evolution can be explained in a stepwise fashion.

There is little doubt that the sodium and potassium gradients preceded the action potential mechanism. Ion gradients of various kinds are found across the membranes of many plant, fungal and bacterial cells, and sodium and potassium gradients like those across the membranes of nerve cells are found in almost all animal cells – most of which, of course, are not excitable and cannot produce action potentials. Furthermore, the 'pump' that is responsible for maintaining the sodium and potassium gradients across the membranes of animal cells resembles, both in its mechanism and its molecular structure, pumps that transport hydrogen and potassium ions in plants, fungi and bacteria.[14] Given a cell with sodium and potassium gradients across its membrane, any inequality in the permeabilities to sodium ions and potassium ions will generate a potential difference across that membrane. And there are quite likely to be inequalities because in aqueous solution potassium ions are smaller than sodium ions. In a cell with a potential difference across its membrane generated in this way, anything that causes a local change in permeability will cause a local change in that potential difference, and electric currents will flow between the changed region and the adjacent regions. These local currents will alter the potential difference across the membrane in the adjacent regions; and *if* that alteration causes an appropriate change in permeability we shall have a propagated action potential. That may seem a big 'if', but action potential mechanisms seem to have evolved more than once. The giant algal cell *Nitella* conducts a rather slow action potential that has been shown to depend on the transient opening of voltage-gated channels; and in the Venus fly-trap (an insectivorous plant) an action potential produced in a similar way causes the sudden springing of the trap that catches the insect.

*   *   *

We have now spent two chapters discussing the machinery that transmits nervous impulses – action potentials – along nerve fibres, without paying any attention to the way in which useful information is carried along the fibre by the stream of impulses. How, we want to know, is the message encoded? The rather surprising answer to that question is discussed in the next chapter.

8 Encoding the Message

The medium is part of *the message*

There is a striking difference between the way in which we receive information by post or telephone and the way in which our brains receive information from our sense organs. The information we derive from a letter or a telephone call is all in the message itself. The same postman brings us different letters, and even if we have more than one telephone line we are not normally concerned to know which line is in use when we receive a call. In contrast, the messages that our brains receive from our sense organs – our eyes, our ears, the sense organs in our skin – all consist of streams of very similar impulses. Because of the all-or-nothing behaviour of nerve fibres, the size of the impulses carries no useful information, so the only information available from the impulses arriving along a given fibre must come from their timing and from identification of the fibre by which they arrive. Somehow then, these features must tell us all we need to know about the initiating stimulus – what sort of stimulus it is, where it is, how intense it is, and how long it lasts.

You might envisage a system, employing all-or-nothing action potentials, in which all the information about the stimulus was encoded in intricate timing patterns – particular rhythms of impulses having particular meanings – but nothing of this kind has ever been found. This is almost certainly because such patterns would not be stable. A nerve fibre that has just conducted a number of impulses in very quick succession conducts more slowly than a nerve fibre that has been at rest. In a rapid burst of impulses there is therefore a tendency for the later impulses to lag progressively, and any code that depended on *precise* timing would not be reliable. In fact, only two straightforward codes appear to be used: a *labelled-line code*, used mainly to convey information about the site and nature of the stimulus, and a simple *frequency code*, used mainly to convey information about its intensity and duration.

Labelled-line code is the unfamiliar jargon phrase used to describe a familiar system. It is the system that allowed the servants in a large Victorian house to know in which room someone was demanding attention – each room having a bell-pull connected by its own wire to one of a row of bells in the servants' quarters. Since, to be sensitive at all, each area of skin (or any other tissue) needs to have sense organs of some kind, with nerve fibres connecting those sense organs to the brain, it is not surprising that the brain should get information about the *site* of a stimulus simply from the identity of the nerve fibres along which impulses are arriving. It is much more surprising that our ability to identify the *nature* of a stimulus – whether it is heat or cold or touch or pressure or vibration or light of a particular colour – depends on information conveyed to the brain by separate nerve fibres connected to sense organs that are selectively sensitive to stimuli of different kinds.

The most striking evidence that the coding is of this type is that if, for some reason, the nerve is stimulated by something other than the normal stimulus, the sensation detected still has the same character. A blow on the eye makes us 'see stars' because, although the stimulus to the retina is mechanical, the impulses arriving at the brain as the result of that mechanical stimulus are treated by the brain as if they were caused by light falling on the retina. Patients who are completely deaf as the result of damage to the cochlea – the organ in the inner ear that is sensitive to sound – hear sound when the nerve from the cochlea is stimulated electrically. The skin contains a variety of separate sense organs individually sensitive to heat, cold, light touch, firm pressure, and vibration. Because they are separate, the sensitivity of the skin to the different modalities varies from point to point, so that you can pick out areas about a millimetre across that are particularly sensitive to warmth or cold. If you find a 'cold spot' and stimulate it, not with something cold but with a small probe heated to about 45°C – a temperature that would be painfully hot if applied over a wide area – the sensation felt is that of cold. Because the brain relies on labelled-line coding to identify the nature of the stimulus, it gets the wrong answer when stimulation is caused by an abnormal stimulus. This was realized by Descartes in the 17th century,[1] and by Johannes Müller in the second quarter of the nineteenth.

Because it is not possible to have a receptor and labelled line for every possible stimulus, recognition of many stimuli depends on the ratio of the extents to which receptors of different kinds are stimulated. We shall see examples of this later in connection with seeing, hearing and smelling.

Frequency coding was first demonstrated in the 1920s by E. D. Adrian and his colleagues, who showed that the frequency of impulses in nerve fibres carrying information from tension receptors in muscles, or from various sense organs in the skin (touch receptors, pressure receptors, hairs), reflected the intensity of stimulation – a stronger stimulus generating more frequent impulses. With the tension receptors in muscle, a maintained stimulus (i.e. a constant pull on the muscle) led to the transmission of impulses at a nearly constant frequency, so that the brain was continuously informed of the strength of the stimulus. With the receptors in the skin, the response to a maintained stimulus – touching the skin, for example – was different. Touching the skin led to a burst of impulses, but the frequency fell quickly to zero (or to some resting level) and then remained there until the touch ceased, when a further burst of impulses was seen. In other words, the touch receptors responded to sudden changes, but they *adapted* to a constant stimulus and initiated no impulses so long as the stimulus was constant. For touch then, the brain is informed of the beginning and the end of the stimulus but is given no information in between. This difference between the tension receptors in muscle and the touch receptors in skin obviously makes sense, since in order to maintain posture or to control movement our brains need continuous information about the pull on relevant muscles, but in general we don't need to be informed continuously about objects touching our skin; it is sufficient, and more economical, for the touch receptors to transmit information only when there is a change. This ability of touch receptors to adapt quickly to a constant stimulus is an example of an important phenomenon that is characteristic of nervous systems: their ability to discard information likely to be redundant.

Although information about the intensity of a stimulus is usually transmitted to the brain by frequency coding, and information about quality by labelled-line coding, there are exceptions. If there are many receptors at a site and they vary in sensitivity, a strong stimulus will excite more nerve fibres than a weak stimulus. Such 'recruitment' of more nerve fibres by a stronger stimulus occurs, for example, in the balancing organs in the ear. And in many situations, making the stimulus stronger may succeed in stimulating receptors further away from the point at which the stimulus is applied. Here, then, labelled-line coding is being used to transmit information about intensity. Conversely, our ability to discriminate between the pitches of low sounds is thought to involve an unusual type of frequency coding.[2]

If stimuli are very strong – intense heat, intense cold, severe pressure – besides their specific sensation, they also cause pain.* This might suggest that pain from, say, intense heat is felt when the brain receives a very large inflow of impulses along the nerves that carry impulses from the heat receptors; that pain from intense cold is felt when the brain receives a large inflow of impulses along the nerves that carry impulses from the cold receptors; and so on. But this turns out to be wrong. It is now known that pain depends on a flow of impulses to the brain along separate nerve fibres whose endings are sensitive to a variety of stimuli that either cause, or are associated with, local damage to the tissues – intense heat, intense cold, mechanical trauma, and the presence of chemical irritants (including certain substances released from damaged tissues). These nerve fibres are of two kinds: small *myelinated* fibres that conduct at between 5 and 30 metres per second, and small *unmyelinated* fibres that conduct at about one metre per second. The endings of the faster fibres tend to be most sensitive to mechanical or thermal stimuli, and their activity is associated with sudden sharp pain. These

---

*Although we talk of the sensation of pain in much the same way as we talk of the sensations of heat or cold or pressure, pain is unlike heat and cold and pressure in not being a stimulus.

endings adapt quickly, so the pain soon fades. The endings of the slower fibres tend to be sensitive also to chemical stimuli, and their activity is associated with a slow burning or aching pain. These endings adapt slowly, if at all, so the pain persists. It is because of the different characteristics of the two kinds of nerve fibre concerned with pain that, when you stub your toe, you get a sudden sharp pain followed about a second later by a persistent aching pain. The evolutionary advantage of having two pathways for information about noxious stimuli, one quick and rapidly adapting, and the other slow and non-adapting, is obvious. The quick pathway makes a rapid withdrawal possible; the slow non-adapting pathway acts as a continuous warning.

9   Interactions Between Nerve Cells

If Galen or Descartes, miraculously resurrected, were to ask us whether we yet understood how messages are transmitted along nerve fibres, we should be able to give them a satisfactory and rather detailed answer. But nerve fibres are merely the extremely elongated portions of nerve cells, and the transmission of messages along them can do no more than ensure that information available at one end of the fibre is also available at the other end. The more interesting things that our nervous systems do all involve interactions between different nerve cells, or between nerve cells and other excitable cells.

If understanding the ways in which excitable cells interact seems a matter of detail, and a long way from important questions about minds, remember that such interactions are massively involved in everything we do or think or feel, and that a large part of the pharmaceutical industry is engaged in developing drugs that affect our thoughts, our feelings and our moods, by affecting these interactions. Drugs whose actions are described – with varying accuracy – as sedative, hypnotic, analgesic, tranquillizing, anxiety-reducing, mood-enhancing, antidepressant, anti-psychotic, hallucinogenic, or psychedelic, produce most of their effects in this way. It turns out, too, that learning and memory also depend on subtle changes in the ways in which nerve cells excite one another.

The most powerful roles of the interactions between nerve cells are found in the workings of central nervous systems (the brain and spinal cord in vertebrates), so it is helpful to begin by looking at the general character of these workings.

It is the job of an animal's central nervous system to ensure that the behaviour of the whole animal, and of each part of it, is appropriate to the current situation – a situation of which the system is made aware by information arriving along sensory nerves. That job is essentially the same as the job of control systems in machines made by human beings. The automatic landing system in an aeroplane uses information about the aeroplane's altitude, tilt, air-speed, engine-speed, position in relation to ground beacons and so on, as well as stored information about the characteristics of the aeroplane, to control the rudder, wing-flaps, engine-speed and whatever, in such a way as to achieve a smooth landing. The computer programmed to make decisions about buying or selling shares uses published information about various stock-market indices, about individual companies, and about the movements of individual share prices, in such a way as to maximize the value of the portfolio. We are not concerned here with the detailed mechanisms of these man-made control systems but it is characteristic of them that they contain 'integrating devices' – that is, devices that allow a single output to be determined by the values of several different inputs – and these devices are often arranged so that the output of one device can be one of the inputs of another. It is arrangements of this kind that give such systems the ability to use information to make appropriate decisions. To obtain the information, the control system must, of course, contain or be suitably connected to devices (sensors) that monitor events and make the information about them available in a usable form; and to make the decisions effective the control system must also be connected to other devices (effectors) that implement the instructions that are issued.

Turning now from machines to central nervous systems, we find essentially the same kind of arrangement. Figure 9.1 shows a typical nerve cell (shaded) and parts of some other nerve cells (unshaded) that interact with it. Each nerve cell (or neuron) consists of a compact cell body bearing two sorts of projection: several highly branched projections called dendrites (from the Greek word for tree), and a much longer one which usually (except in the cerebral cortex) has branches only at its far end. This single long projection – which may be up to a metre in length – is called the nerve fibre or *axon* (from the Greek word for axle), and it is the conduction of impulses along such axons that we have been considering. In Figure 9.1, the fine branches at the end of the axon of the shaded nerve cell can be seen to expand into small knobs which are in contact with the dendrites or cell body of other nerve cells. Similarly, the cell body and dendrites of the shaded nerve cell are in contact with knobs at the ends of terminal branches of axons of nerve cells which are outside the Figure. These points of contact are called *synapses* (from the Greek word for connection), and it is events at synapses on the dendrites and cell body of a nerve cell that determine whether that cell initiates an impulse which then passes along its axon. I shall say more about these events in a moment, but what *causes* an 'event' at a synapse is the arrival of an action potential at the synaptic knob. Since there are many synapses but only a single axon, the arrangement ensures that the single output of the shaded cell (the stream of impulses that pass along its axon) will be affected by many inputs (the impulses that arrive at the numerous synaptic knobs in contact with the dendrites or cell body). And since both outputs and inputs are simply nerve impulses – all-or-nothing action potentials – any output can be used as an input in another part of the system provided that the geometry of the nerve network allows it to get there.

Despite the resemblance between the arrangements of inputs and outputs in nerve cells and the formal structure that provides the basis of many man-made control systems, there are obvious differences between central nervous systems

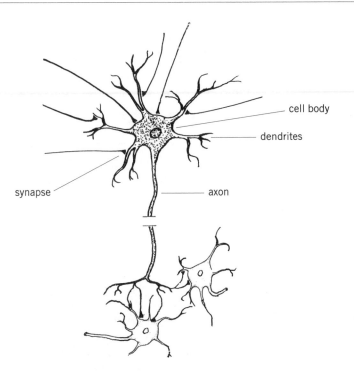

FIGURE 9.1 A nerve cell (or neuron) as an integrating device. The output of the shaded nerve cell – the number of impulses passing down its axon – depends on the inputs to it arriving along axons from other nerve cells.

and man-made systems. In the first place, the basic aims of central nervous systems are set by the evolutionary history of the animal rather than by an inventor; they are the outcome of natural selection. Secondly, in many animals, the control exercised by the central nervous system appears to be much more sophisticated than anything achieved by machines; and in the highest animals the exercise of that control is sometimes associated with consciousness. Both great sophistication and consciousness, however, seem to be associated only with central nervous systems that contain very large numbers of nerve cells, connected in very complicated ways. The human brain contains at least 10^{11} nerve cells, and since, on average, each is involved in more than 1000 synapses, there must be something like 10^{14} synapses. So it is not surprising that the human brain can control behaviour that is very sophisticated indeed. Whether an artificial control system of similar complexity, and connected to suitable sensors, could also be conscious is a question we shall return to; for the moment I want to concentrate on the nature and role of the events at individual synapses.

It turns out that events at a synapse may either increase or decrease the chance that an impulse will be initiated by the nerve cell whose dendrites or cell body form part of the synapse. A given synapse, though, always acts in the same way, so we can talk about excitatory and inhibitory synapses. What determines whether the shaded nerve cell in Figure 9.1 initiates an impulse or not – whether the cell 'fires' as the neurophysiologists say – therefore depends not only on the frequency of impulses along each of the axons whose terminal branches form synapses with that cell, but also *on whether each of those synapses is excitatory or inhibitory.* Remarkably, this notion that the behaviour of a nerve cell reflects the sum of excitatory and inhibitory effects on it was proposed by Charles Sherrington, on the basis of his studies of reflex actions, long before anything was known about the physiology or the detailed anatomy of synapses.

THE MECHANISMS OF SYNAPTIC TRANSMISSION

In discussing the ways in which synapses work, it is convenient to refer to the cell along whose axon the action potential reaches the synapse as the *presynaptic cell,* and the other cell forming the synapse – the cell whose excitability is being increased or decreased – as the *postsynaptic cell.* At the synapse, these two cells are separated by a very narrow gap – the *synaptic cleft* – that is usually about 20 millionths of a millimetre wide. The crucial question is: How does the arrival of an action potential at a nerve terminal of the presynaptic cell increase or decrease the likelihood that an action potential will start in the cell body of the postsynaptic cell?

This question was first asked – and some eighty years later first answered – about a synapse that is so atypical that it is not usually referred to as a synapse at all but as a *neuromuscular junction.* This 'junction' is the point at which a terminal branch of a motor nerve fibre lies very close to the surface of the muscle fibre that it excites; and it is peculiar as a synapse because the arrival of a *single* impulse along the nerve fibre is sufficient to generate an action potential in the muscle fibre. (It is the action potential in the muscle fibre that activates the contractile machinery so that the muscle contracts.) In 1877, Du Bois-Reymond wondered how the excitation of a muscle by its nerve might be explained, and he suggested two alternatives. *Either* the muscle might be excited by local electric currents caused by the action potential in the nerve; *or* the nerve terminal might release some substance that excited the muscle. The discovery, early in the following century, that the actions of certain nerves could be mimicked by adrenaline,* and those of certain other nerves by muscarine (a substance obtained from *Amanita muscaria,* the familiar scarlet, or in north America orange-yellow, mushroom with white spots) supported the notion that nerves may produce their effects by chemical rather than electrical means, but did not prove it.

---

*Known as epinephrine in the USA.

The first proof that transmission at a nerve ending could be chemical came from a famous experiment by Otto Loewi, working in the Austrian town of Graz, in 1921.[1] Years later, Loewi recalled that he had had the idea for this experiment in a dream – indeed in two dreams, for although he had written down the idea when he had woken in the middle of the night, he had been unable to decipher his writing in the morning. Fortunately, the dream recurred. What he did in his experiment was to stimulate the branch of the vagus nerve going to a steadily beating frog's heart through whose chambers a suitable salt solution was being slowly circulated. One of the actions of the vagus nerve is to slow the heart, and, as he expected, the heart slowed. He then transferred to a second heart the salt solution present in the first, and he found that the second heart also slowed. As the only connection between the two hearts was the transferred fluid, it followed that stimulation of the vagus nerve to the first heart must have released a substance capable of slowing frog hearts. Loewi was not able to isolate and analyse the active substance, so he simply called it Vagus-Stoff (vagus stuff). It was later shown to be acetylcholine.

Although Loewi's experiment showed that the vagus nerve slowed the heart by releasing an active substance, the terminal branches of the vagus nerve do not form synapses in the heart, so the experiment did not prove that transmission *at synapses* could be chemical. Proof of that came some years later from experiments by Henry Dale and his colleagues on muscles that, unlike heart muscle, do not contract spontaneously but only when stimulated through their nerves.[2] They showed that stimulation of the motor nerves to the tongue of an anaesthetized cat or to the leg muscle of an anaesthetized dog was accompanied by the release, in the muscles, of something with properties indistinguishable from those of acetylcholine. What is more, injection of a small quantity of acetylcholine into the artery supplying a small muscle made the muscle twitch. Acetylcholine is quickly destroyed by an enzyme present in many tissues, including muscle, and Dale and his colleagues found that inhibition of this enzyme by a suitable drug increased the size of the muscle twitch; and this was true whether the twitch was caused by the arrival of an impulse along the nerve or by the injection of acetylcholine into the artery. These experiments provided strong evidence that the release of acetylcholine was not some side effect but was itself the cause of excitation of the muscle.

Dale's clinching evidence came from experiments with the paralytic drug curare, the active ingredient in the arrow poison used by South American Indians. Nearly a century earlier, the distinguished French physiologist Claude Bernard had found that, after paralysis had been induced with curare, the supposedly paralysed muscle still contracted if it itself was given an electric shock; he also found that the nerve supplying the muscle was still able to conduct action potentials. It seemed, then, that the effect of curare was not on the nerve itself or the muscle itself but was on events at the neuromuscular junction. Dale and his colleagues showed that, as well as blocking neuromuscular transmission, curare prevented the contraction of a muscle that follows the injection of acetylcholine into its artery – just what would be expected if the release of acetylcholine from the nerve terminal is what normally excites the muscle.[3]

These experiments of Dale and his colleagues not only answered Du Bois-Reymond's question – establishing that transmission at the neuromuscular junction is chemical rather than electrical, and that acetylcholine is the transmitter – they also established the kind of criteria that have been used ever since to identify synaptic transmitters at other kinds of synapse.

Identifying the transmitter is, of course, only the first step in elucidating how a chemical synapse works. Two further questions need to be answered: How does the arrival of the nerve impulse at the nerve terminal of the presynaptic cell cause the release of transmitter into the synaptic cleft? And how does the presence of transmitter in the synaptic cleft affect the postsynaptic cell? It was work on the neuromuscular junction which first provided answers to these questions and led the way to an understanding of other synapses.

Exciting the muscle fibre

At the neuromuscular junctions in the muscles of vertebrates, the motor nerve fibre divides into very fine terminal branches which lie in grooves in the cell membrane of the muscle fibre. The area of the muscle cell membrane in which the fine terminal branches of the nerve are embedded – referred to as the *motor end-plate* – has special properties. It contains protein molecules which can act as 'gated' ion channels (see p. 117), the 'key' to the gate being acetylcholine. Normally the channels are closed, but when molecules of acetylcholine are present in the synaptic cleft and bind to special sites on the outer face of the protein molecules, the channels open. While a channel is open, sodium, potassium and calcium ions can all cross the membrane in either direction, but the main effect of channel opening is a rapid entry of sodium ions.[4] The result of the opening of the acetylcholine-gated channels is therefore an inward movement of positive charge, and a consequent reduction of the potential difference across the membrane in the region of the motor end-plate – a reduction known, somewhat illogically, as the *end-plate potential*. How much the potential difference is reduced depends on how many acetylcholine-gated channels are opened. If enough channels are opened the fall in the potential difference will cause local currents big enough to excite the muscle fibre.

Within the last fifteen years, the acetylcholine-gated channels at motor end-plates have been isolated, and their approximate structure has been worked out.[5] Just as the isolation of the voltage-gated sodium channels from nerves was achieved with the help of toxins from the Japanese puffer fish and from certain marine algae, so the isolation of the acetylcholine-gated channels from motor

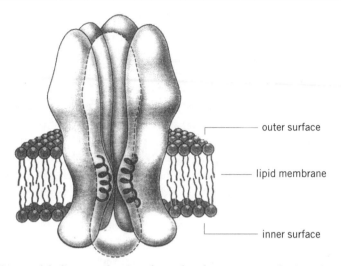

outer surface

lipid membrane

inner surface

FIGURE 9.2 The acetylcholine-gated cation channel at the neuromuscular junction –
the best studied of all transmitter-gated channels. (From N. Unwin.)

end-plates was achieved by taking advantage of their ability to bind very tightly
to toxins extracted from cobra venom or from the venom of a smaller Indian
snake, the banded krait. And to add to the witches' cauldron, the best source of
acetylcholine-gated channels proved to be the electric organ of the *Torpedo* or
stingray – the same fish that Meno accused Socrates of resembling because of the
numbing effect of his arguments. It turns out that each channel consists of five
protein subunits arranged in a circle to form a pore, which is the pathway for the
ions (Figure 9.2). Negatively charged groups projecting into the pore repel all
negatively charged ions and make it selectively permeable to ions carrying positive charges. Part of the pore is narrower than the rest, and this is thought to be
where closing of the channel occurs – probably by a concerted slight twisting of
the subunits. The binding of two acetylcholine molecules to sites on the outer
surface of two of the subunits reverses this twisting and opens the pore.

Releasing the transmitter

In 1952, Paul Fatt and Bernard Katz, working with frog muscles at University
College, London, noticed something very surprising.[6] They were measuring the
potential difference across a motor end-plate, when they saw that, from time to
time, there were tiny transient fluctuations in the potential difference *even when
no impulses were arriving along the nerve fibre*. The fluctuations took the form of
sudden drops (nearly always of about 1 millivolt) in the potential difference
followed by a slower return to the original level. Apart from their size, they were

reminiscent of the much larger drops in the potential difference across the membrane that occur when nerve impulses arrive at a motor end-plate – the *end-plate potentials* – so Fatt and Katz called them *miniature end-plate potentials*. Like end-plate potentials, the miniature end-plate potentials were prevented by curare and they increased in size in the presence of drugs that inhibit the enzyme that destroys acetylcholine. It looked, then, as though the miniature end-plate potentials were caused by the release of acetylcholine, and to account for their uniform size it was necessary to suppose that packets of acetylcholine of roughly uniform size were being released spontaneously and randomly from the nerve terminal.

There were, then, two crucial questions. What were these packets that contained the acetylcholine? And could the full-size end-plate potential – the event that actually excites the muscle fibre – be caused by the simultaneous release of many of these packets?

The first question was answered by electron microscope studies, which showed that nerve terminals at the neuromuscular junction contain large numbers of small vesicles of uniform size – vesicles which were later shown to contain acetylcholine. The second question was largely answered by further experiments on the frog neuromuscular junction by Katz and his colleagues.[7] They showed that *reducing the electrical potential difference across the membrane of the nerve terminal* (by passing current between electrodes placed on the nerve near the terminal) increased the frequency of the miniature end-plate potentials. By reducing the potential difference sufficiently they could increase the frequency over a hundredfold until the individual miniature end-plate potentials could no longer be distinguished, and the response looked like a full-size end-plate potential.

One link in the chain of events was still unexplained. How does reducing the potential difference across the membrane of the nerve terminal increase the frequency with which the vesicles release their contents to the exterior? The clue to the solution of this problem was the requirement for calcium ions in the bathing solution. Motor nerve terminals have in their membranes a large number of voltage-gated channels that are selective for calcium ions. Reduction of the potential difference across the membrane opens these channels, and this leads to a sudden inflow of calcium ions. In ways which are still not fully understood, but which involve several specialized proteins in the vesicle membrane and in the membrane of the nerve terminal, calcium ions lead to the fusion of these two membranes, followed by the formation of a pore which then widens releasing the vesicle contents to the exterior.[8] Tetanus toxin, botulinus toxin, and the toxin of the black widow spider all owe their toxicity to their effects on the processes involved in the release of vesicle contents at synapses.

Despite the number of links in the chain of events that leads from the arrival of a nerve action potential at the motor nerve terminals to the contraction of the muscle fibre, the *synaptic delay* – the time between the arrival of the nerve action potential at the nerve terminals and the beginning of the fall in the potential difference across the muscle membrane at the motor end-plate – is less than one millisecond.

SYNAPSES IN THE CENTRAL NERVOUS SYSTEM

I have discussed the events at the vertebrate neuromuscular junction at length because it was through the investigation of these events that synaptic transmission was first understood; and the neuromuscular junction is still the best understood of all synapses. Its job, though, is quite different from that of most other synapses; it is simply to ensure that each impulse arriving along the nerve fibre excites the muscle fibre. Much more interesting are the synapses in the central nervous system, whose job it is to form part of integrating devices each of which allows many inputs to influence a single output.

The striking feature of these synapses is that they may be excitatory or inhibitory, and in the early 1950s John Eccles and his colleagues in Dunedin and Canberra decided to see what determined whether the arrival of an impulse at a particular synapse tended to excite or inhibit the post-synaptic cell. To do this they looked at the behaviour of motor neurons in the spinal cord of the anaesthetized cat during reflex actions in which the motor neuron was either excited or inhibited.[9] (Motor neurons are the nerve cells whose axons excite muscles at neuromuscular junctions.)

What they found was that the arrival of a single impulse along the presynaptic fibre going to a synapse on the motor neuron (the postsynaptic cell) caused a small and transient swing in the electrical potential difference across the membrane of the motor neuron, *but the direction of the swing depended on whether the synapse was excitatory or inhibitory*. At excitatory synapses, the arrival of an impulse caused a small transient fall in the potential difference across the membrane of the motor neuron; at inhibitory synapses there was a small transient rise. Although the fall caused by the arrival of an impulse at an excitatory synapse (unlike the fall caused by the arrival of an impulse at the neuromuscular junction) was itself much too small to excite the motor neuron, the simultaneous or nearly simultaneous arrival of enough impulses did change the potential difference across the motor-neuron membrane sufficiently for the neuron to fire. Arrival of impulses at inhibitory synapses made the motor neuron less excitable by moving the potential difference across the cell membrane even further from the threshold for firing.

It seems then, that at each moment, whether a motor neuron in the spinal cord fires or not depends on the magnitudes and directions of the transient swings in the potential difference across the cell membrane caused by nerve impulses arriving at synapses on the cell body and dendrites. The probability of firing is also affected by the geometrical distribution of the active synapses, because those that are situated on remote dendritic branches of the motor neuron have less influence.

Since the classical work of Eccles and others in the 1950s a great deal of work has been done on the properties of synapses in the central nervous system, and a picture of extraordinary richness has emerged, and is, indeed, still emerging. This is not the place for a detailed examination of that picture, but I want to say enough to give an idea of the power and variety of the units which, when 10^{14} of them are suitably arranged and suitably supported, seem to provide the necessary – and perhaps the sufficient – conditions for our mental life. But don't worry if you can't remember the details. What matters is the different patterns of behaviour that synapses can display.

A wealth of chemical synapses

Chemical synapses differ from one another in their roles and in their mechanisms. The most straightforward excitatory and inhibitory synapses in the central nervous system resemble neuromuscular junctions in that channels in the cell membrane of the postsynaptic cell are opened transiently by a transmitter which is released from the presynaptic cell and then promptly removed; the channels are transmitter-gated. In excitatory synapses the channels are selectively permeable to small positively charged ions, and, as in the neuromuscular junction, the main effect of their opening is a rapid inflow of sodium ions, which causes a reduction in the electrical potential difference across the membrane. In contrast, at inhibitory synapses the channels are selectively permeable *either* to small negatively charged ions – which, in effect, means chloride ions – or to potassium ions. In either case, the effect is to increase the potential difference across the nerve cell membrane and to make the nerve cell more difficult to excite.[10]

The excitatory and inhibitory synapses differ not only in the selectivity of the channels in the postsynaptic membrane but also in the transmitter that is released from the terminal of the presynaptic cell and that opens these channels. Despite the similarity between the mechanism of the main excitatory synapses in the central nervous system and the mechanism of the neuromuscular junctions, the usual excitatory transmitter in the central nervous system is not acetylcholine but *glutamate* – the charged form of the amino acid glutamic acid. The usual inhibitory transmitters are also amino acids: *gamma-aminobutyric acid* (GABA, pronounced 'gabba', for short) – one of the most surprising compounds found in the Murchison meteorite (see pages 69–70) – and *glycine*, the simplest of all the amino-acids.

All the transmitter-gated channels we have been talking about are proteins, and they show a strong family resemblance in their structure. This resemblance may be partly the result of constraints imposed by similarity of function, but a detailed comparison of amino-acid sequences points to a common evolutionary origin. Within each channel type there are variations, and the different varieties may behave in ways which allow them to have different roles – not just in stimulating or inhibiting, but also in providing the crucial machinery for making decisions, for remembering and for learning.

'Channel power'

Most transmitter-gated channels behave like gates with locks that are opened by the appropriate transmitter; whether the channel is open or closed depends only on whether that transmitter has recently been released into the synaptic cleft. The great majority of the channels responsible for rapid synaptic transmission at excitatory synapses in the central nervous system work in this way, with glutamate being the effective transmitter. There is, though, another kind of glutamate-gated channel – the so-called NMDA receptor – whose behaviour is more subtle and has had momentous consequences.

NMDA is the abbreviation of N-methyl-D-aspartate. It is a substance that does not occur naturally, so none of the NMDA receptors in your brain or mine have ever met, or are ever likely to meet, a molecule of NMDA. To call such channels NMDA receptors is therefore peculiar, but not as crazy as it looks. If you want to study one variety of glutamate-gated channels in a bit of brain that contains several varieties, you either have to separate them, or inhibit all but the kind you want to study, or try and find a substance slightly different from glutamate which opens only the kind of channel you are interested in. And it turns out that, for this last approach, NMDA fits the bill perfectly: it has no effect on the more humdrum glutamate-gated channels but opens the more interesting ones.

Why are they interesting? Their first peculiarity is that, at the normal resting potential, the channel is plugged by a magnesium ion, so glutamate has little effect until the magnesium ion is released by a lowering of the potential difference across the membrane in the neighbourhood of the channel. This means that the channel behaves as though it has two gates in series, one opened by a change in the potential difference across the membrane and the other by the presence of glutamate. A synapse involving a channel of this type *will therefore respond to glutamate only if the potential difference across the membrane is lowered by the simultaneous activity of other excitatory synapses on the same neuron.* Theoretically, at least, this implies that neurons with synapses involving such NMDA receptors could serve as 'logic gates' – opening of the gate occurring if, and only if, event 'A' and event 'B' occur simultaneously or almost simultaneously. In other words, neurons containing synapses of this kind could make formal 'decisions' of the type: 'If A *and* B, then C'.

By linking together enough decision-making devices of such simple kinds, it is theoretically possible to create a mechanism capable of controlling very complicated behaviour. But the behaviour of human beings and of many animals is not merely complicated; it demonstrates memory and the ability to learn. To explain these features, you have to postulate the existence of mechanisms that produce long-term changes in the properties of the control system, and again the peculiar properties of NMDA receptors seem to provide an answer. Their channels are not only permeable to sodium and potassium ions but are also extremely permeable to calcium ions. The binding of glutamate to NMDA receptors sitting in a membrane that has had the potential difference across it sufficiently reduced to relieve the magnesium block, therefore leads to an entry of calcium ions; these, by activating different enzymes, produce a variety of effects in the post-synaptic cell including long-lasting changes in synaptic responsiveness – just what is necessary for memory and learning. Later we shall see that there is direct evidence that NMDA receptors are involved in such processes (Chapter 19).

A puzzling but important feature of all the transmitter-gated channels we have been discussing is that they often have, as well as the receptor sites for the transmitter, other sites that bind drugs specifically and with a high affinity. For example, the GABA-gated channels at inhibitory synapses in the brain have on their outer surface, not only the receptor sites for GABA, but also sites that can bind barbiturates and different sites that can bind benzodiazepines – 'anti-anxiety' agents such as Valium and Librium, and the sleep-inducing drug Mogadon.* The binding of either kind of drug apparently increases the affinity of the GABA-binding sites for GABA, and so increases the inhibitory effect of any GABA that may be present. The usefulness of barbiturates and of benzodiazepines is thought to be related, at least in part, to these effects, but neither barbiturates nor benzodiazepines are normally found in the body; so what, if anything, normally binds to these sites? Are there physiological analogues of barbiturates, or of substance like Valium and Librium, which have a normal role in increasing the inhibitory effect of GABA? In contrast to the barbiturates and benzodiazepines, which increase the affinity of the GABA-gated channels for GABA, strychnine reduces the affinity of the glycine-gated channels for glycine. The normal effect of glycine at inhibitory synapses at which glycine is the transmitter is therefore reduced, and this is probably why strychnine causes convulsions. Synapses at which glycine is the transmitter are also involved in the convulsions caused by tetanus toxin, but here the toxin is thought to act by preventing the release of glycine from the presynaptic cell.

All these effects of drugs on transmitter-gated channels involve inhibitory synapses, but drugs can also bind to and affect channels at excitatory synapses. A

---

*The capital letters indicate that Valium, Librium and Mogadon are trade names.

dramatic example is the inhibition of NMDA receptors by the drug phencyclidine, better known by its street name 'angel dust'. Those who take this drug get
auditory and visual hallucinations, a sense of being disconnected from
their environment, and delusions of being controlled by external agents – all
features also found in schizophrenics. It is uncertain, though, whether these
'psychotomimetic' actions of phencyclidine are related to its NMDA-receptor-
blocking effect, since the drug also binds to some of the receptor sites at
which opioids (morphine-like substances) that occur naturally in our bodies
normally act, and it also increases the concentration of the neurotransmitter
dopamine at nerve terminals in the forebrain.[11]

Neuromodulators

Synapses in which transmitters act directly on ion channels usually act very
rapidly and very briefly – the NMDA receptors are an exception – and they are
responsible for most of the rapid synaptic transmission that occurs in the brain
and spinal cord. There is, however, a large group of synapses whose function is
different. At these synapses, the role of the transmitter is to cause longer-lasting
change in the way in which the post-synaptic cell responds to excitation or
inhibition. The transmitters that act at synapses of this kind are often known as
neuromodulators, and they are usually small peptides (short chains of amino-
acids bonded together as they are in proteins), though some are amino-acids or
simple derivatives of amino-acids. They act by binding to receptors, in the membrane of the post-synaptic cell, that respond by initiating a 'cascade' of enzymic
reactions – a series of steps in which the enzyme catalysing each step is activated
by a product of the preceding step. The cascade usually leads to the production
of a so-called 'second messenger' – the transmitter, of course, being the 'first
messenger'. Second messengers are small soluble molecules that diffuse rapidly
in the post-synaptic cell, where they can have a variety of effects. They may themselves bind directly to ion channels in the cell membrane, opening or closing
them and producing corresponding changes in excitability. Or they may initiate
further enzymic cascades. In this way they can open or close ion channels indirectly, they can alter the sensitivity of receptors in the cell membrane to other
transmitters, and they can regulate the synthesis of new proteins, including the
synthesis of new receptors.

   Whereas the transmitters that bind to receptors that are themselves transmitter-gated channels produce changes in excitability in something like a
millisecond, those that act by initiating cascades of enzymic reactions take
times of the order of hundreds of milliseconds to produce their effects. Those
effects also tend to last much longer – seconds or minutes rather than tens of
milliseconds; and days or still longer periods where the synthesis of new proteins
is involved. These longer-lasting effects (like the long-lasting effects caused by the

action of glutamate on NMDA receptors) are thought to be involved in learning and memory.

Although there are more than fifty neuromodulators which bind to receptors that act through the production of second messengers, all such receptors – or at any rate all such receptors whose structure has been analysed – seem to be similar; and detailed studies suggest that they all belong to related gene families. What is more, the first step in the cascade of enzymic reactions that leads to the production of the second messenger is always similar, involving one or other of a set of about a dozen proteins, which have different actions but strong structural resemblances showing that they all belong to one family.[12] It seems then that the enormous richness and variety of chemical synapses in the brain and spinal cord has been made possible by the evolution of many diverse forms from a small number of ancestral membrane proteins. Since related proteins and peptide transmitters are involved in the signalling that sometimes occurs between adjacent *non-excitable* cells,* it looks as though it was the advantages of possessing mechanisms that allowed adjacent cells to influence one another both subtly and reliably that provided the original evolutionary drive for the development of these chemical signalling systems. If that is right, we can think of the synapses we have been discussing as devices that have evolved because of the benefits conferred by coupling sophisticated *local* cell-signalling systems with the rapid *long-distance* transmission of information along nerve axons. It is the powerful combination of these two systems that has made possible the evolution of both brains and minds.

Electrical synapses

One result of the success of work on chemical transmission at synapses was that the possibility that some synapses might be electrical was largely ignored. There was therefore some surprise when, in the late 1950s, it was discovered that transmission at a giant synapse in the crayfish was electrical.[13] Since then electrical synapses have been shown to be quite common in invertebrates, and to occur too, though more rarely, in vertebrates, including mammals. For an action potential in one cell to act directly to reduce the membrane potential of an adjacent cell sufficiently to excite it there must be an electrical connection between the interiors of the two cells. This connection is provided by so-called gap junctions, hexagonal arrays of protein particles each of which forms an aqueous channel linking the cells.

Why should organisms have evolved both electrical and chemical synapses? Gap junctions are not restricted to excitable cells but occur very widely, allowing small molecules as well as ions to pass from cell to cell. Their function in synapses

---

*That is, all the cells in the body other than nerve cells, muscle cells, and the receptor cells of sense organs.

is therefore simply one example of their more general function, and this suggests that electrical synapses might well have evolved first. But electrical synapses, though they act almost instantaneously, have major drawbacks. In the first place they are necessarily excitatory.[14] Secondly, they lack the advantages provided by the existence of different transmitters and of receptors that respond to those transmitters in different ways. And thirdly, in order for the presynaptic cell to deliver sufficient current to excite the postsynaptic cell, the presynaptic cell must be the bigger of the two. To try to depolarize a large cell with current generated by a small synaptic knob would be 'like trying to heat a cannon ball with a hot knitting needle.'[15]

Given these disadvantages, why should animals have retained electrical synapses at all, once chemical synapses had evolved? One obvious advantage is speed. The giant synapse of the crayfish, in which electrical transmission was first demonstrated, is involved in the production of the violent forward flip of the tail that enables the crayfish to make a sudden retreat. It may not be a coincidence that, with quite different neural arrangements, electrical synapses are also involved in producing the violent sideways flip of the tail that bony fish use to evade predators. A second advantage of electrical synapses is that, by coupling a group of cells together electrically, they can ensure that if those cells fire they will fire almost simultaneously. This is how, when it is disturbed, the marine mollusc known as a 'sea hare' is able to release a sudden cloud of Tyrian purple.

The role of electrical synapses in mammals is obscure. Though such synapses have been found in several sites in the mammalian brainstem, the overwhelming majority of synapses in our brains are chemical. This is not surprising, given the ability of chemical synapses to be excitatory or inhibitory, to be selective for different transmitters, and to produce more or less lasting changes by the use of a great variety of neuromodulators. We sometimes need speed but we nearly always need flexibility.

10 'The Doors of Perception'*

those five portals by which we can alone apprehend
—E. M. Forster, *The Machine Stops*

In his book *On the Soul* Aristotle pointed out that humans and the higher animals have five senses: sight, hearing, smell, taste and touch – the last including the senses of heat and cold. He had an ingenious, though entirely spurious, argument that there could not be more than five, and we are still tempted to attribute any inexplicable awareness to an occult 'sixth sense'. If we ignore the rather peculiar senses of balance, acceleration, and awareness of the position of the parts of our own body, the traditional five seem adequate for us, but it is clear that some animals have senses that we do not: bats avoid obstacles by echo-location, migrating birds navigate partly by using their sensitivity to the earth's magnetic field, some fish find their prey in murky waters by detecting small changes in electrical resistance, and the duck-billed platypus finds insects and molluscs buried in the sand by detecting their electrical activity.

The existence of these unfamiliar senses raises the question: Are such unfamiliar senses accompanied by correspondingly unfamiliar sensations, and if so what might such sensations be like? Thomas Nagel's famous essay 'What is it like to be a bat?'[1] discusses questions of this kind, and such questions can be asked about the sensations of any supposedly sentient creature that differs from ourselves (including, of course, other people). Leaving aside for now the philosophical problems associated with sensations, I want to tackle a question that is less obscure: How do our sense organs work? And, in particular, how do the individual sensory receptors – the entities within sense organs that actually respond to adequate stimuli – work?

In choosing to discuss sense organs rather than sensations, I may be accused of giving too much attention to tedious concerns about mechanism, so let me try to justify this choice. First, in a human being it is the sensory receptors that provide the overwhelmingly predominant input to the entire nervous system – that is to say, to the system that, almost alone, controls our thoughts and our feelings as well as our actions. Secondly, though the occurrence of events in our sensory

*The phrase was Blake's before it was appropriated by Aldous Huxley.

receptors is not a *sufficient* condition for us to experience sensations caused by changes in our surroundings, it is, in normal circumstances, a *necessary* condition. And thirdly, though knowledge of sensory receptors cannot account for the intrinsic qualities of sensations – the redness of red, the painfulness of pain – it can often account for our ability (or our inability) to detect different stimuli, and to discriminate between them. It can explain, for example, how it is that we can distinguish between an oboe and a flute playing the same note. It can explain why – difficult though it is to believe – we are unable to distinguish between the colour of a white screen illuminated with yellow light and the same screen illuminated with an appropriate mixture of red and green light; in both cases the screen looks yellow to anyone with normal vision. And it can often explain striking differences in the ability to discriminate that are found by comparing different individuals, or the same individual in different circumstances.

There is also a more general point. Without consideration of the behaviour of sensory receptors, and of the nervous system of which they form part, philosophical speculations can go wildly wrong. In his essay 'Doors of Perception', which gives a marvellous account of the effects of taking mescalin, Aldous Huxley accepts the suggestion of the Cambridge philosopher C. D. Broad (following Henri Bergson):

> that the function of the brain and nervous system and sense organs is in the main *eliminative* and not productive. Each person is at each moment capable of remembering all that has ever happened to him and of perceiving everything that is happening everywhere in the universe. The function of the brain and nervous system is to protect us from being overwhelmed and confused by this mass of largely useless and irrelevant knowledge, by shutting out what we should otherwise perceive or remember at any moment, and leaving only that small and special selection which is likely to be practically useful

'According to such a theory', Huxley claims,

> each one of us is potentially Mind at Large. But . . . to make biological survival possible, Mind at Large has to be funnelled through the reducing valve of the brain and nervous system. What comes out at the other end is a measly trickle of the kind of consciousness which will help us stay alive . . .

Although it is certainly true that rejection of 'useless and irrelevant' information is an important part of the information processing that goes on in the nervous system, the notion that the function of the brain and sense organs is 'in the main eliminative and not productive' is unhelpful. And the picture of the brain as a reducing valve, reducing to a measly trickle of consciousness what would otherwise be a flood from the rest of the universe, is absurd. But to prove the one unhelpful and the other absurd you have to look at what the brain and sense organs actually do.

When we looked at the coding of information in sensory nerves (Chapter 8), we saw that the impulses that pass along such nerves are always rather similar, and that what enables them to convey information is the use of two types of code: a labelled-line code and a frequency code. For the labelled-line code to work, the individual sensory receptors that feed information into the labelled lines must respond only to stimuli of a particular type – pressure, acceleration, a rise or a fall in temperature, sound within a particular range of pitches, light within a partic-ular range of wavelengths, the presence of molecules of a certain character. For the frequency code to work, the sensory receptors must be able to adjust the fre-quency of the nerve impulses to the strength of the stimulus. So we want to know how sensory receptors transmit nerve impulses in response to appropriate stimuli of adequate strength; what makes individual sensory receptors respond only to stimuli of a particular type; and how the relation between the intensity of stimulation and the frequency of nerve impulses is achieved.

Rather surprisingly, despite the enormous variety of different individual sensory receptors, the behaviour of all of them fits into a fairly simple overall pattern. What all sensory receptors cells have in common is that in the presence of the appropriate stimulus the cell membrane alters its permeability to small positively charged ions. This causes a change in the electrical potential difference (or voltage) across the cell membrane, and this change – the *receptor potential* – provided it exceeds some threshold level, starts off one or more nerve impulses in the nerve fibre that links the receptor to the central nervous system. Stronger stimulation causes bigger receptor potentials, which give rise to more frequent impulses. The relation between the strength of the stimulus, the initial* fre-quency of the impulses it causes and, so far as one can judge, the subjective strength of the resulting sensation, shows a characteristic pattern for all sense organs. If the strength of the stimulus falls below a certain threshold, the stimu-lus is ineffective. If it exceeds a certain level, the (initial) frequency of impulses and the strength of sensation both saturate. In between threshold and saturation, an increase in the strength of the stimulus leads to an increase in the frequency of impulses and the subjective strength of sensation. For most of this range, the smallest *increase* in stimulus strength that can be detected is roughly propor-tional to the initial strength of the stimulus.** This makes good sense, because it means that sensitivity is nicely adjusted to the task in hand. Carrying a kilogram, you can just detect an increment in weight of about 30 grams, but carrying only 100 grams, you can detect an increment of as little as 3 grams.

The whole process from stimulation of a sensory receptor to the initiation of nerve impulses is known as *sensory transduction*, and it is often misrepresented

*It is necessary to specify the initial frequency because in sense organs that show adaptation, such as touch receptors in the skin, the frequency falls even if the stimulus is maintained at constant strength.

**This is the well-known Weber-Fechner law, named after its 19th-century discoverers.

in textbooks of physiology as the conversion of the energy of the stimulus – whether mechanical, thermal, electromagnetic or chemical – into the energy of nerve impulses. In fact the energy for the nerve impulses comes from metabolic processes within the nerve fibre, and what is significant is not the transfer of energy – which at best is tiny and may be non-existent (suppose being plunged into darkness is the stimulus) – but the transfer of information. This is not the place for a systematic survey of the behaviour of sensory receptors, but, by choosing a few examples, I want to illustrate their striking characteristics: their selectivity, their sensitivity, and the way in which different sensory receptors can combine their outputs to yield information of a more sophisticated kind.

THE MECHANORECEPTORS

The crucial feature of a sensory receptor that makes it sensitive to a stimulus of one kind rather than another is the nature of the machinery that generates a receptor potential when an appropriate stimulus is present. Perhaps the most straightforward machinery, at least in principle, is found in the so-called mechanoreceptors – the sensory receptors that enable us to know not only when we are being touched, the tension and stretch of our muscles, and the position of our joints, but also the direction of the force of gravity, whether we are being subjected to linear or rotational acceleration and, most impressively, the nature and quality of sounds in our environment.

What makes it possible for receptors of this kind to serve such a variety of ends is partly differences in their position and local anatomical arrangements, and partly more subtle differences in the transduction machinery. For example, our sense of balance, our sense of acceleration and our sense of hearing all depend on cells in the inner ear which carry at one end a bundle of stiff hairs of graded lengths linked at their tips by fine connecting strands (see Figure 10.1). Bending of the bundle in the direction of the tallest hair stretches the links and this opens (mechanically gated) channels permeable to small positively charged ions, so reducing the potential difference across the membrane of the hair cell. Bending of the bundle in the opposite direction slackens the links and leads to an increase in the potential difference. The magnitude of the potential difference controls the rate at which the hair cell releases an excitatory transmitter at a synapse between itself and the terminal branch of a sensory nerve, and so controls the frequency of nerve impulses passing to the brain along that nerve. With the hairs unbent, the sensory fibres carry impulses at a low frequency, and this frequency can be increased or decreased depending on the direction in which the hairs are bent – literally a push–pull system.

Whether the hair cells are sensitive to gravity or linear acceleration or angular acceleration or sound depends on how they are arranged. If they are arranged in

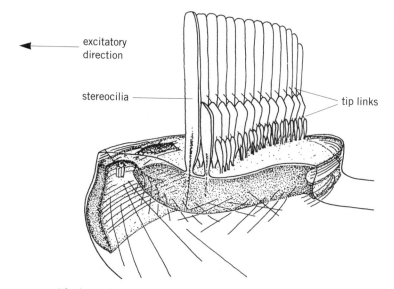

FIGURE 10.1 The hairs (labelled stereocilia) at the top of a hair cell. The hairs are arranged in rows of different height, and the tip links always connect cells in different rows. (From C. M. Hackney & D. N. Furness.)

a patch in the wall of a liquid-filled container, with the hairs projecting into the liquid, and the tips of the hairs embedded in a blob of jelly loaded with heavy crystals of calcite, any tilting of the structure will bend the hair cells; we therefore have an organ sensitive to changes in position relative to the direction of the force of gravity. Because of the inertia of the heavy blob of jelly, the organ will also be sensitive to linear acceleration. Organs of precisely this kind are found in the inner ears of all vertebrates and also in many invertebrates.

But now suppose that the hair cells are arranged differently. Suppose that they are arranged in a patch in the wall of a liquid-filled semicircular canal and the tips of the hairs are embedded in a blob of *unweighted* jelly that projects into the lumen of the canal. If the organ is suddenly rotated in the plane of the semi-circular canal, what happens is rather similar to what happens if you suddenly rotate a cup of tea. Initially, because of inertia, the liquid lags behind the canal, pushing on the blob of jelly and bending the hair cells. If the rotation is contin-ued, the viscosity of the liquid will ensure that before long it rotates at the same speed as the canal and the hair cells will return to their equilibrium position. If the rotation is then stopped, the inertia of the liquid will make it continue to move for a time, and the hair cells will be bent in the opposite direction, until the viscosity of the liquid brings it to rest. With three semicircular canals arranged in three planes at right angles to one another, we therefore have an organ sensitive to rotational acceleration in any plane. Like the organs sensitive to gravity and linear acceleration, such sets of semicircular canals are found in the inner ears of all vertebrates, and similar structures are found in a few invertebrates.

Hearing

The part of our inner ear that is concerned with hearing is thought to have evolved from the lateral-line organ in fishes, a series of canals that run along each side of the fish, opening to the surface at intervals. These canals contain hair cells, which are sensitive to low-frequency vibrations in the water, including those produced by other fish. The astonishing ability of a school of fish to execute complicated manoeuvres with almost perfect synchrony is said to depend on information from the lateral-line organs rather than the eyes.[2]

The hair cells that enable us to hear are arranged along the whole length of the *basilar membrane* – a strip of stiff membrane about 3 cm long that forms a partition along the *cochlea* (Latin for snail) – a narrow spiral bony tube in the inner ear. The cochlea is filled with fluid, which is vibrated by sound waves through the action of a system of levers that connects the eardrum with a flexible membrane at one end of the cochlea. Under the microscope the basilar membrane shows cross striations – about 24000 of them in the human basilar membrane – suggesting the presence of transverse fibres. Helmholtz suggested that these fibres might be made to vibrate individually by sounds of appropriate pitch, just as the individual strings of a piano (with the sustaining pedal depressed) show sympathetic resonance – that is to say, they vibrate in the presence of sounds of the same pitch as that produced when the string is struck. Since the width of the basilar membrane increases progressively along its length, it seemed likely to Helmholtz that notes of different pitch would cause different bits of basilar membrane to vibrate, so stimulating different nerve fibres. It was this arrangement, he believed, that made pitch detection possible.

For many years this 'resonance' theory, was accepted much as Helmholtz had proposed it, but eventually it became clear that the analogy with the strings of a piano does not quite work. In the 1930s, the Hungarian-born American physiologist Georg von Békésy managed to look directly at the basilar membrane *in situ*, and he found that exposure to a note of a fixed pitch did not cause vibration at a particular point (as in a piano) but instead set up travelling waves in the basilar membrane, rather like waves in a rhythmically flicked rope of which the far end is fixed. And though the point of *maximal* displacement of the membrane did vary with pitch as Helmholtz's theory required, the 'tuning' did not appear nearly sharp enough to account for our observed ability to discriminate between notes of different pitch. At first, the unexpectedly good discrimination was therefore attributed to analysis in the cerebral cortex, but that theory had to be abandoned when recordings from individual nerve fibres in the auditory nerve showed that the hair cells stimulating each nerve fibre must themselves be highly tuned. How can hair cells from adjacent regions of the basilar membrane distinguish so well between slightly different pitches when the amplitude of the displacement of the basilar membrane at the two adjacent regions is practically the same? The answer seems to be different in reptiles and in mammals.

In reptiles, each hair cell possesses a mechanism that makes the electrical potential difference across the cell membrane tend to oscillate (that is, wax and wane regularly) at a particular frequency, and this makes the hair cell much more sensitive to vibration of its hairs at that frequency. The oscillation is caused by the interplay of two kinds of ion channel in the membrane – voltage-gated calcium channels and calcium-gated potassium channels.[3] The two kinds of channel act alternately in a strange *pas de deux*, each causing changes that activate the other. In mammals some of the energy stored in the electrical potential difference across the hair-cell membrane can be used to move the hair cell – though the mechanism responsible is not understood.[4] As the basilar membrane vibrates, the fluctuating potential difference across the hair-cell membrane causes a fluctuating force to act on the hair cell, thus reinforcing the motion caused by the sound. What determines the tuning of each hair cell is not clear.

The sensitivity and selectivity of the mammalian cochlea are both astonishing. In humans, the threshold of hearing – that is to say, the smallest amount of energy that has to fall on our ears for us to hear anything – depends very greatly on pitch; in the region of greatest sensitivity – the middle of the top octave of the piano – we can hear notes so soft that the variation in pressure, with each compression and rarefaction, is less than one ten thousandth of one millionth of atmospheric pressure, and the magnitude of the displacement of air molecules is less than the diameter of a hydrogen molecule. The ear is much less sensitive to sounds of pitches outside this range; the sensitivity to middle C, for example, is about 150 times less, and to the lowest C on the piano about twenty-five million times less.*

The evolutionary advantage of great sensitivity to sound is obvious, but why should we, and so many other animals, have evolved such elaborate machinery for discriminating pitch? After all, several groups of insects – grasshoppers, crickets and cicadas – use sounds for communication, particularly for attracting mates, but their hearing organs are relatively insensitive to pitch, and information seems to be transmitted from insect to insect solely by changes in intensity and rhythm. Nobody knows the answer to this evolutionary question, but a plausible suggestion is that pitch discrimination first developed because of its role in the localization of the source of a sound – a matter of great importance whether

*This enormous difference in sensitivity to notes at the top and bottom end of the piano is not obvious when you listen to the instrument being played because, when the notes are struck with equal force, the fraction of mechanical energy that is converted to sound energy is very much greater for notes at the bottom end.

you are trying to find prey or to avoid a predator. To see why it is has this role, and how pitch discrimination works, we need to look a little more closely at the nature of sound waves.

When we are in a room listening to a violin playing middle C, what is happening is that the vibration of the string, caused by the passage of the bow over it, is causing minute oscillations in the pressure of the air adjacent to the string. These oscillations in pressure – sound waves – spread in all directions and when they reach our ear-drums, either directly or after reflection from walls or objects, they make the drums vibrate at the same frequency as the vibrating string (261 vibrations per second) and we hear middle C.

A graph of air pressure at any one point in the room plotted against time would show a regular succession of identical waves of rather complicated shape occurring at the rate of 261 per second. If, instead of listening to a violin, we listen to a trumpet or a tuning fork producing the same note, the graph of air pressure against time would show waves with the same frequency, but of different shape. If we bow the violin more firmly, blow more vigorously into the trumpet, or strike the tuning fork harder, the amplitude of the sound waves will increase, but the frequency will remain the same, and though the waveforms will change to some extent the characteristic differences between the three instruments will be maintained. It is the amplitude of the sound waves that determines the loudness of the sound, their frequency that determines the pitch, and the shape of their waveform that partly determines the quality – what musicians call the timbre. Timbre also depends on other factors, such as the way individual notes start and stop, and the presence and character of vibrato.[5]

If you imagine a beetle crawling at constant speed round and round the rim of a clockface, a graph of the vertical distance between the beetle and the line joining 9 o'clock and 3 o'clock is called a sine wave (see Figure 10.2a). (It is the simplest of all regularly repetitive waveforms, and it is also the waveform of the sound produced by a tuning fork.)[6] In 1822, the French mathematician Fourier proved that any regularly repetitive waveform can be built up by adding together sine waves whose frequencies are equal to or multiples of the fundamental frequency (that is, the frequency at which the pattern is repeated) (see Figure 10.2b). And just as it is possible to build up any regularly repetitive waveform from sine waves with frequencies limited to the fundamental frequency and multiples of it, so it is possible, mathematically, to analyse any complex regularly repetitive waveform into its component sine waves. 'Fourier analysis' is the counterpart of 'Fourier synthesis'.

Fourier's discovery was a discovery in pure mathematics, but for sound waves the different multiples of the fundamental frequency are frequencies of the different harmonics; and Helmholtz realized that the ear recognizes sound quality by carrying out a Fourier analysis. This does not mean that when you recognize a French horn as a French horn you are doing mathematics in your head. It simply

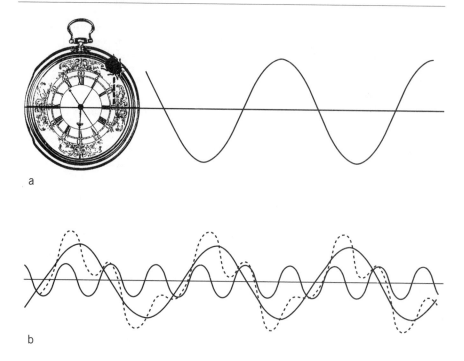

a

b

FIGURE 10.2 (a) Shows the way in which the steady movement of a beetle round the rim of a clockface generates a sine wave. (b) The dotted curve shows the effect of adding together two sine waves with frequencies in the ratio 1:3. By adding more and more frequencies, and choosing the amplitudes (and phases) appropriately, any waveform can be synthesized from sine waves.

means that a French horn playing, say, middle C will stimulate not only your hair cells tuned to middle C, but also those tuned to the harmonics of middle C that are present in the French horn sound; and the intensity of stimulation of each hair cell will reflect the fraction of the sound energy that is in the corresponding harmonic.* In this way, a battery of sensory receptors, each maximally sensitive to a particular and rather simple stimulus, is able to convey information about the environment of a much more sophisticated kind.

It is this ability to break down complex, and possibly rapidly changing, waveforms that facilitates the localization of the source of a sound. For sounds of pitch up to the B nearly three octaves above middle C, we determine where (in the horizontal plane) a sound comes from mainly by comparing the times of

---

*The intensity of stimulation of the different hair cells will not give any information about the relative phases of the different harmonics, and Helmholtz was able to show that the ear is indeed insensitive to the phases of pure tones contributing to a sound. For this reason it is, strictly speaking, the harmonic content rather than the waveform that determines the timbre.

arrival of the sound at our two ears.* At the moment the sound starts this presents no difficulty, but for complex and continuing sounds the comparison can only be made by comparing the information received from the two ears along nerves 'tuned' to similar frequencies.

The extent to which an ability to distinguish between the timbres of different musical sounds has been advantageous in evolutionary history is difficult to assess, except perhaps in songbirds and humpback whales. But the ability of the hair cells to discriminate between different frequencies is also useful in distinguishing the qualities of non-musical sounds. Such sounds do not consist of a fundamental and harmonics but of an almost continuous range of frequencies, and the character of the sound may change from moment to moment. The ability of the ear to measure, each moment, the relative contributions of vibrations of different frequency makes it possible for the owner of the ear to discriminate between patterns of sound that are only subtly different – provided, of course, that the ear is backed up by a well-developed auditory cortex. In any event, the sophistication of the machinery in the cochleas of reptiles, birds and mammals suggests that, in the evolutionary history of the higher vertebrates, the survival value of the abilities conferred by that machinery has been important. And speech, of course, could not have developed without it.

'THE VISION THING'

The human eye is the most sophisticated, the most fascinating and, happily, the best understood of all sense organs. Like a camera it consists of an optical device that produces an image of part of the outside world on a light-sensitive sheet. In the camera, the light-sensitive sheet is the photographic film; in the eye it is the retina. In both, each point on the sheet receives light from just one tiny area of the visual field, and in both the changes produced at each point on the sheet depend on the intensity and colour of the light falling on it. The retina, though, has a tougher job. Unlike photographic film, which is used only once, the retina has to respond continuously; and it must not only record the information in the image but also transmit it. In the need to respond continuously and to transmit, the retina is, of course, like other sense organs, but it differs from them in an interesting way. Information transmitted along the sensory nerves from other sense organs is usually restricted to information about the degree of stimulation of the individual sensory receptors (but cf**). The retina is more subtle. Along with the layer of photoreceptors –

*To locate sounds of higher pitch we depend on the difference in the intensity of the sound at the two ears caused by the blocking effect of the head.

**An exception is the transmission of information about low frequency sounds heard by the ear – see endnote 2, Chapter 8. Also, where one sensory nerve serves several receptors there may be some averaging.

receptors that respond to light – it contains a network of nerve cells which acts as a small computer. The network takes the information provided by the individual receptor cells and does a certain amount of processing on it before transmitting the result along the fibres of the optic nerve to the brain. The nature of this processing, and the purpose it serves, we will come to later; for the moment I just want to consider the transduction machinery in the photoreceptors.

In 1876, Franz Boll, in Rome, reported that if frogs were kept in the dark their retinas appeared purple. A year or two later Willy Kühne, who had followed Helmholtz as Professor of Physiology at Heidelberg, succeeded in extracting the purple substance, and he showed that it turned first yellow and then white when exposed to light. In the course of the following century, this substance (*visual purple* or *rhodopsin*), or substances very like it, were shown to be present in the eyes of all vertebrates or invertebrates, and also in the light-sensitive pigmented spots that are found in unicellular and other primitive animals, and even in some mobile algae. Rhodopsin, and the related visual pigments, always consist of two parts: a colourless protein – a member of a class of proteins called *opsins* – and, deep within the opsin and joined to it, a derivative of vitamin A called retinal. The retinal, within the opsin, can exist in two shapes, 'kinked' and 'straight'. In the dark, it is kinked, but if the rhodopsin is illuminated and the retinal absorbs light, it straightens. In doing so it delivers a micro-kick to the opsin part of the molecule, converting it momentarily into an active enzyme. The essence of visual transduction is a kick-start.[7]

What does the opsin do during the brief period it is an active enzyme? We have seen (p. 135) that there is a class of membrane receptors which are found at synapses and which, when triggered by the appropriate transmitter, set off a cascade of enzyme reactions inside the cell. We have also seen that receptors of this class all belong to the same family of proteins, as shown by similarities in their structure. The structure of the opsins shows that they too belong to this family; and they too set off a cascade of enzyme reactions when they are activated. The end result of this cascade is to *close* sodium channels in the cell membrane, so *increasing* the potential difference across that membrane and *suppressing* the release of a transmitter at a synapse between the photoreceptor cell and a nerve cell.

It may seem surprising that light should suppress rather than stimulate the release of transmitter by a photoreceptor, and it is certainly the opposite of what is usually expected of a sense receptor in the presence of its normal stimulus. But whether the stimulus increases or decreases the rate of release of transmitter is immaterial. The response of a given photoreceptor depends on the illumination, but we need photoreceptors to detect changes both from dark to light and from light to dark. The sudden shadow is as likely to be significant as the sudden light. One way of meeting this need would be to have separate 'light' and 'shade' receptors, with different transduction mechanisms – just as we have separate receptors for heat and cold in the skin. A simpler way, and the way the eye actually works, is to have a single receptor, and then, if necessary, to reverse the sign of the signal at the next stage. Light causes a decrease in the rate of release of transmitter at the synapse between the photoreceptor cell and a nerve cell. If the synapse is inhibitory, reducing the release of transmitter will tend to stimulate the nerve cell. If the synapse is stimulatory, reducing the release of transmitter will tend to inhibit the nerve cell. In fact, both kinds of synapse are found, and they are found on different nerve cells, so increasing the light falling on a photoreceptor will stimulate one nerve cell, and decreasing the light will stimulate a different nerve cell.

Sensitivity: the need for adjustment

A major difficulty faced by the retinal photoreceptors is that the intensity of sunlight is about ten billion times that of starlight, yet we need to be able to see on moonless nights as well as on sunny days. What we are usually interested in, though, is not the absolute brightness of the different parts of the scene – the white teeth, the black panther – but their relative brightness. And *in a uniformly lit scene*, the lightest part of the scene before us at any moment (excluding the light source) is not likely to be more than twenty times as bright as the darkest that we need to distinguish. What we need, then, is a mechanism that allows us to differentiate brightnesses over something like a twenty-fold range, but whose absolute sensitivity can be adjusted over an enormously wider range – becoming greater when more sensitivity is needed because the light is weak, and smaller when less sensitivity is needed because the light is strong.

That just such a change in sensitivity with illumination occurs is, of course, familiar to all of us. When we go from bright sunlight into a dimly lit room we see very little until our eyes adapt to the dark. When we go from a dimly lit room into bright sunlight we are dazzled until our eyes adapt to the light. Our pupils narrow in the light and widen in the dark, but they cannot change the fraction of light entering the eye by more than a factor of ten, so the main adjustment must occur in the retina.

The retina achieves the necessary changes in sensitivity in two ways. First, the problem is made slightly easier by the existence of two sorts of photoreceptor: one sort (cylindrical and known as *rods*) suitable for dim light – roughly speaking, moonlight and darker – and the other (conical and known as *cones*) for brighter light. But even with this division of labour, as lighting conditions change, the sensitivities of both rods and cones need to vary by many orders of magnitude. This variation is brought about automatically, partly by variation in the concentration of rhodopsin itself, and partly by changes in the calcium concentration in the photoreceptor which lead to changes in the concentrations of key reactants in the enzyme cascade, and so alter (appropriately) the sensitivity of the transduction machinery.[8]

'The dark is light enough'

As remarkable as the ability of the retina to change its sensitivity is its ability to detect light when there is scarcely any light. During the second world war, Selig Hecht and his colleagues in New York made the first accurate measurements of the minimum amount of light that could be detected by a subject who was fully dark-adapted.[9] They flashed small spots of very dim light very briefly on to the retina, and asked the subject to report when a flash was seen. By using light of the wavelength which rhodopsin absorbs best, and varying the brightness and duration of the flashes, they were able to show that a sensation was possible when as little as five quanta of light were absorbed by rhodopsin molecules.* Since the area of the illuminated spot was many times greater than the area of a rod, there was little chance that more than one quantum was absorbed by rhodopsin molecules in any one rod; the experiment therefore implied, first, that a rod could be excited by a single quantum, and secondly that several rods had to be excited simultaneously for a sensation of light to be experienced.**

The first of these conclusions means that the rod (in the fully dark-adapted eye) approaches the theoretical limit of sensitivity. What makes this almost incredible sensitivity possible is the amplification that occurs at some of the steps in the enzyme cascade. ('Cascade' is not, in fact, a very good metaphor for describing the events in enzyme cascades, for in a real cascade the amount of water falling through each step is the same.) As a result of this amplification the absorption of a single quantum of light (or photon) can cause the closure of hundreds of sodium channels, and increase the potential difference across the rod membrane enough to affect significantly the release of transmitter at the synapse between the rod and the adjacent nerve cell.

The requirement that a small number of rods be stimulated simultaneously for the flash to be seen is thought to be a device for ensuring that a sensation of light does not often occur simply as a result of the thermal agitation of the rhodopsin molecules. Activation of any particular rhodopsin molecule by thermal agitation is a highly improbable event, but since each rod contains about a hundred million such molecules, it does occur at a significant rate leading to the stimulation of individual rods even when no light is present. The need for several rods to be stimulated simultaneously greatly reduces the chance that these infrequent events are associated with any visual sensation, though it does not

---

*Although in many ways light behaves as waves, in others it behaves as a stream of particles, known as quanta of light or photons. The individual quantum is the smallest quantity of light that can exist.

**More recent work by Hecht and his colleagues suggests that a more reliable figure for the minimum number of quanta needed is about a dozen. The general conclusions from their experiments are not affected.

reduce it to zero. A fully dark-adapted subject sitting in a totally dark room sees, not blackness, but a very dark slightly shimmering grey.

Seeing colours

In moonlight, we see no colours. This is generally explained by saying that in moonlight we are using only our rods, which are colour blind, and that it is only in brighter light, when we can use our cones, that we see colours. The explanation is correct, but also misleading, for it tends to suggest that individual cones are not colour blind – that they can transmit information about the colour of the light falling on them. In fact, no single photoreceptor can provide information about both the intensity and the colour of light falling on it, because, though its sensitivity to light will doubtless vary with the colour, it has no way of distinguishing between a less intense light of a colour to which it is more sensitive, and a more intense light of a colour to which it is less sensitive. Each cone is as colour blind as each rod. What makes our cones special – apart, of course, from their ability to function in bright light – is that there are different kinds, with slightly different visual pigments, and that the eye has the ability to compare the responses of cones of the different kinds.

These cone pigments are similar to rhodopsin, but they have slightly different opsins which alter their relative abilities to absorb light of different wavelengths, and therefore their sensitivity to light of different wavelengths. One type of cone is most sensitive to light of rather short wavelengths (peak sensitivity in the violet), one is most sensitive to light of medium wavelengths (peak sensitivity in the green) and one is most sensitive to light of longer wavelengths (peak sensitivity in the yellowish green). The peaks are not sharp and each type of cone is sensitive, to some extent, to a rather broad range of wavelengths. Together the three types of cone cover the whole of the range of wavelengths that we can see, and, except towards the ends of the range, light of any wavelength will stimulate at least two types of cone. Despite the position of the peaks, and the breadth of the curves, the three types of cone – short wavelength, medium wavelength and long wavelength – are traditionally (and confusingly) known as 'blue-sensitive', 'green-sensitive' and 'red-sensitive', and sometimes simply as 'blue cones', 'green cones' and 'red cones'.

This system of determining colour using just three kinds of 'broadly tuned' receptor is very different from the system used by the cochlea to determine pitch, which, as we saw, depends on a very large number of hair cells each tuned to a narrow range of pitches. A system like that in the cochlea would not work for vision because, if we are to make use of the information in the retinal image, the cone system needs to provide information about both colour and intensity for each area of retina, and, in the central part of the retina – the part used for detailed vision – at nearly each point in the retina. Indeed, it was Thomas Young's realization, in 1801,[10] that there would not be sufficient room for a large number of sharply tuned receptors, that led him to suggest that there were only three kinds, corresponding to 'the three principal colours, red, yellow and blue'.

It is because our sensations of colour depend primarily (though, as we shall see later, not simply) on the relative responses of only three broadly tuned receptors that we cannot distinguish between a white screen illuminated with yellow light (i.e. light whose wavelength is about 580 nanometres) and the same screen illuminated with an appropriate mixture of red light and green light. The difficulty arises because the blue-sensitive cones are sensitive to violet and blue light, and slightly to blue-green light, but not to light of longer wavelengths. It follows that discrimination by the eye between red light, orange light, yellow light and green (excluding blue-green) light depends solely on the ratio of the responses of the red-sensitive and the green-sensitive cones. By changing the proportions of red light and green light in the mixture progressively from nearly all red to nearly all green, we can therefore produce in turn mixtures indistinguishable from each shade of colour in the spectrum from red, through orange and yellow, to mid-green. This is very different from the situation in the ear: a musician with perfect pitch can pick out the individual notes in a chord, and all of us can tell that a chord is a chord, and can distinguish between the pure tone produced by a tuning fork and notes of the same pitch containing harmonics, such as might be produced by a violin or a trumpet.

The failure of the eye to distinguish between pure colours and appropriate mixtures has been put to practical use. By the early part of the 18th century, the rules for producing 'all the colours of nature and of art' by mixing red, yellow and blue paints had been codified, and in the 1720s Jacob Christoph Le Blon published a slim book describing his method of making colour prints by superimposing the impressions of three mezzotint plates each printing a single colour.[11] The distinction between additive mixing (the mixing of lights of different colours) and subtractive mixing (the mixing of pigments of different colours, each pigment absorbing light of some wavelengths so that what is reflected from the paper is white light less what each pigment has absorbed) was appreciated by Le Blon, though not fully understood till much later. It is additive mixing, of red, green and blue light, that is used in colour television and in modern colour printing.* More subtle additive mixing is used by pointilliste painters.

The 'trichromatic theory' also provides very satisfying explanations of coloured after-images and of colour blindness. If you look at an indigo square on a white background steadily for half a minute (keeping your eyes fixed on one spot) and then transfer your gaze to a white wall, you see a square of yellow, which gradually fades. If you repeat the experiment with a magenta square, the

---

*Unlike the traditional colour printer, who superimposed colours, the modern printer makes the picture out of discrete dots in three colours – yellow, magenta and cyan – together with black, the dots being too small to be resolved at the normal viewing distance.

after-image is green. The explanation of such after-images is that the more each type of cone is stimulated by the original stimulus, the less responsive it becomes. When the gaze is transferred to the white wall, the white light reflected from the wall will therefore not produce an equal response from all three types of cone; the response will be weakest from the cones that were previously stimulated the most. In the first example, the blue-sensitive cones will be much the most exhausted, so the white wall will stimulate mainly the red-sensitive and green-sensitive cones, giving a sensation of yellow.

Complete colour blindness – the inability to distinguish any colours – is extremely rare,* but about eight per cent of Caucasian males have abnormal colour vision, and in about a quarter of these the abnormality is more than trivial.[12] The members of this more seriously affected group lack either green-sensitive or red-sensitive cones, and are said to be *red-green blind*. Curiously, although an explanation of colour vision and colour blindness on these general lines was first suggested as long ago as 1781 – by George Palmer, a London glass merchant[13] – and although the trichromatic theory fits experimental observations on both normal and colour-blind subjects so well that it has long been almost universally accepted, it was not until the late 1950s and early 1960s that direct proof was obtained. William Rushton, in Cambridge, managed to measure the absorption spectrum of cone pigments in the living human eye, by shining lights into the eye and measuring the light reflected back.[14] Although the difference in composition between the incoming and reflected light does not all represent absorption by the visual pigment, the part of it that does can be differentiated by bleaching the pigment away with a very strong light and seeing how the results are affected. In this way, Rushton and his colleagues were able to measure the absorption spectra of the pigments in the 'red-sensitive' and 'green-sensitive' cones, and to detect abnormal pigments in colour-blind subjects. Shortly afterwards, Paul Brown and George Wald at Harvard, and, independently, Edward MacNichol and his colleagues at Johns Hopkins University succeeded in distinguishing the cone pigments by shining microscopic beams of light through the outer parts of single cones obtained from monkey or human retinas.[15]

Because the blue-sensitive cones scarcely absorb light beyond the mid-green region of the spectrum, anyone lacking either green-sensitive or red-sensitive cones has only one set of cones that respond to light in the red, orange and yellow regions, and the less bluish part of the green region. All these colours will therefore be indistinguishable. The English chemist John Dalton – the originator of

*Except in the island of Pingelap, in Micronesia, where about 10 per cent of the population are totally colour blind, having rods but no cones. This is thought to be the result of the spread, by genetic drift, of a mutation in a single individual centuries ago when the population was extremely small. (See Cavalli-Sforza, L. L. and Cavalli-Sforza, F. (1995) *The Great Human Diasporas*, p. 99, Addison-Wesley, Reading, Mass.)

the atomic theory, and the first person to give a clear account of his own colour blindness – claimed to have been unable to distinguish between the colour of red sealing wax and the colour of a laurel leaf.[16] And George III told Fanny Burney that 'The Duke of Marlborough[17] actually cannot tell scarlet from green!'. ('How unfortunate', Fanny commented in her diary, '. . . that such an eye should possess objects worthy the most discerning – the treasures of Blenheim!')

The 6 per cent of Caucasian males who have less serious defects of colour vision, do not lack either red-sensitive or green-sensitive cones, but the pigment in one or other type is abnormal, so that the absorption curves of the two types are much closer than normal. Because those who suffer from this kind of colour anomaly – and I write as one of them – have three types of cone, they resemble normal subjects in needing three 'primary' colours to match the full range of colours in the spectrum; but they cannot distinguish as many hues as normal subjects, and they sometimes confuse browns and olive greens. They also tend not to notice small areas of red against a background of greens or browns – berries in a rowan tree, poppies in a hedgerow, a line of red in what seemed to be a sober Harris tweed – though once the red areas are noticed they appear quite red. Whether the red-sensitive or the green-sensitive cones are abnormal can be decided by a dramatic test introduced by the physicist, Lord Rayleigh over a century ago. The subject is asked to match a patch of yellow light projected onto a white screen from one projector, by adjusting the composition of a mixture of red and green light projected onto an adjacent area of the screen from two other projectors. To get a match, the subject with the anomalous red-sensitive cones turns up the red light until the colour of the mixed patch looks bright pink to a normal subject. The subject with the anomalous green-sensitive cones turns up the green light until the colour of the mixed patch is bright green. Despite the startling results of this test, the disability caused by anomalous red-sensitive or green-sensitive cones is not serious. The only time I found it embarrassing was during an experiment that involved recognizing a particular colour change in a solution. I was working with an American colleague and we couldn't agree on what the colour was. It turned out that my colleague, too, had defective colour vision, but whereas my abnormality was in the red-sensitive cones, his was in the green-sensitive ones.

In 1986, Jeremy Nathans and his colleagues at Stanford determined the structure of the genes for the three human cone pigments.[18] They showed that the sequence of amino-acids in all three cone pigments is about 40 per cent identical with the sequence in rhodopsin; but whereas the green-sensitive and red-sensitive pigments differ from each other in only about 4 per cent of their amino-acids, there are much bigger differences between them and the blue-sensitive pigment. This suggests that the separation of red-sensitive and green-sensitive pigments occurred relatively recently in our evolutionary history. And as, outside humans, trichromatic vision like ours is found only in apes and Old

World monkeys,[19] it looks as though the separation must have occurred after the Old World and the New World monkeys diverged about thirty million years ago.

Why did it occur? As we saw earlier, what the possession of both red-sensitive and green-sensitive cones confers is the ability to discriminate colours in the part of the spectrum between mid-green and far-red. The Cambridge psychologist John Mollon has suggested that this ability was particularly valuable to our primate ancestors because it allowed them to pick out fruits that are yellow or orange when ripe, against the background of foliage; and he points out that there is evidence for the parallel evolution of monkey colour vision and of trees with fruits that are of this kind and are too big to be eaten by birds. As he puts it: 'With only a little exaggeration, one could say that our trichromatic colour vision . . . is a device invented by certain fruiting trees in order to propagate themselves.'[20]

Because the absorption spectrum of the pigment in the red-sensitive cones is only a little different from the absorption spectrum of the pigment in the green-sensitive cones, the light exciting each type of cone can be brought to a focus at almost the same distance from the lens. In contrast, the shorter-wavelength light that best excites the blue-sensitive cones is much more strongly refracted and cannot be brought to a focus in the same plane; colour discrimination made possible by having blue-sensitive and green-sensitive cones is therefore obtained at the cost of some blurring of the image. In the small central region of the retina that is used for the most detailed vision, such blurring would not be tolerable, and, in fact, very few blue-sensitive cones are found. As a result, this part of the retina is almost 'blue-blind', as you can see by looking at two spots that are of the same size but whose colours – say red and purple – differ in such a way that detection of the difference depends on stimulation of blue-sensitive cones. As the spots are moved further and further from the eye, a point is reached at which the retinal image of the spots has shrunk sufficiently to fall wholly on the part of the retina lacking these cones, and the colour difference is no longer detectable.

What the retina tells the brain

We have seen that the retina, unlike all our other sense organs, does some processing of the information it collects before transmitting it to the brain. This arrangement may have evolved simply because it greatly reduces the amount of information to be transmitted along the optic nerve, but another possibility is that it reflects a peculiarity in development: the retina is the only sense organ that originates as an outgrowth from the developing brain. It can, therefore, be regarded as an archipelago of the brain, connected to the rest by a nerve tract called the optic nerve. We should not be surprised, then, that the retina, like other parts of the brain, is involved in information processing.

If this way of regarding the retina is correct, investigation of the working of the retina should tell us not only about vision but also something about the sorts of way in which a part of the brain handles information. And, for investigation, the retina has three great advantages. It is a sheet less than a quarter of a milli-metre thick, so all the nerve cells in it are accessible. In many animals, its detailed microscopic anatomy – the wiring diagram of the nerve network – is fairly well understood. And because the receptors are not at some remote site but are spread out on the surface of the network, and are sensitive to light, it is easy to control the input of information to each part of the system very precisely.

Figure 10.3 shows, in a simplified form, the microscopic structure of the human retina. At the top of the diagram is a layer of closely packed rods and cones. At the bottom is a layer of cells called *ganglion cells*, whose axons – carrying information from the retina to the brain – form the optic nerve. In between these two layers is a layer consisting mostly of cells called *bipolar cells* because each has two branches, one synapsing with rods or cones, and the other with one or more ganglion cells. This arrangement obviously provides a direct route for informa-tion from the photoreceptors, via the bipolar cells, to the ganglion cells. There are, though, two complications.

First, since there are about 125 million rods or cones in the human eye but only about one million ganglion cells, there must be great convergence. Convergence has advantages and disadvantages. The greater the number of photoreceptors that feed information to a ganglion cell, the more likely that cell is to fire in dim

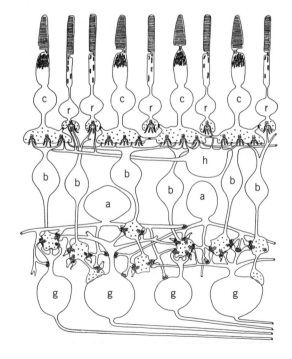

FIGURE 10.3 The structure of the mammalian retina. **r** and **c** denote rods and cones; **b** bipolar cells, **g** ganglion cells, **h** horizontal cells and **a** amacrine cells. Information flows 'vertically' from rods and cones, via bipolars, to ganglion cells. The horizontal cells and amacrine cells allow the activity of a ganglion cell to be affected by activity in adjacent regions of the retina. (From J. E. Dowling and B. B. Boycott, slightly simplified.)

light, so the greater the sensitivity. But because convergence destroys knowledge of the precise location of the photoreceptor that has been stimulated, that greater sensitivity is bought at the expense of a loss of the ability to discriminate fine detail. Evolution has coped nicely with this problem. We obviously need both high sensitivity and accuracy, but we do not usually need both at the same time. For rod vision in dim light, and for both rod and cone vision near the periphery of the visual field, we can afford to sacrifice accuracy for greater sensitivity; for cone vision in bright light, particularly in the central part of the retina that we use for detailed vision, we can sacrifice sensitivity for accuracy. What we find, then, is just what you might expect to find: great convergence in the periphery of the retina, less in the centre.

The second, and more interesting, complication is that as well as the direct route for information there are indirect routes, which allow information to be transferred laterally in the retina, i.e. parallel to its surface. The layer of the retina that contains bipolar cells also contains two other types of cell, both with long branches running parallel to the surface of the retina. The branches of the cells of one type form synapses with the photoreceptors and also with the branches of the bipolar cells that extend towards the photoreceptors. The branches of the cells of the other type form synapses with the ganglion cells and also with the branches of the bipolar cells that extend towards the ganglion cells.

The existence of these indirect routes, making use of nerve cells with branches running parallel to the surface of the retina, implies that the activity of a ganglion cell depends not only on the degree of stimulation of photoreceptor cells connected to it directly by bipolar cells, but also on the activity of photoreceptors in neighbouring regions of the retina. This has important consequences. It turns out that, through lateral connections, activity in each small area of the retina tends to *inhibit* activity in adjacent areas, and the result of this is that boundaries between light and dark areas of the visual field are sharpened. Such *lateral inhibition*, as it is called, is a common feature in sensory pathways; but although the anatomy was worked out by Santiago Ramón y Cajal, in Madrid, in about 1900, it was not until fifty years later that direct evidence for *lateral inhibition* by ganglion cells was obtained, and its significance understood.

Ganglion cells are the easiest of all retinal nerve cells to record from, and their behaviour, when the retina is illuminated, is of great interest because it is only through their axons that information passes from the retina to the brain. Yet the results of early experiments to study the effects of illumination were extremely puzzling. Unexpectedly, even in the dark, most ganglion cells were found to fire irregularly at a rate of 2–20 impulses per second. Even more unexpectedly, flooding the retina with light caused only a small increase in the rate of firing, and sometimes had no effect at all. How could the sensory nerves from the eye show so little response when the eye was flooded with light?

The solution to this unexpected problem was as unexpected as the problem.

It was discovered by Stephen Kuffler, in Baltimore, who worked with the eyes of anaesthetized cats, and who showed that, though *uniform* illumination of the retina had only rather small effects on the ganglion cells, exploration of the retina with a small spot of light produced dramatic effects.[21] He found that for each ganglion cell there was a small roughly circular area of retina – the *receptive field* of that particular ganglion cell – in which illumination with a small spot of light caused large changes in firing rate. Each receptive field can be thought of as the area of retina that a particular ganglion cell monitors, and, as one might expect, receptive fields are small in the part of the retina that is used for seeing detail, and larger elsewhere.

More detailed exploration of cat ganglion cells by Kuffler showed that whether the firing rate increased or decreased depended on where, in the receptive field, the spot of light fell – the effect of light in a circular central area being the opposite of the effect of light in the surrounding ring-shaped area. Most ganglion cells fell into one or other of two classes that behaved in opposite ways. In *on-centre cells* illumination in the central area increased the firing rate; in *off-centre cells* illumination in the central area decreased the firing rate.[22] The centre/surround antagonism within each receptive field was generally poised so that the effect of illuminating the whole of the peripheral ring-shaped area nearly balanced the effect of illuminating the whole of the central area. This, of course, explains why changing the illumination uniformly over the whole retina produced such undramatic responses. We can, nevertheless, easily detect changes in the uniform illumination of a blank screen because a minority of ganglion cells do not show centre/surround antagonism but respond to the overall brightness in their receptive fields.

Kuffler found that ganglion cells with *on-centres* and *off-surrounds* were about as common as ganglion cells with *off-centres* and *on-surrounds*. The receptive fields of the on-centre cells form a mosaic covering the whole retina, and a similar mosaic is formed by the receptive fields of the off-centre cells. The rough equality in numbers of the on-centre and off-centre cells, and the relative ineffectiveness of uniform illumination, both make sense if the main role of the retina is to measure contrast between adjacent areas rather than to measure absolute levels of illumination in any area. And it makes sense that measurement of contrast *should* be the main role, first, because, as we saw before, it is the relative brightness of different parts of a scene rather than the absolute brightness that gives the more important information – it is more important to see that what you are looking at is shaped like a tiger and has stripes like a tiger than to notice how brightly it is lit – and, secondly, because it is obviously economical for the retina to send detailed information where that information is relevant to significant features in the retinal image – edges, angles, gradations of tone – and less information where the retinal image is uniform.

But though the ganglion cells in the mammalian retina transmit to the brain the results of comparisons between the centres and the surrounds of their receptive fields, the retinal circuitry in mammals does not generally make it possible for individual ganglion cells to identify and locate such things as edges and angles. Paradoxically, the frog retina has greater analytic power.

In 1953, the year in which Kuffler reported centre/surround antagonism in the ganglion cells of the cat retina, Horace Barlow reported surprising observations on the frog retina.[23] In experiments to explore the responsiveness of individual ganglion cells, he found that one particular type of ganglion cell was 'most effectively driven by something like a black disc, subtending a degree or so, moved rapidly to and fro within the unit's receptive field'. What is more, when the same stimulus was presented to intact frogs they tended to 'turn towards the target and make repeated feeding responses consisting of a snap and a jump'. Barlow concluded that ganglion cells of this type, together with the neurons feeding information to them, acted as 'bug detectors'; and this implied that, in frogs, analysis to extract significant features from the visual scene – analysis of a kind that everyone had imagined could occur only in mysterious 'centres' in the brain – can occur to an important extent in the much simpler circuitry of the retina.

Later work on frogs showed that, in each area of the retina, ganglion cells of five different types send information about five different features of the retinal image, including the existence of edges, both moving and stationary.[24] By concentrating on a few useful features in this way, the frog's retina provides what is essential, and at the same time reduces the flow of information along the optic nerve to an amount that can be handled by the small amount of the frog's brain that is devoted to vision. But that reduction means that a great deal of information in the visual image is lost. Frogs and toads are extraordinarily good at catching flies and avoiding predators, but for them the visual scene must be very poor – certainly no real life *Wind in the Willows*. In contrast, in mammals, less analysis is done in the retina, and much more information is sent to the visual areas of the brain. What is sent, though, is neither a point by point description of the retinal image, nor a summary of crucial features in that image; it is simply the results of the first steps towards an analysis of the retinal image aimed at revealing such features, an analysis which will be taken further by the parts of the brain concerned with vision.

What about colour? In primates there are three main classes of ganglion cells that receive information from cones and that are therefore responsible for transmitting all the information we use in daylight vision. Since there are three types of cone, you might guess that each class of ganglion cell handles information from cones of only one type, but that is not how the retina evolved. Rather surprisingly, an important class of ganglion cells that receive information from cones are not concerned with colour discrimination at all. They have circular receptive fields with centre/surround antagonism, but the information from the red-sensitive cones and the green-sensitive cones is added together, so the cell simply compares the brightness of the centre and the surround.* These cells are therefore concerned with detailed daytime vision but without colour. Ganglion cells of another class also have circular receptive fields with centre/surround antagonism; but only cones of one type, say red-sensitive, send information to the ganglion cell from the centre of the receptive field, and only cones of a different type, say green-sensitive, send information from the ring-shaped surround. Such ganglion cells therefore compare red and green in adjacent areas. Ganglion cells of a third class have circular receptive fields but lack centre/surround antagonism. Instead, over the whole of the cell's receptive field, signals from red-sensitive and from green-sensitive cones have similar effects (say stimulation) whereas signals from blue-sensitive cones have the opposite effect (say inhibition). These cells, in effect, add red and green together and oppose it to blue. Since we have seen that yellow light stimulates both red-sensitive and green-sensitive cones, we can think of these ganglion cells as balancing blue against yellow.

In any event, it is clear that the information provided by any one ganglion cell gives highly ambiguous information about the illumination of its receptive field. These ambiguities can only be resolved by the brain, using information from several ganglion cells, reporting from the same area of retina** – which is, presumably, why there is so much overlap between the receptive fields.

Finally, what about movement? Detecting movement in the visual field is obviously extremely important for any animal – if only to catch prey or avoid predators – but until the 1960s it was widely assumed that, in mammals and birds, movement was a feature that would be extracted by the brain from the changing pattern of information transmitted from the eyes. Then in the early 1960s work on rabbits[25] and on pigeons[26] showed that some retinal ganglion cells behaved in an unexpected way: they responded to movement and were directionally sensitive. Anywhere within the cell's receptive field, a movement in one direction would cause a burst of impulses; a movement in the opposite direction would not. Different ganglion cells of this type had different preferred directions, even in the same area of retina. And the stimulus could be a moving spot of light on a dark background or a moving dark spot on a light background. In other words, these cells ignored the nature of the contrast, and the position of the moving object within the receptive field and were concerned only with movement and its direction. In the rabbit, some cells of this kind were also selective for the speed of movement, responding only to very fast or to very slow movements.

Ganglion cells sensitive to movement in a particular direction have not been

*These cells are thought not to receive information from blue-sensitive cones.

**We shall see later (Chapter 14) that the colour actually seen also depends on information reaching the brain from other parts of the retina, and that this dependency helps the brain to identify colours despite changes in the illuminating light.

found in primates, but their existence in some retinas is significant because, as Horace Barlow first pointed out in connection with his experiments on frogs, it shows that relatively simple synaptic circuitry can achieve a moderately sophisticated [unconscious] perceptual task.

Figure 10.4 shows a scheme suggested by Horace Barlow and William Levick to explain how they think this is done.[27] The three boxes in the top row of the diagram represent receptor cells; each, when stimulated, excites an intermediate nerve cell immediately below it, and also – and this is the crux of the scheme – inhibits the intermediate nerve cell of its left-hand neighbour but not of its right-hand neighbour. (The plus and minus signs indicate the excitatory and inhibitory synapses.) Provided that the inhibitory effect is strong enough to prevent the intermediate cell from firing, and that it lasts long enough for a boundary between light and dark areas in the image on the retina to pass from one receptor to its neighbour, a boundary passing from right to left on the retina will not excite the intermediate cells, but a boundary moving from left to right will.

A striking observation by Barlow and Levick in favour of this hypothesis is that, if the boundary is moved continuously through the receptive field in the null direction, there is no stimulation of the ganglion cell, but if, while the boundary is within the receptive field, its movement is stopped for a moment, there is a transient burst of impulses from the ganglion cell just when the movement resumes. Presumably, while the boundary is stationary, the inhibitory effect

FIGURE 10.4   Barlow and Levick's scheme (explained in the text) to account for the directional sensitivity of ganglion cells in the rabbit's retina. (Modified from H. B. Barlow and W. R. Levick.)

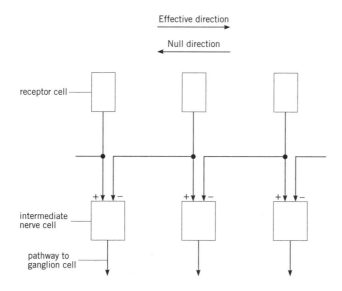

wears off, so that when the movement is first resumed and the boundary reaches new receptors the signals from them get through to the ganglion cell.

TASTE AND SMELL

If you eat an apple, while pinching your nose with your fingers so that no air passes through it, the apple has rather little taste. It may taste sweet or sour, but it does not taste much like an apple. The reason is that the cells on the surface of the tongue that are concerned with taste are able to discriminate between only five main qualities: sweet, sour, salt, bitter and 'umami' – a Japanese word indicating the taste of glutamic acid, something like chicken broth or the taste of meat extracts.[28] All more subtle tastes, and all smells, depend on the stimulating effects of volatile substances on receptor cells in the membrane that lines the air passages above and behind the nose. Somehow, these olfactory receptor cells, which in humans occupy an area about the size of a postage stamp, provide information that enables us to discriminate between an enormous number of smells, including the smells of artificial substances that neither we nor our ancestors have ever met before. And dogs, rabbits and opossums, with far more olfactory receptor cells, do far better.

Recognition of (and response to) different chemical substances is not, of course, a function limited to receptor cells involved in smelling and tasting. As we have seen, it occurs in connection with various kinds of signalling in the body. And although in smelling, and in tasting sweet or bitter substances, the substances to be recognized are different from the neuromodulators that act at synapses, similar transduction mechanisms are involved.[29] The binding of an appropriate molecule to the surface of an olfactory receptor cell sets off a cascade of reactions that leads to a reduction in the electrical potential difference across the cell membrane. What is more, the protein molecules that act as receptors for the volatile substances that we smell, and for the sweet or bitter substances that we taste, belong to the same group of related gene families as the protein molecules that act as receptors at neuromodulator-activated synapses and the opsin molecules in photoreceptors. It seems, then, that in the evolution of these very different systems – receptors at neuromodulator-activated synapses, photoreceptors, olfactory receptors and taste receptors – natural selection has been playing a number of variations on the same theme. And it has been playing for a long time. When a yeast cell is ready to mate, it releases a small peptide that makes potential partners stop proliferating and get ready to conjugate. The receptor, on the surface of the partner, that binds the peptide and starts this change in behaviour is a member of that same group of related gene families.[30]

It is said in highly respectable textbooks that humans can distinguish more than 10,000 different smells. Whether or not that figure is right, and whether or not one believes the more extreme claims of tea tasters and wine buffs, the figure is certainly large. Is it possible that there are as many kinds of olfactory receptor cells as there are distinguishable smells, and that each cell is specific for one smell? Since there are perhaps ten million olfactory receptor cells, the hypothesis that there are, say, 10,000 different kinds is not impossible, but it would imply that about a tenth of all human genes code for olfactory receptor proteins, so it is not very likely. There is, though, now good evidence in rodents for the existence of something like a thousand distinct (though related) receptor proteins, only one of which is present on the surface of each receptor cell.[31]

If there are far more distinguishable smells than there are different kinds of receptor protein, and each receptor cells carries only one kind of receptor protein, it looks as though each receptor cell is likely to be involved in the detection of more than one smell. And there is independent evidence for that. It is not easy to record the electrical activity of individual olfactory receptor cells, but such recordings have been made, and what they show is that each receptor cell has a broad spectrum of sensitivity, responding strongly to some odorous molecules, less to others, and not at all to yet others.[32] We therefore have a huge population of receptor cells of perhaps a thousand different kinds, each kind with its own apparently idiosyncratic pattern of preferences, and each cell sending impulses along a particular axon to the brain. Can useful information be transmitted in this way?

The answer is that it can, provided each receptor cell behaves consistently. The way the system could work is nicely illustrated by an analogy used by my colleague Roger Carpenter.[33] He compares the collection of olfactory receptors to a nursery full of spoilt children, each child having its own strong and arbitrary food fads. He imagines these children seated at a dinner table and provided with push buttons with which they can register approval or disapproval of what is set in front of them. If the push buttons are connected to an array of lights, then, although any individual child's preferences may appear idiosyncratic, any particular dish will result in a characteristic and reproducible pattern of lights by which it may be recognized.

These curious arrangements for identifying smells are different both from the orderly array of a very large number of narrowly tuned receptors that identifies pitch, and the trio of more broadly tuned receptors that identifies colour. And this difference has a very striking consequence. It is possible to predict accurately the result of mixing sounds of different pitch or lights of different colour; but because there is no continuously variable quantity characteristic of smells that corresponds to the frequencies or wavelengths characteristic of sounds and lights, and because the rules determining preferences in the olfactory receptor cells are not known, there are, notoriously, no simple rules for predicting the results of mixing smells.[34]

But the quincunx of heaven runs low, and 'tis time to close the five ports of knowledge.
—Sir Thomas Browne *The Garden of Cyrus*

11   A Cook's Tour of the Brain

When Thomas Cook started his grand circular tours, they were designed so that a traveller, without too much expenditure of time or effort, could get a general idea of the area to be toured, hurry through the dull parts, linger in those that were more interesting, and emerge knowing a little of the history and features of the area and much better equipped to make more detailed visits later. That is, roughly speaking, the sort of tour of the human brain that I want to conduct in this chapter.

AN OVERALL VIEW

The easiest way to make sense of the structure and organization of the human brain is to look at the way the brain develops in the embryo, and at its evolutionary history. In vertebrates, including ourselves, the nervous system starts as a midline groove in the surface layer of cells on the back of the embryo. This groove becomes deeper, and soon forms a thick-walled tube which separates from the surface, and is destined to form the brain and spinal cord. As the embryo develops, the front end of the tube, which is closed, swells into three connected vesicles which will form the *forebrain*, the *midbrain* and the *hindbrain*, respectively. Later the forebrain divides into an expanded *endbrain*, and a *between-brain* that lies between the endbrain and the midbrain. The way these different regions develop, and the functions they have is different in the different classes of vertebrate but there is a common overall pattern.

You can get some idea of the evolution of the brain – from fish, through amphibia and reptiles, to mammals – by comparing the brains of animals living today. The striking thing in this evolutionary series is the progressive enlargement of the endbrain. In all four classes it develops to form two cerebral hemispheres, but these are small and fused in the fish, larger in amphibia and reptiles, and very large in mammals, particularly in primates. In humans, the cerebral hemispheres are so large that they fill most of the space in the skull. This great increase in size is accompanied by the takeover of roles that in lower vertebrates are performed by other parts of the brain, and by the appearance of behaviour of a complexity not seen in lower vertebrates.

The human forebrain

Figure 11.1 shows the entire human brain viewed from the left side. The very large cerebral hemisphere dwarfs the rest of the brain, parts of which can be seen peeping out below. The surface of the cerebral hemisphere is a crumpled sheet of neurons and supporting cells from 2 to 5 mm thick. This sheet is the *cerebral cortex*, and the many folds and fissures increase the effective area nearly threefold. Underlying the cortex are masses of axons, which, being mainly myelinated (p. 115), look white in contrast to the 'grey matter' of the cortex. A very large bundle of axons – the *corpus callosum* – crosses from one hemisphere to the other, and provides the main pathway for the transfer of information between the two hemispheres. Deep within the white matter of each hemisphere are three further collections of neurons and supporting cells, the *basal ganglia*, the *hippocampus* (from a fanciful resemblance of its shape, in cross-section, to a 'sea horse' – *hippokampos* in Greek) and the almond-shaped *amygdala* (from the Greek word for almond). The basal ganglia are largely involved in the control of movement – it is their malfunctioning that causes the rigidity and tremor in Parkinson's disease. The hippocampus and amygdala, together with other structures play a vital part in memory and emotion.

FIGURE 11.1 The brain from the left side, showing the lobes of the left cerebral hemisphere and the primary motor and sensory areas of the cortex.

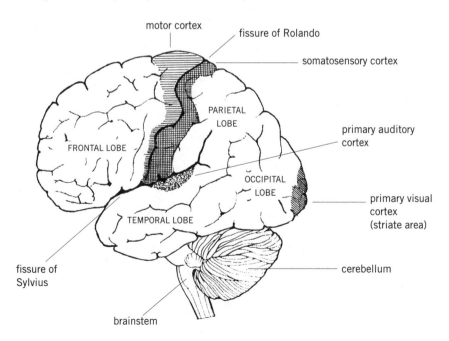

The total surface area of the cortex is about a quarter of a square metre – a little larger than a large pocket handkerchief – and it contains something like 100 billion neurons. It is almost certainly to this extraordinary structure, more than to any other part of the brain, that we as a species, owe our remarkable intellectual abilities.

The between-brain shows nothing like the same expansion in the course of evolution. In all vertebrates, during the embryological development of the between-brain, an outgrowth on each side develops into the retina of the eye and the optic nerve. A conspicuous feature of the mammalian between-brain is the presence, in each side wall, of a large mass of neurons called the *thalamus* – the Latin form of a Greek word meaning 'inner room'. A consequence of the takeover of functions by the cerebral cortex is that, in mammals, information about all sensations has to be carried to the cortex. Some information about smell passes directly from the olfactory organs to a part of the cortex, but information from all the other sense organs (and also information from other parts of the brain) reaches the cortex almost exclusively via one or other thalamus. Each thalamus therefore acts as a great relay station, but this cannot be its sole function as there are even more nerve fibres carrying information from the cortex to the thalamus than there are carrying information from the thalamus to the cortex. The role of these back connections is not known, but a fashionable hypothesis is that they make it possible for the cortex to use representations of information it has just received, to select signals from the thalamus that are most likely to be useful for subsequent cortical processing.[1]

In the floor of the between-brain are several collections of neurons that together forms the *hypothalamus* – *hypo* being Greek for below. The hypothalamus is tiny but by controlling the pituitary gland, which secretes hormones that influence other hormone-secreting glands, it dominates the entire hormonal system in the body, and has important effects on metabolism, growth and various processes involved in reproduction. It also acts through the *autonomic nervous system* – a discrete part of the nervous system that, as its name suggests, controls events in the body that occur more or less automatically, though not necessarily unconsciously. Of this part of the nervous system, one division (the *parasympathetic nervous system*) is concerned with 'housekeeping' functions such as appetite, thirst, salt and water balance, body temperature, the movements of the gut and the emptying of the bladder. The other division (the *sympathetic nervous system*) is continuously concerned with the control of blood pressure, but it is particularly active when the body has to be prepared for vigorous action. As generations of medical students have been taught, it is the system for 'fright, flight and fight'. Yet another role of the hypothalamus is to act with other parts of the brain in controlling sleep and wakefulness, and in producing some of the physical changes in the body that are normally associated with emotions such as fear, anger or pleasure.

During embryological development, an outgrowth from the roof of the between-brain forms the pineal gland – Descartes' 'seat of the soul'. The evolutionary origin of this gland seems almost as unlikely as Descartes' hypothesis, but it is believed to represent the vestige of a third eye (situated at the top of the head), an eye that is found in the fossils of certain extinct fish, amphibia and reptiles, and that still exists in one living reptile – the tuatara, a spiny, lizard-like creature found on some offshore islands in New Zealand. In those mammals in which the onset of the breeding season is controlled by the duration of daylight, there is evidence that secretion of a hormone by the pineal is involved. In humans, it is not clear that the pineal has any normal role, though because it secretes a substance (melatonin) during the dark hours but not much during the day, a role in maintaining daily rhythms has been suggested. There have even been attempts to use melatonin to control jetlag. A particular kind of pineal tumour is associated with (among other disorders) extraordinary sexual precocity that can lead to a child of three or four having external genital organs of nearly adult size.

The midbrain

In all vertebrates except mammals, the roof of the midbrain swells into a pair of domes, one on each side. In fish, amphibia and reptiles, these are the main sites for handling visual information and are therefore called optic lobes. In mammals the roof of the midbrain swells into four domes. One pair correspond to the optic lobes in lower vertebrates, but (the handling of visual information having been largely taken over by the cerebral hemispheres) they are mainly concerned with eye movements. The other pair are concerned with handling auditory information on its way to the cerebral hemispheres, and their development is thought to be associated with the evolution of the *cochlea*. They are particularly large in the bat, where they play an important part in the echo-location system. The midbrain also contains collections of nerve cells that, together with nerve cells in the forebrain above and in the hind brain below, are concerned with movement, sleep and arousal.

The hind brain

The hind brain is a tapering tube linking the midbrain and the spinal cord. The roof adjacent to the midbrain is expanded into a large highly folded structure, the cerebellum, which is concerned with balance and posture, with the fine control of movement, and in particular with the learning of complex patterns of movement. The flautist who completes a fast and intricate passage without a thought of the fingering is believed to have stored the necessary instructions in her cerebellum. In all vertebrates the hind brain is concerned with the control of the circulation and of respiration. In lower vertebrates, it is also concerned with vibration sense (or hearing) and taste, and with the initiation of voluntary movements, but in mammals these functions have largely been taken over by the cerebral cortex.

That completes the rapid part of our itinerary, and I want to spend the rest of this chapter looking in a more leisurely way at the cerebral cortex.

THE CEREBRAL CORTEX: WHAT HAPPENS WHERE

The convolutions

The strikingly convoluted appearance of the surface of the human cerebral cortex was noted in an Egyptian papyrus of about 1700 BC, which compared it with the film and corrugations seen on the surface of molten copper as it cools.[2] Few of us nowadays are familiar with the appearance of cooling molten copper, but the comparison is, anyway, misleading for it suggests that the convolutions are arbitrary and inconstant. In fact they are sufficiently similar in different brains to be used as landmarks; and in exploring the working of the cortex we need landmarks.

In Figure 11.1 the left cerebral hemisphere is seen from the side. Two striking fissures – the fissure of Rolando, running downwards and forwards from just behind the highest part of the brain, and the fissure of Sylvius, running backwards and slightly upwards from the notch near the front of the brain – divide the hemisphere into frontal, parietal and temporal lobes. (The names of the fissures celebrate Italian and Dutch anatomists of the 18th and 17th centuries; the names of the lobes come from the different bones of the skull with which they are in contact.) A fourth lobe, the occipital lobe at the back end of the brain, is not clearly demarcated from the parietal and temporal lobes on the outward-facing surface of the hemisphere, but has a clear boundary on the surface (not visible in the Figure) that faces the opposite hemisphere. The surface of each lobe is subdivided by numerous smaller fissures into convoluted flattened areas called *gyri* (singular *gyrus*, Latin for circuit).

The idea that different parts of the cerebral hemispheres have different functions is now so familiar that it seems obvious, but it has had a long and chequered history. Hippocrates warned against making incisions in the brain because of the danger of causing convulsions on the *opposite* side of the body. By the 18th century it was well known that injury to one side of the brain often caused paralysis of limbs on the opposite side, and in 1778 Nicolas Saucerotte, an army surgeon, described extremely unpleasant experiments in which he damaged selected areas of the cerebral cortex of dogs by trephining through the skull; he showed that, depending on where he placed the trephine, he could cause paralysis of the opposite forelimb or of the opposite hindlimb. In the early 19th century many attempts were made to identify the functions of different parts of the brain by destroying them, or by stimulating them mechanically, chemically or electrically, but the results were confused and difficult to interpret.[3] The main problem was that, without anaesthetics or precautions against infection, the surgical procedures were so traumatic that the animals tended not to survive long enough to recover fully from the shock of the operation. Even if they did, it was unsafe to argue that, because damage to an area caused loss of a particular function, performance of that function was the normal role of the area; the effect might be an indirect one. Conversely, a genuine loss of function might be concealed by compensatory effects of other parts of the brain. When electrical stimulation was used, the results were complicated by the spread of current to surrounding structures. And the whole question was complicated by the enormous and unsavoury red herring of phrenology.

The bumps

At the end of the 18th century, Franz Josef Gall, a physician practising in Vienna, had developed a new system of psychology. As a boy, he had noticed, or thought he had noticed, that people with particularly good memories had prominent eyes. Regarding the brain as the organ of the mind, he therefore wondered whether a variety of mental characteristics might not be reflected in physical features of the brain, and hence of the skull. His system was based on a number of crucial assumptions.[4] He assumed that human mental powers can be analysed into a definite number (initially twenty-seven) of independent 'faculties'; that each of these faculties has its seat in a definite region of the surface of the brain; that the size of each region in an individual reflects the strength of the corresponding faculty; and that, because the skull fits the brain like a glove, it is possible to judge the sizes of the different regions by measuring the bumps on the surface of the head. If all Gall's assumptions had been justified, and if his allocation of faculties to regions had been correct, he would have invented a marvellously objective way of assessing character; and in this lay the enormous popular appeal of his system – an appeal that was increased when, in 1802, the Austrian government, influenced by the ecclesiastical authorities, interdicted his lectures as being dangerous to religion. In 1807 he settled in Paris where he made many converts and a great deal of money. His pupil and colleague, J. C. Spurzheim, was so successful in lecturing on phrenology in Great Britain and America that, by 1832, there were twenty-nine phrenological societies in Great Britain, and several English-language phrenological journals.

Phrenology was, almost from the start, both fashionable and a subject of ridicule. The trouble with it was, in the first place, that not all the basic assumptions were justified, and secondly, that, although the allocation of faculties to regions was supposed to be based on empirical evidence from the study of subjects with known character (Gall had been particularly conscientious in visiting gaols and lunatic asylums), it was largely the product of guesswork. Gradually the influence of phrenology declined, though it lingered on into the 20th century. Francis Crick recalls being taken to a phrenologist by his mother when he was a boy;[5] the last British phrenological society was disbanded only in 1967; and porcelain heads with the classical faculties marked on them can still be found in British antique shops. What is curious amid all the quackery is that Gall was a highly competent anatomist who did important work sorting out the pathways of the motor nerve fibres that carry information from the cerebral cortex to the spinal cord; and his emphasis on the role of the grey matter of the cerebral cortex in intellectual processes was, of course, correct.*

A prediction confirmed

Gall had associated speech with the frontal lobes of the brain. As always, the evidence was flimsy, but Gall's view was accepted by Jean Baptiste Bouillaud, later Professor of Medicine at the Charité hospital in Paris, whose clinical studies on patients with loss of speech appeared to support it.[6] Despite successive attempts to persuade them, the members of the Académie de Médicine remained unconvinced, however, and in a debate in the stormy political climate of 1848 the exasperated Bouillaud made an offer: he would pay 500 francs to anyone who would provide him with an example of a deep lesion of the frontal lobes of the brain without a lesion of speech. He also made a prediction: 'At this very moment', he said, 'there is a patient at [the Hospital] Bicêtre who has all the freedom of his intelligence and his movements . . . but he cannot speak . . . I am not afraid to affirm that this man carries a deep lesion in the anterior lobules of his brain.' Thirteen years later, at a meeting of the Société d'Anthropologie, Paul Broca, a surgeon at Bicêtre, presented the brain of a patient who had died the previous day.[7] The patient was a Monsieur Leborgne, a maker of cobbler's lasts by trade. Epileptic since his youth, he had been admitted to the hospital twenty-one years earlier, at the age of thirty, when he had become unable to speak though still able to use his tongue and lips normally in every other way. During his long stay in the hospital, he was able to understand speech, and to make himself understood by

*Jerry Fodor has argued that Gall also deserves credit for the way in which, in slicing the cognitive powers of the mind into its component faculties, he rejected the usual 'horizontal' divisions into memory, imagination, attention, sensibility, perception, and so on – faculties that are independent of the subject matter and whose operations cross 'content domains' – and instead chose 'vertical' divisions into faculties that are distinguished by reference to their subject matter – amatory propensity, love of children, friendliness, combativeness, acquisitiveness, cautiousness, vanity, benevolence, religious sentiment, wit, and so on. (The words in quotes are those used by Fodor.) See Fodor, J. A. (1983) *The Modularity of Mind*, MIT Press, Cambridge, Mass. Whether Fodor is right in claiming that 'the notion of a vertical faculty is among the great historical contributions to the development of theoretical psychology' seems to me doubtful.

gestures, but his only answer to any question was 'tan tan', or, when he became angry, 'Sacré nom de Dieu'. It is not clear whether this unfortunate man was the man Bouillaud had been referring to thirteen years earlier, but it seems likely. In any event, the frontal lobe of his left cerebral hemisphere showed a fluid-filled cavity about the size of a hen's egg.

Inability to speak 'in individuals who are neither paralysed nor idiots' seemed to Broca to deserve a special name, and eventually it became known as *aphasia* (from the Greek *aphatos* meaning speechless).

Within two years of Leborgne's death, Broca had collected eight cases of aphasia associated with a lesion in the same part of the frontal cortex; and because in many of them the lesion was much more circumscribed, Broca could argue convincingly that the critical area was the posterior part of what is now known as the inferior frontal gyrus (see Figure 16.1, p. 262). In all these cases the lesion was on the left side, but the nearly perfect bilateral symmetry of the brain's anatomy made Broca, and others, reluctant to accept that this was anything more than coincidence. Three events changed this view. The first was the steady accumulation of cases in all of which the lesion was on the left; by March 1864 Broca had collected twenty cases. The second was a demonstration to the Société Anatomique of a case in which a lesion of the inferior frontal gyrus *on the right side* was *not* associated with any speech disorder. The third was the acquisition and publication by the Académie de Médicine of a memoir by a general practitioner from Montpellier called Marc Dax. Dax had died twenty-six years earlier, but in the last year of his life he had given a paper to a congress of physicians in southern France describing forty cases of patients who had lost the power of speech; in all of them, though there were no post-mortem findings, there was evidence (usually from the sidedness of the associated paralysis) that the lesion was on the left side of the brain. By 1864, Broca was convinced: 'Nous parlons avec l'hémisphère gauche'.

At about the same time as Broca, in Paris, was demonstrating a connection of speech with a specific area of the cortex, John Hughlings Jackson, at the Hospital for the Paralysed and Epileptic in Queen Square, London, was beginning the famous series of clinical observations that led him to suggest that particular areas of cortex could cause movements of particular parts of the body.[8] He was especially interested in those less common epileptic fits – still called 'Jacksonian' – in which the convulsions begin unilaterally; and he demonstrated that, in different patients, the jerking would begin in the thumb or the big toe, or the angle of the mouth, and would then spread in a reproducible manner. Believing that the cause of epilepsy was the 'occasional sudden, excessive, rapid and local discharge of the grey matter' he suggested that the site of the initial jerking was determined by the site of the initial nervous activity, and that the spread of the seizure reflected the spread of this activity over regions of the brain in which movements of parts of the body were represented. The reproducibility of the spread of the seizure suggested that there was an orderly representation of the parts of the body.

Frau Hitzig's dressing table

Within a decade, there was strong support for Hughlings Jackson's hypothesis from the experiments of Gustav Fritsch and Eduard Hitzig in Berlin.[9] There is a story that, while dressing a wound of the brain during the Prussian–Danish war, Fritsch had noticed that irritation of the surface of the brain on one side caused twitching of the body on the opposite side. In any event, he and Hitzig determined to investigate the effect of stimulating the surface of the brains of anaesthetized dogs, using weak electric currents. At that time there were no facilities at the Physiological Institute in Berlin for working on warm-blooded animals so the first experiments were done in Hitzig's home, and indeed on Frau Hitzig's dressing table. Using a pair of platinum wires stuck through a cork, and currents just strong enough to be felt on the tongue, Fritsch and Hitzig explored the surface of the cerebral hemispheres, and found that stimulation of certain areas in the front half of the brain caused contractions of groups of muscles, producing discrete movements of one or other limb on the opposite side of the body. In later experiments Hitzig defined the limits of these 'motor areas' in both the dog and the monkey.

At about the same time, David Ferrier, physiologist and physician at King's College, London, but working in a laboratory attached to the West Riding Lunatic Asylum in Yorkshire, made a detailed map of the motor areas in the monkey.[10] By electrical stimulation using very weak currents and electrodes only a millimetre apart, he could produce tiny normal-looking movements – the flick of an eyelid, the twitch of an ear. In this way he was able to identify particular areas of cortex with particular parts of the body, and he then showed that removal of an identified area led to paralysis of the corresponding part. Even before Ferrier's work (but following the publication of Fritsch and Hitzig's paper in 1870), Jean Martin Charcot, at the Salpêtrière Hospital* in Paris, had made careful comparisons of clinical observations and of post-mortem examinations of his patients' brains, and had concluded that damage to the gyrus just in front of the fissure of Rolando (Figure 11.1) led to disturbances of movement. Whether these disturbances affected the arm or leg or head depended on which part of the gyrus was damaged.

A second speech centre

All these examples of localization of function in the cerebral cortex have been concerned with motor activity. What about sensory activity? In 1874, thirteen years after Broca had demonstrated the fluid-filled cavity in the brain of

---

*Built as an arsenal by Louis XIII, its name comes from the saltpetre which had once been manufactured there. In Charcot's time it had about 5000 inhabitants.

Monsieur Leborgne, Carl Wernicke, a twenty-six-year-old physician in Breslau, published a short monograph drawing attention to cases of aphasia that differed in several ways from those described by Broca.[11] First, the patients, unlike those of Broca, did have difficulty in understanding speech. Secondly, their difficulty in speaking was not in finding words, but in using them correctly. Instead of the sparse, telegram-like style characteristic of Broca's aphasia, these patients tended to be fluent, sometimes very fluent; but the words were not always the right words or even real words, and the fluent sentences did not always convey much meaning, so that the physician could easily misdiagnose the patient as confused. And thirdly, the brain lesion was not in Broca's area but in the left temporal lobe adjacent to the fissure of Sylvius (see Figure 16.1, p. 262). Modern American examples of the speech of patients with Broca's aphasia and Wernicke's aphasia are given in Chapter 16 (pp. 262–3).

Wernicke attempted to make sense of these differences.[12] In Broca's aphasia, the lesion was in the area which, when stimulated by Hitzig in monkeys, had caused movement of the mouth and tongue. Broca's area, Wernicke argued, was therefore probably the centre for representation of movements made in speaking, and damage to it would cause loss of the ability to speak, without any loss of comprehension. In contrast, the lesion in what was to become known as Wernicke's area was close to part of the cortex that was thought to be concerned with hearing. Wernicke's area was, then, presumably the centre for representation of sound patterns, and damage to it would cause some loss of the ability to understand, or even simply repeat, the spoken word. The predictions that followed from this picture of the organization of speech in the brain, and their dramatic confirmation in some patients, is an interesting story that we will come back to when discussing language (Chapter 16). What is significant in the present context is that, in the 1870s, the division of aphasias into primarily motor (Broca's aphasia) and primarily sensory (Wernicke's aphasia) suggested that particular areas or 'centres' in the cerebral cortex were concerned with particular motor or particular sensory activities. And, judging from the two 'speech centres' it seemed that the levels of activity that were controlled by the 'centres' were intermediate between simple movements or sensations and the highly complex activities – veneration, filial love and so on – that were envisaged by the phrenologists.

Maps in the brain

An even more striking example of the way in which individual areas of the cortex have individual roles came from the work of Salomon Henschen, the Professor of Medicine at Upsala.[13] In 1888 he published the first of many articles in which he correlated loss of vision in parts of the visual field with damage to parts of the so-called striate cortex in the hindermost part of the occipital lobe. (The name 'striate' comes from a stripe which is found in this part of the cortex when it is cut across.) Henschen found, first, that loss of the striate cortex on one side was always associated with blindness in the opposite half of the visual field in both eyes.* (The explanation, he showed later, is that all the fibres in the optic nerve that come from the left half of *both* retinas connect with the striate cortex in the left occipital lobe, and all those from the right half of *both* retinas connect with the striate cortex in the right occipital lobe (see Figure 11.2). Since we use the left halves of our retinas to see objects to the right of the mid line, and vice versa, loss of the striate cortex on one side leads to blindness in the opposite half field.) This was impressive, but even more remarkable results came from looking at the effects of small lesions in this part of the cortex. Damage to a small area in, say, the left striate cortex gave rise to a small area of blindness – a *scotoma* – at a *corresponding point* in the right half-fields of both eyes.** This correlation led Henschen to conclude that damage to adjacent areas of the striate cortex caused scotomas in adjacent areas of the visual field; in other words, for an observer whose eyes are still, the visual world seemed to be mapped point by point onto the striate cortex, just as the real world is mapped onto the pages in our atlases. This was localization with a vengeance. And because what we see depends on what is in the images on our retinas, it implied that the fibres in the optic nerves must connect particular points in the retina to corresponding points in the striate cortex. It is worth noting, though, that this does not mean that equal areas of retina are handled by equal areas of striate cortex. In fact, as you might suspect, the areas of striate cortex that are involved in handling information from the small central areas of the retina concerned with detailed vision are much more extensive than those concerned with information from the peripheral parts of the retina.

Shocking the brain

Speech and vision are rather specialized operations so it was, perhaps, not surprising to find areas of the cortex devoted to them. What about more mundane tasks such as the handling of information from touch receptors in the skin or from stretch receptors in the muscles and joints? Answering this question required different approaches depending on whether the owner of the cortex was human or animal. In humans the most successful approach was to stimulate different areas of the cortex, using weak electric shocks, and to ask the patient to report any sensation that followed. This kind of procedure requires, of course,

*A similar effect had been found by Herman Munk, eleven years earlier, following removal of one of the occipital lobes in a monkey.

**The patient is not usually aware of such scotomas, any more than the normal subject is aware of the normal 'blind spot' (the part of the visual field corresponding to the insensitive area of the retina where the optic nerve enters), but they can be detected by asking the patient to fixate on a particular point and then seeing whether a small spot of light is noticed when it is placed at different points in the field of view.

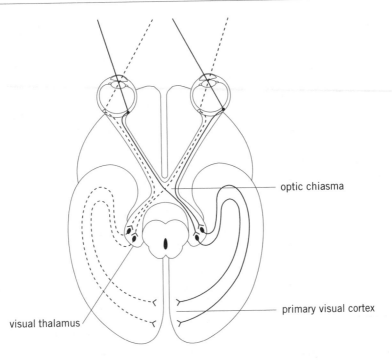

optic chiasma

primary visual cortex

visual thalamus

FIGURE 11.2 The optic pathways. Because optic nerve fibres from the nasal half of each retina cross at the optic chiasma, all information about the *left* half of the visual field ends up in the *right* primary visual cortex (black route), and all information about the *right* half of the visual field ends up in the *left* primary visual cortex (dashed route). The two cortical areas are connected by fibres in the corpus callosum (not shown).

that the cortex be accessible and the patient be conscious. There also needs to be a strong medical justification for exposing the patient to a procedure that inevitably carries some risk.

In the first recorded case in which the human cortex was stimulated, this third condition was not satisfied and the outcome was catastrophic. In 1874, Roberts Bartholow, professor of materia medica at the Medical College of Ohio, in Cincinnati, had as a patient a thirty-year-old housemaid, Mary Rafferty, the top of whose skull had been eroded over an area about two inches in diameter by a malignant ulcer.[14] The pulsating brain was visible, covered only by a thin membranous sheet. As the brain had already been deeply penetrated by incisions made for the escape of pus, Bartholow argued that fine needles could be inserted into it without risk of material injury. He showed that such insertion caused no pain. A succession of brief electric shocks to the left side of the brain caused muscles to contract in both the right arm and right leg, and this was accompanied by 'strong and unpleasant tingling' in both these limbs. There was also slight contraction of the muscle in the left eyelid and widening of both pupils. Similar stimulation on the right side of the brain caused corresponding results. With the needle still in the right side, Bartholow increased the strength of the shocks, but the patient became very distressed, lost consciousness and was violently convulsed for five minutes. Twenty minutes after the beginning of the attack she regained consciousness and complained of some weakness and vertigo. Two days later the unfortunate patient was found to be extremely ill and partially paralysed on the right side. While being questioned she became unconscious, and the following day she had a convulsion and lapsed into a coma from which she did not recover. A post-mortem examination of the brain suggested that the cause of death was thrombosis in the main vessel draining venous blood from the brain, the course of this vessel lying right in the centre of the ulcer. Bartholow's paper caused a storm of criticism, which he seems to have weathered. His name appears as one of the founder members of the American Neurological Association, which had its first meeting the following year.[15]

As surgical techniques improved, electrical stimulation of the exposed cortex was to become a useful tool in operating on patients with brain tumours or with epileptic foci that were causing intolerable seizures. Before cutting into the cortex in the neighbourhood of a tumour it was only sensible to try to identify the function of that part of the cortex by looking at the response to weak electrical stimulation. In patients with focal epilepsy, stimulation might reveal the site of the focus by causing the premonitory signs of an attack.

In the 1880s and 1890s there were half a dozen reports of the effects of cortical stimulation, which confirmed that very small electric shocks in the region of the Rolandic fissure caused movements on the opposite side of the body. Sometimes, too, there were tingling sensations, but it was not until 1908 that Harvey Cushing succeeded in producing sensation without movement.[16] This was in the course of investigating two patients, both suffering from distressing epileptic attacks, inaugurated in each instance by strange feelings in the right hand. In each case the cortex was exposed under anaesthesia, but the stimulation was done with the patient fully conscious so that the surgeon might be guided by any sensations that were produced. (The previous year, in an operation to excise a cortical tumour, Cushing had shown that, when the patient had recovered consciousness after the first part of the operation, cutting the cortex and removing the tumour 'not only occasioned no discomfort, but was attended on the part of the patient by a lively and helpful interest in the performance'.) Stimulation of the gyrus just behind the fissure of Rolando caused no movements but definite sensory impressions in the opposite hand and arm – in one patient a sensation of numbness, particularly in the little finger, in the other a sensation 'as though someone had touched or stroked the index finger' or, when the point of stimulation was lower down on the gyrus, as though the back of the hand had been stroked.

The most extensive studies of the effects of stimulating the cortex of conscious patients were made by Wilder Penfield,[17] and his colleagues at Montreal, in a huge series of cases extending from the 1930s to the 1950s. The justification for operating was usually that the patients had tumours needing removal, or epileptic foci causing intolerable seizures. These studies confirmed that stimulation of the gyrus just in front of the fissure of Rolando generally caused movement, and stimulation of the gyrus just behind generally caused sensations. The former gyrus therefore became known as the *motor cortex,* and the latter as the *somatosensory cortex* (*soma* is the Greek for body).*

The studies also showed that the different parts of the body were represented in an orderly fashion on the surfaces of the two gyri. Figure 11.3 summarizes the patterns of representation – in other words, the system of mapping – and it shows several striking features. First, the arrangement is similar in both gyri so the motor cortex and the sensory cortex dealing with each part of the body are close together across the Rolandic fissure. Secondly, the area of cortex devoted to each part of the body is wildly disproportionate to the size of that part; it relates more to the part's sensitivity and discriminatory ability or the complexity of its behaviour. The areas devoted to the trunk, hips and legs are small, and those devoted to the fingers, thumb, lips and tongue comparatively huge. Penfield's artist has illustrated this by draping two 'homunculi' over the surfaces of the cerebral hemispheres sliced in the plane of the fissure of Rolando. One of these homunculi is sensory, the other motor. Both are upside down, and both show not only gross distortion from the normal human proportions, but also curious discontinuities – the face area, for example, is unconnected with the head or neck and appears (the right way up) below the hands. There are other subtleties, too. Both for movement and sensation, combined activity of the hand and fingers seems to be represented separately from activity of the fingers alone.

Unlike humans, animals cannot describe their sensations, so cortical stimulation followed by interrogation cannot be used to identify the areas of cortex that are concerned with sensation. One way of identifying these areas is to remove bits of cortex and to see whether the response to stimulation of sense organs in different regions of the body is affected. Another way is to take advantage of the electrical activity – the so-called 'evoked potentials' – that can be detected in an area of sensory cortex when sense receptors in the part of the body represented in that area are stimulated. Both methods have been used extensively and show an organization not basically different from that found in humans so long as one allows for the different relative importance of different activities. In rats, for example, there is a relatively enormous area of cortex devoted to the whiskers, in pigs to the snout, and in raccoons (which show great manual dexterity) to the fingers.

*Aldous Huxley's *soma,* which gave such pleasure in his *Brave New World,* is presumably derived from the identical (though unrelated) word in Sanskrit, which refers to an intoxicating drink used in Vedic rituals.

It is interesting that the *movements* caused by stimulation of the human or animal motor cortex – opening or closing of the hand, flexion or extension of the leg, masticatory movements, swallowing – involve the coordinated actions of several muscles, but they are never complex learned movements. They are, as Penfield pointed out, the sort of movements that a baby is able to do at birth or shortly afterwards. The *sensations* caused by stimulating the human somatosensory cortex tend to be tingling or numbness. Even the feeling of being touched or stroked described by one of Cushing's patients is unusual. Clearly, whether motor or sensory, the activities represented in the gyri bordering the fissure of Rolando seem to be rather elementary. This is not true for all areas.

Some of the most startling effects of cortical stimulation were reported by Penfield in operations in which he stimulated the temporal cortex in patients who habitually suffered from severe epileptic seizures of a peculiar kind. The peculiarity lay not in the convulsions themselves – they might even be absent – but in the initial stages of the attacks, which always involved a change in psychological state. The patient entered either a dream-like or hallucinatory state, or a state of altered perception of the environment with perhaps a sense of *déjà vu* or detachment or remoteness or unreality; things might seem to grow larger or smaller, or sounds to be abnormally loud or totally stilled. (The curious 'turns' of the governess in Henry James's *The Turn of the Screw* – the 'intense hush', the 'seeing with a stranger sharpness', the sense of dread – are an example of attacks of this kind that never proceeded beyond the initial stage.)*

In 1938 Penfield operated on a fourteen-year-old girl who had had a single convulsion followed by coma and transient paralysis in infancy, and then periodic seizures from the age of eleven:[18]

> Her attacks were characterized by sudden fright and screaming. She then held on to people about her for protection. This was followed by falling and occasionally by a major convulsion. On careful questioning it was learned that during the preliminary period of fright she invariably saw herself in a scene that she remembered to have occurred at the age of 7 years.
>
> The scene is as follows: A little girl was walking through a field where the grass was high. It was a lovely day and her brothers were walking ahead of her. A man came up behind her and said: 'How would you like to get into this bag with the snakes?' She was very frightened and screamed to her brothers, and they all ran home, where she told her mother about the event. The mother remembers the fright and the story, and the brothers still recall the occasion and remember seeing the man.

*In a fascinating discussion of the governess as a case of temporal lobe epilepsy, J. Purdon Martin (*British Medical Journal*, 22 Dec 1973, 717–721) points out that Henry James's friend and London publisher, Frederick Macmillan, knew Hughlings Jackson and published the journal *Brain*, of which Hughlings Jackson was an editor and which included accounts of temporal lobe epilepsy.

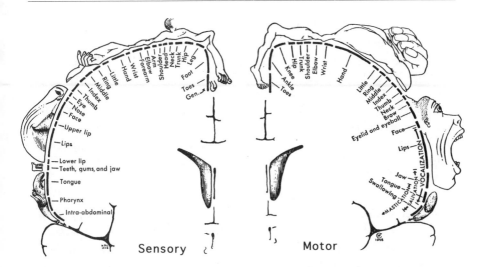

FIGURE 11.3 Penfield's homunculi. Representation of the different parts of the body in the somatosensory cortex and the motor cortex. (From W. Penfield and T. Rasmussen.)

After that, she occasionally had nightmares during her sleep and in the dream the scene was re-enacted. Three or four years later . . . it was recognized that she had attacks by day in which she habitually complained that she saw the scene of her fright. There was a little girl, whom she identified with herself, in the now familiar surroundings. During the attack she was conscious of the actual environment and called those present by name; yet she also saw herself as a little girl with such distinctness that she was filled with terror lest she should be struck or smothered from behind. She seemed to be thinking with two minds.

When Penfield examined the brain he found evidence of an old haemorrhage which, he thought, probably explained the convulsions in infancy. Stimulation far back in the right occipital lobe caused the patient to see coloured stars in the left half of the visual field, presumably because the electrode was in the striate area. But the significant result occurred when he stimulated the outward-facing surface of the right temporal lobe. Depending on the placing of the electrode, he found he could produce in the patient different portions of her 'dream'.

Penfield's interpretation of his observations, and the history, is that the bleeding at the surface of the brain in infancy caused some damage; the serious fright at the age of seven caused periodic nightmares in which the pattern of nervous activity that recorded the frightening events was repeatedly reproduced in proper sequence; that at the age of eleven (as a late consequence of the early damage) the patient began to get seizures; that the point of origin of the epileptic discharge was in the right hemisphere near the junction of the temporal and occipital lobes; and that 'the discharge followed the well-worn synaptic pattern that was capable of waking the childhood memory'. He points out that the temporal cortex must be the repository of many other patterns but supposes that conduction along the neuronal links associated with the childhood memory had been facilitated first by the repeated nightmares and later by the repeated epileptic attacks. The results of electrical stimulation suggested that the neuronal pattern could be activated from different points, initiating different stages of the dream. If the electrode was held in place, the hallucination progressed like a story unfolding, and it included not just the visual and auditory components but also the terror. In Penfield's words, 'She could hear what the man said . . . she could sense the fright that possessed her, feel her body running.'

Although the attacks did not always proceed to convulsions, they were so wrecking the girl's life that Penfield decided to excise a large part of the right temporal cortex. (Because it was the right cortex there was no risk of causing aphasia.) After the operation the girl 'no longer had hallucinations. But when she was asked about the experience, she could still remember it – the meadow, the man, and her fright'. This implies that the memory was also stored somewhere else in the brain, possibly, Penfield suggested, in the opposite temporal cortex.[19]

The 'black reaction' and after

A completely different approach to understanding the cerebral cortex is to look at its microscopic structure. In the course of the 19th century there were great advances in the design of microscopes and in the methods of preparing tissues for microscopy. There were new methods for preserving the tissues, for staining them, and for embedding them in materials (tallow, soap, paraffin wax, celloidin) so that slices could be cut thin enough to be examined by shining light through them. By the end of the 1860s the structure of most tissues was fairly well known, but the structure of nervous tissues, and particularly of the grey matter of the brain, remained puzzling. The reason was straightforward. In most tissues the cells had simple shapes and in suitably stained sections of the tissue their arrangement was not too difficult to decipher. Nerve cells had very complicated shapes with long highly branched processes, and in grey matter they were packed tightly together. They were also, apart from their nuclei, very difficult to stain. It was not even clear that the grey matter was made up of individual cells. Around 1867, Joseph von Gerlach had looked at teased preparations of cortex stained with gold salts, and had concluded that the fine nerve processes in the grey matter joined together to form a continuous network. Gerlach's *reticular theory* – *reticulum* is Latin for net – was to be a matter of contention for the next forty years.

What was needed to sort out the tangle was a method of staining that would pick out an individual cell in its entirety without staining the adjacent cells. In 1873, Camillo Golgi, then aged thirty, was working as resident physician at the Pia

Casa degli Incurabili in the little Lombardy town of Abbiategrasso. Here, in a rudimentary laboratory set up in his kitchen, and working mostly at night, he developed a method – the *reazione nera* (black reaction) – which did just this. Fragments of brain that had been hardened for several days in a solution of potassium dichromate, were soaked for several more days in a dilute solution of silver nitrate. The method was temperamental but when it worked properly the results were astonishing. Under the microscope, thin sections of tissue looked mostly yellow (from the dichromate), but the occasional cell down to its finest ramifications would be jet black 'with the sharpness and brilliance of an indian ink drawing on a pale yellow background'[20] – see Figure 11.4. The black material was silver chromate, but why it appeared in only a small fraction of the nerve cells was (and, more than a century later, still is) a puzzle.

Golgi's method revolutionized knowledge of the fine structure of the nervous system. Golgi himself was the first to differentiate between nerve cells and supporting cells (neuroglia), and he was able to distinguish two important classes of nerve cells in the cortex: pyramidal cells, whose axons left the cortex to enter the white matter, and large star-shaped cells whose axons ramified locally in the cortex. In the hippocampus he found an elegant and regular architecture of nerve cells of different types. On the general question of whether nerve cells were separate entities or formed a network, Golgi was less successful. Despite having invented the method that was to solve the problem, he got the wrong answer. He persuaded himself that the fine terminal branches of axons did fuse with the branches of other axons to form a network. (The dendrites he thought were not involved, and he assumed, again wrongly, that they were merely nutritive in function, without any role in conduction.)

What was needed to resolve the problem was a clear demonstration of the way axon terminals did in fact end in the brain. This demonstration was to be provided by Santiago Ramón y Cajal.

Cajal's life has been splendidly described in the autobiography he wrote in his last years.[21] Born in 1852, the son of a poor surgeon in a tiny village in Navarre, he survived being kicked unconscious by a horse at the age of four, and at seven he coped with his father's letters about patients while his father was away. His parents saw a brilliant future for him, but at school he was a disaster – shy, loathing Latin, Greek and mathematics, and forever being punished. He was keen on games, fighting, bird-nesting and the countryside, and skilled at drawing, which his father thought a waste of time. After brief apprenticeships to a barber and a shoemaker and a further attempt at schooling, he was taken in hand by his father, who decided to teach him about human anatomy, starting with bones he acquired at night from the local cemetery. (This was not *quite* grave robbing. In his memoirs, Cajal describes skeletal remains piled in a hollow 'derived ... from those wholesale exhumations ... which the living impose upon the dead ... under pretext of scarcity of space'.) Interested at last, Cajal seems to have been trans-

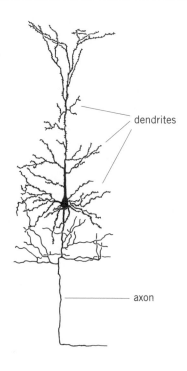

dendrites

axon

FIGURE 11.4 A nerve cell from the cortex of a mouse, stained (using the Golgi method) and drawn by Ramón y Cajal.

formed. He impressed his father with his anatomical drawings; he studied physics, chemistry, and natural history, and began a course of medical studies at Zaragosa, where his father now had an appointment. When he was twenty-one he qualified as a doctor and, being conscripted into the army, joined the Army Medical Service and was sent to Cuba. Here he learnt about corruption and military incompetence, and nearly died of malaria and tuberculosis. Invalided out of the army, he returned to Spain, gradually regained his health, read Helmholtz's *Physiological Optics,* together with the works of Lamarck, Darwin and Spencer, and just before his twenty-fifth birthday was appointed to a temporary assistantship in anatomy at Zaragosa. From then on his career was a succession of successes, culminating in the award of a Nobel prize for Medicine and Physiology in 1906.

At the age of thirty-two, while visiting Madrid, Cajal was introduced by a colleague to Golgi's *reazione nera,* and was amazed at its power and its capriciousness. After much labour he managed to make it less capricious, but his success in demolishing the reticular theory, by showing that the terminal branches of axons did not fuse with parts of other nerve cells, depended less on improvements in method than on his choice of material to be examined. Finding the grey matter of the adult mammalian brain an impenetrable thicket, he decided to look at embryonic brains. He discovered that by choosing a stage of development before the myelin sheaths were formed, and when the nerve cells were still relatively small, he could obtain clear pictures of the terminal branches of axons, and they were perfectly free.

This work, and work that followed from it, not only demolished the reticular theory but also established that the contacts between axon terminals and parts of adjacent nerve cells were far from random and sometimes followed a tightly controlled pattern. Ironically, Ramón y Cajal was never able to convince Golgi that the reticular theory was wrong, and in 1906, when they were jointly awarded the Nobel Prize for Physiology and Medicine, Golgi embarrassed everyone by using his lecture to attack Ramón y Cajal and defend the theory.

Another crucial problem tackled by Cajal was the direction of conduction within nerve cells. The many contacts between axon terminals and the dendrites of adjacent nerve cells made Golgi's notion that the dendrites were purely nutritive seem very unlikely. But if dendrites were involved in the interactions between nerve cells, in which direction did they conduct? Cajal solved this problem elegantly by looking at the detailed anatomy of the retina. Knowing that the light stimulus acts on the rods and cones, and that the message concerning the stimulus is eventually carried by the optic nerve to the brain, he was able to conclude from the arrangements in the intervening layers of the retina that the direction of conduction within each neuron must be from dendrites to cell body to axon. He drew the same conclusion from his studies of the olfactory bulb.

During the 1890s, Cajal and others studied the fine structure of the cerebral cortex, work that led on to the very detailed studies of Korbinian Brodmann and Oscar Vogt in Germany during the first decade of the 20th century. On the basis of these studies, Brodmann divided the human cerebral cortex into fifty-two discrete areas. Some of these areas correlated well with areas of known function – Brodmann's area 4, for example, fitted well with the motor area, Brodmann's areas 1, 2 and 3 with the somatosensory area, Brodmann's areas 41 and 42 with the auditory area. Despite the time that has elapsed since Brodmann's work, and despite the idiosyncratic numbering system he used – he seems to have allotted numbers according to the order in which he studied the different areas[22] – his fifty-two areas remain a standard for reference, and one of Brodmann's diagrams (Figure 11.5) still tends to appear in current textbooks of physiology.

Although Brodmann's fifty-two areas look different, their structures are all variations on the same basic six-layered pattern. Roughly speaking, the cells in one of the middle layers receive most of the sensory information coming into the cortex; the cells in the inner layers send information to distant regions of the cortex or to other parts of the brain; the cells in the outer layers mainly make local connections with cells in adjacent cortex.

The persistence of Brodmann's diagram in current textbooks does not mean that there have been no advances since 1910. In particular, ingenious methods have been developed to follow the course of individual nerve fibres within the brain. If radioactive amino-acids or sugars are injected in a region containing the

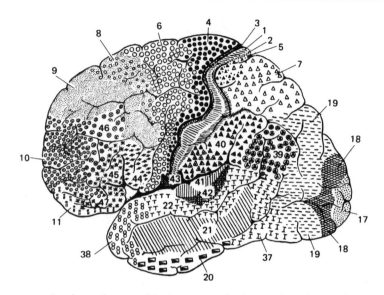

FIGURE 11.5 Brodmann's map of the human cerebral cortex based on its fine structure. What was significant about Brodmann's work was that areas defined by their microscopic appearance often correlated with areas defined by their function. (The numbers have no significance and are merely for reference.)

cell bodies of nerve cells, they are taken up by the cells, incorporated into proteins or glycoproteins and transported to the nerve terminals. If very thin slices of the brain tissue are placed against photographic film, the presence of the radioactive material shows up as a blackened area of film. Conversely, the location of cell bodies belonging to particular nerve terminals can be identified by exploiting the ability of nerve terminals to take up radioactively labelled proteins or small particles in the neighbourhood of the terminal and to transport them back to the cell body. Even more ingeniously, viruses that can cross synapses have been used to trace chains of functionally connected nerve cells.

The 'association areas'

We have seen that the effects of electrical stimulation of different areas of cortex, or the effects of disease or damage in particular areas, led to the identification of the motor cortex, the somatosensory cortex, and the visual cortex. There are also well defined areas of cortex concerned with hearing and with smell. For a long time, what the remaining areas of the cortex did was not clear, and because electrical stimulation caused little in the way of motor or sensory responses they were called 'silent areas'.

A striking difference between these silent areas and the areas with well-defined functions, was pointed out by Paul Emil Flechsig, the professor of psychiatry in Leipzig, as the result of a long series of investigations beginning in the 1890s. By examining the brains of premature, full-term and early post-natal infants he could divide the cerebral cortex into areas in which the myelin sheaths were laid down before birth and those in which they were laid down later. He found that the areas in which myelination occurred before birth corresponded to the motor cortex, and the various sensory cortices. The areas in which myelination was delayed until after birth corresponded to the intervening silent areas. As myelination is necessary for proper functioning, the delayed myelination of these intervening areas suggested that they only started functioning as the organism began to acquire experience. Flechsig called them *association areas* and he supposed that they not only brought together the information from the different sensory areas, but were also responsible for the higher intellectual functions that develop after birth.

In recent years, the distinction between the sensory areas and the association areas has become blurred, with the realization that fibres bringing sensory information from the thalamus reach all portions of the cortex. And the area of 'association cortex' has shrunk, as more careful investigation has revealed that some areas previously labelled association areas are, in fact, secondary areas for processing motor or sensory information of a particular kind. For example, next to the primary visual area (striate area) there are areas concerned specifically with colour or with movement. But even after the shrinkage, in humans, about half of the total area of cortex remains association cortex (see Figure 11.6).

The proportion of the cortex that is occupied by the association areas is much greater in humans than in other primates, and much greater in primates than in non-primate mammals. This is compatible with Flechsig's view that this part of the cortex is responsible for higher mental functions. If that view is right, it is important to know just what goes on in these areas. Unfortunately, their 'silence' means that the most straightforward method of investigating function – seeing the effects of electrical stimulation – is useless; and attempts to elicit electrical activity in areas of association cortex by applying normal stimuli to the skin or the sense organs have proved unrewarding. Until the development of modern scanning methods – which we shall look at later – elucidation of the functions of different parts of the association cortex therefore depended on the study of the effects of local damage caused by disease, accident or surgical interference.

Studies of this kind led to vigorous controversy about whether there was any localization of function within the association cortex.[23] Obviously, areas concerned with the integration of information of different kinds could not be isolated from each other, but did that imply that the association cortex worked as a single unit? And if it did, did different parts of it have different functions, or did they all act in the same way? This latter alternative was strongly advocated by the American neurophysiologist Karl S. Lashley, who supported his case with

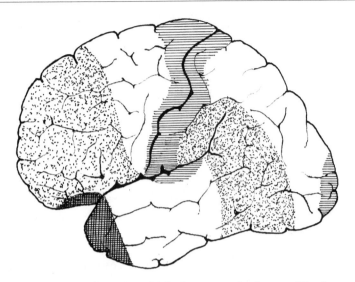

FIGURE 11.6 The association areas in the human cerebral cortex. The three association areas are shown either stippled or cross-hatched. Primary motor or sensory areas are shaded with horizontal lines. The white portion of the frontal lobe is a secondary motor area; the white portions of the other lobes include secondary or higher-order sensory areas.

some surprising observations on rats.[24] He found that the effect of damage to the association areas on a rat's ability to learn mazes, (or to cope with mazes learnt before the damage was inflicted) *depended solely on the volume of cortex damaged, and not at all on its location.* Lashley's findings are puzzling and have never been satisfactorily explained, but one possibility is that the tasks he set his rats involved so many different areas that damage to most areas caused a falling off in performance. In any event, Lashley's view eventually had to be abandoned because of overwhelming clinical evidence.

By the middle of the century it was clear that the effects of disease, trauma, or surgical interference on the various areas of the association cortex were very different depending on which area was damaged. Here I only want to give the briefest summary of the pattern that is suggested by those differences. Roughly speaking, the part of the association cortex that lies wholly in the frontal lobe (see Figure 11.6) appears to be concerned with planning, especially the planning of complex patterns of movements, including those involved in speech – in the left hemisphere it includes Broca's area. The part that straddles the parietal, occipital and temporal lobes is concerned with bringing together somatosensory, visual and auditory information, to promote the creation and use of complex perceptions, including shapes and words – in the left hemisphere it includes Wernicke's area. The part that is divided between the lower part of the frontal lobe and the front part of the temporal lobe contains areas which, when damaged, cause abnormalities in emotion and memory.

. . . LA VISITE EST TERMINÉE

That completes our tour, and it also completes this section of the book. We have seen what sort of message nerves carry and how they are able to carry them. We have seen in some detail how nerve cells interact with one another, and in the sketchiest outline how networks of such interacting cells might form the basis of control systems capable of deciding, of remembering, and of learning. We have seen how sense organs are able to detect changes in their environment and to code the information they acquire in a form suitable for transmission along nerves to the brain. And we have looked at the overall organization of the human brain and had a glimpse of its evolutionary history. At each stage of the story I have tried to say something about the observations, experiments and arguments that have led to current views, so that those views do not seem like the towers of Toledo at sunrise – ravishing but insubstantial, and resting on nothing but the mist over the Tagus.

III LOOKING AT SEEING

We find certain things about seeing puzzling, because we do not find the whole business of seeing puzzling enough.
—Wittgenstein, *Philosophical Investigations*, IIxi

'What does it mean, to see?' David Marr asked in the first sentence of his cele-brated book on vision.[1] 'The plain man's answer (and Aristotle's too)', he contin-ued, 'would be, to know what is where by looking.' But this won't do. The plain man who does not know the meaning of *seeing* is unlikely to know the meaning of *looking*. Marr himself, I suspect, was not happy with his definition, for he promptly offered another. 'In other words, vision is the *process* of discovering from images what is present in the world, and where it is.' This definition, too, may not help the plain man who does not know what seeing is, but it does say something useful to the vast majority of people, who do know. The images Marr refers to are the images on the retina, and it is the processing of the information in those images that is the problem we are concerned with in understanding sight. In Chapter 10 we considered the question: What does the retina tell the brain? We now need to look at the astonishing ways in which the brain makes use of that information to enable its owner to 'know what is where by looking'. More specifically, we want to understand how it is that we are able to distinguish shape, colour, movement and depth, to recognize objects almost irrespective of their orientation, their distance from us, and how they are lit, and to store a vast amount of visual information so that it is available for future use.

In thinking about the role of the human brain in vision, it is easy to slip imper-ceptibly into the common-sense dualist view of the interaction between mind and brain. We reject the notion that our pineal glands contain souls, or that our heads contain homunculi looking at private neural television screens; but because the information that our retinas send to the brain relates to a scene, we do tend to assume that whatever in the brain is handling that information must be looking at some kind of representation of the scene. And the obvious candi-date for the entity doing the looking is the mind. The temptation to think in this way is even greater if the scene we are considering is in the memory. 'I can see it so clearly in my mind's eye', we say of some well-remembered room; and we often can, right down to the pattern of the wallpaper and the graining on the wood of the Victorian dresser. We speak in the same way of imaginary scenes and halluci-nations. 'In my mind's eye, Horatio' is Hamlet's reply when asked where he saw his father's ghost.

Of course, the mind's eye is a metaphor, but it is a metaphor that makes sense only because, however sophisticated our philosophy may be, in everyday life we are all common-sense dualists. Macbeth is explicitly dualistic in trying to under-stand the dagger that he can see but cannot clutch: '... or art thou but a dagger of the mind, a false creation, proceeding from the heat-oppressèd brain?' There is nothing reprehensible about dualism in everyday life – think of the circumlocu-tions that would be necessary if it were given up – but it isn't helpful in trying to understand how the brain uses the information it receives from the eyes. So what approaches are helpful?

There are four, of which two have been used for over a century, and two are much more recent. The most obvious approach is to treat the whole of the visual system as a 'black box'; in other words to act as if the machinery responsible is both inaccessible and invisible, so that information about it can only be deduced from the system's overall behaviour – the relations between various stimuli and the sensations they cause. This is the approach that led Thomas Young to the trichromatic theory of vision, a century and a half before cones with three dif-ferent pigments were identified directly. It is often called *psychophysical*, from a term introduced by Fechner in 1860, and it includes the study of optical illusions and of the kind of phenomena that interested the Gestalt psychologists just before the first world war and between the wars.

The other long-established approach is *neurological* – the study of the effects on vision of injury or disease. This is not quite a black-box approach because in the operating theatre and the post-mortem room the investigator has some access to the inside of the box.

Of the two more recent approaches, the older can conveniently be called *neurophysiological*, since it involves studying the responses of nerve cells in the brain – we have already looked at the responses of nerve cells in the retina – to stimulation of the receptors in the eye. Most of this work has involved recording from microelectrodes placed close to, or inserted into, the nerve cells, and has necessarily been done on animals, but the introduction of modern scanning techniques has made it possible to detect activity in different areas of the conscious human brain.

The last and most recent approach may be called *computational*. It is concerned more with the kind of information processing that is done in the visual part of the nervous system, than with the precise nature of the machinery that does it.

We shall look at the four approaches in turn.

12   Illusions

A psychophysical approach to vision

Knowing 'what is where by looking' is so effortless, immediate and effective, that it is only recently that the enormous complexity of the job done by the visual system has been appreciated. Consciousness, of course, has always been a mystery but, leaving consciousness aside, the problem of converting the information in the pattern of light on the retina into information about objects in front of the eye used not to seem too difficult. Just how wrong that view was became apparent as soon as people tried to make machines that could see – see in the sense that they could control their behaviour to deal appropriately with objects in front of them when the only source of information about the objects was visual. As David Marr puts it:

> in the 1960s almost no one realized that machine vision was difficult. The field had to go through the same experience as the machine translation field did in its fiascos of the 1950s before it was at last realized that here were some problems that had to be taken seriously. The reason for this misperception is that we humans are ourselves so good at vision.[1]

There were, though, much earlier indications that the translation of patterns of light on the retina into knowledge of objects in the world was a more complicated affair than the ease of seeing suggested. These came most obviously from observations and experiments in which the perceptions of scenes by normal people showed unexpected and sometimes illusory features; they also came from the study of those disorders of seeing in which there is nothing wrong with the eyes themselves, or with the nerves leading from the eyes to the brain. It is information derived from some of the unexpected or illusory features of perceptions by normal people that I want to discuss in this chapter. (Further unexpected features and illusions will come later.)

In his great *Treatise on Physiological Optics*, and in later writings,[2] Helmholtz pointed out that the formation of perceptions involves a good deal of unconscious inference – inference which he thought was of an inductive kind and based largely on experiences in early childhood. Inductive inference is not wholly reliable, and occasionally we form illusory visual perceptions which mislead us about the real world. We are prone to such illusions, Helmholtz argued, when for any reason the particular pattern of stimulation on the retina is formed in an unusual way; and it is characteristic of them that what we think we see is what would usually produce that pattern. The primary example Helmholtz gives is that a blow on the outer corner of the eye gives the impression of light somewhere in the direction of the bridge of the nose, because 'if the region of the retina in the outer corner of the eye is to be stimulated [by light], the light actually has to enter the eye from the direction of the bridge of the nose'.

A more unusual example is an illusion to which I was introduced by the experimental psychologist Richard Gregory. He sat me in a very dimly lit room in front of a table on which was a large white cube, and told me to look at the cube. After I had become accustomed to the dim light, he fired a flashgun so as to illuminate, momentarily and intensely, the cube I was looking at. After the flash I still felt that I was seeing the cube, but of course the vigorous activity in my retinas was the aftermath of the intense activity that had been caused by the image of the cube in its moment of glory – I was seeing an after-image. Gregory asked me to keep looking at the cube while I slowly leaned forwards and backwards. Had I really been seeing the dimly lit cube, as I leant forward the retinal image would have got bigger, and as I leant backwards it would have got smaller. In fact, because the size of the after-image did not change, what I seemed to see as I leant forwards and backwards was the cube strangely receding and advancing, but always keeping the same distance from my face. This illusory motion fits Helmholtz's prediction nicely, since real motion of this kind is just what would cause the pattern of retinal stimulation that I was experiencing.

Gregory then asked me to repeat the experiment but, instead of looking at a white cube, I was to look at my hand held out in front of me; and after the flash, instead of rocking backwards and forwards, I was to move my hand alternately towards and away from my face. Because information from my muscles and joints made me aware of the real changes in the position of my hand, the explanation of the unexpected constancy of the size of the image that had worked for the cube was not available. What I seemed to see was that, as I moved my hand towards me it shrank alarmingly; as I moved it away it grew. Again this fits nicely with Helmholtz's prediction, since only by shrinking and growing can an object moving alternately towards and away from the eyes keep the size of its image on the retina constant. There was no question, though, of conscious argument – of weighing the available hypotheses in the two situations. In both experiments the inferences were immediate and effortless; it was only after the illusions had been experienced that I was able to rationalize them. Finally, Gregory completed the second experiment by asking me to keep looking at my hand while I twisted it. There was a creepy moment as the hand seemed to come out of its skin, and then the illusion that I was looking at my hand was destroyed. What I was seeing was an after-image of my hand; and the real hand was moving unseen in the dim light.

Many well-known visual illusions can be interpreted along the lines

Helmholtz suggested. The ghostly white triangle in Kanizsa's illusion (Figure 12.1a) is inferred because that is the obvious way of accounting for the missing sectors in the three black discs. The apparent inequality of the two equal horizontal lines in the Ponzo illusion (Figure 12.1b) is inferred because the near-vertical converging lines tend to be interpreted as parallel lines receding into the distance, like a railway track. The startling variations in the sizes of objects or people placed at different positions in the Ames 'Distorting Room' (Figure 12.2) are inferred because the room is assumed to be rectangular. The spherical bumps and hollows in Ramachandran's photograph (Figure 12.3)[3] interchange when the photograph is inverted, because the inference of shape from shading is ambiguous, depending on the direction of the light, and we tend to assume that light comes from above. (Curiously, most of us are quite unaware of this tendency, so we find Ramachandran's trick all the more startling.) Of course, explanations of this kind do not tell us how the brain manages to make the inferences though they provide a clue to the kind of information processing that may be involved.

Helmholtz's suggestion that the inferences are based largely on experiences in early childhood is part of an empiricist tradition that goes back to John Locke's belief that the mind is initially 'white paper, void of all characters', but Helmholtz was not relying on the empiricist tradition; he had an ingenious argument to support his suggestion.

Everyone would accept that a person wandering about a familiar room in the evening twilight, with objects scarcely discernible, is able to find his way without mistakes only by making use of knowledge from earlier visual impressions. Yet, Helmholtz points out, even when we look round a room flooded with sunshine, 'a large part of our perceptual image may be due to factors of memory and experience . . . Looking at the room with one eye shut, we think we see it just as distinctly and definitely as with both eyes. And yet we should get exactly the same view if every point in the room were shifted arbitrarily to a different distance from the eye, provided they all remained on the same line of sight'. Figure 12.4 illustrates Helmholtz's argument (which is of course the basis of the Ames Distorting Room). If we keep still and use only one eye, the images on our retinas are compatible with a host of differently shaped objects. The picture on the wall opposite me and somewhat to my left, as I write, is an oblong picture flat on the wall, but an identical image on my retina could be produced by an unusually shaped picture of appropriate size hanging in mid air. The lamp shade is circular, but the image of its rim on my retina is oval, and such an image could also be produced by an oval lampshade of appropriate size and held at the correct angle. That I see the picture as flat on the wall and the lampshade as circular can only be the result of my being able (unconsciously) to choose between alternative interpretations and to use earlier experiences to reject those that are unlikely. And that implies that some kind of neural representation of previous experiences (or

FIGURE 12.1 (a) The Kanizsa
illusion – the ghostly white
triangle; (b) the Ponzo illusion –
the upper horizontal bar looks
longer than the lower.

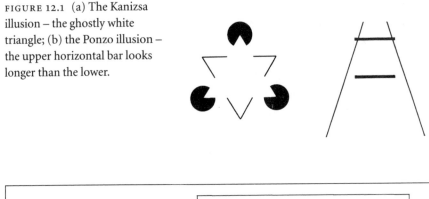

FIGURE 12.2 The Adelbert Ames distorting room. The room is constructed so that,
when it is viewed through the eyehole with one eye, it produces an image on the retina
identical with that of a rectangular room of uniform height. In fact – see (a) in Figure –
the far wall recedes and both the floor and the ceiling slope. When people are placed in
the far corners of the room – see (b) in Figure – their size is judged in relation to the
dimensions of the room *assuming it to be rectangular*. ((a) is from W. H. Ittelson.)

FIGURE 12.3 Ramachandran's
bumps and hollows. Invert the page
and the bumps become hollows, the
hollows bumps. (From V. S.
Ramachandran.)

of generalizations produced from them) serves not only for conscious memory but also for the formation of current visual perceptions.

'Filling in'

There is another kind of interesting visual effect that any of us can experience and that provides yet another example of the subtlety of visual perception. It is rather like an illusion in its unexpectedness, but the result is not necessarily illusory. Figure 12.5a shows a cross and a circle. Shut your left eye and hold the page about ten inches from your face. Focus on the cross and bring the page slowly towards your right eye, continuing to look at the cross. At some point the circle will disappear. What has happened is that the image of the circle is now on your blind spot – the part of the retina where the optic nerve arises and there are no rods or cones. As you would expect, in the absence of any photoreceptors you cannot see the circle; what is more surprising is that, somehow, the visual system seems to have filled in the gap making it the same as the surroundings. You may wonder whether that is the right way to describe what has happened; might it just be that you don't notice a gap because you don't see anything at all in the region of the circle? Now repeat the experiment looking at Figure 12.5b, in which a white cross and a white disc appear on a patterned background. Using your right eye, focus

FIGURE 12.4 Drawing to illustrate Helmholtz's argument. To the venerable figure looking with only one eye, the curiously shaped object in the foreground is indistinguishable from a cube because all the identifiable features are on the same line of sight as they would be if the object were a cube. (From a drawing by O. J. Braddick based partly on an 18th century drawing.)

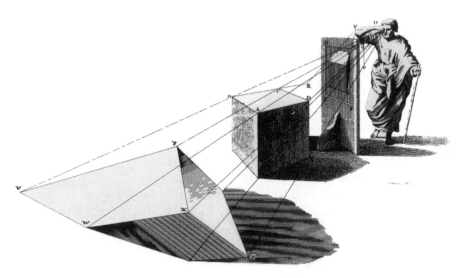

on the white cross and bring the page towards you, continuing to look at the cross. What you will probably see is that the disc disappears, but the pattern seems to be continuous across the space where the disc was. Here, this 'filling in' is illusory, but it is because of filling in that you do not see a black or grey area corresponding to your blind spot when you look at a large white screen with one eye. For the same reason, patients with retinal damage causing small islands of blindness (scotomas) in their visual fields are usually not aware of them unless the central part of the retina – the part which is used for detailed vision – is affected. Interesting examples of illusory 'filling in' have been described by V. S. Ramachandran.[4]

THE GESTALT PSYCHOLOGISTS

Starting in Germany, a few years before the First World War, a novel twist to the psychophysical approach to visual perception was taken by the *Gestalt psychologists*.[5] Like other psychophysicists, they hoped to discover how perceptions are formed by looking for *laws* that relate what we feel we are seeing to what we are actually looking at. But they were not happy with the notion that what is involved is merely the mechanical putting together of independently existing sensations. *Gestalt* is the German for *form* or *configuration*, and what particularly impressed the members of the Gestalt school was the way in which grasping the overall configuration could change the whole picture. Edgar Rubin's famous drawing of a vase or a pair of faces (depending on what is taken as figure and what as background), and Jastrow's duck/rabbit are the classical examples of this phenomenon – see Figure 12.6. It is conventional to say that the view of perception taken by the Gestalt psychologists was opposed to the classical view of empiricists such as Locke and Helmholtz, yet the two interpretations of Jastrow's duck/rabbit fit very well with Helmholtz's idea that past experience makes an important contribution to the formation of visual perceptions.

Figure 12.7 illustrates three of the laws discovered by the Gestalt psychologists: the *law of proximity*; the *law of similarity*, and the *law of good continuation*. Rubin's drawing (Figure 12.6a) illustrates a fourth law, the law of *enclosedness*: it is much easier to see the vase than the two faces, because enclosed regions tend to be seen as 'figure' rather than 'ground'. (This tendency, too, could be an example of experience-based inference: you are more likely to see the entire boundary of an object if it is in the front of the scene.) But though the Gestalt psychologists emphasized the creative aspects of visual perception, and the laws they described drew attention to interesting features of perception that might otherwise have been ignored, the laws themselves were never explained and attempts to link them to physiological models, admittedly of a rather implausible kind, were a total failure. By the time realistic physiological models of the

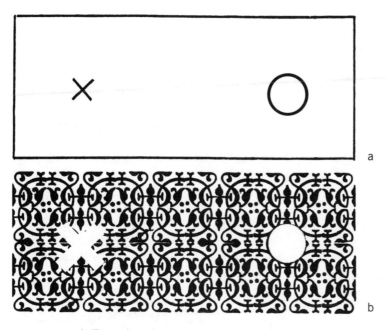

FIGURE 12.5 'Filling in'. For instructions see p. 197 of text.

FIGURE 12.6 Visually ambiguous figures. (a) Edgar Rubin's vase or two faces; (b) Jastrow's duck-rabbit.

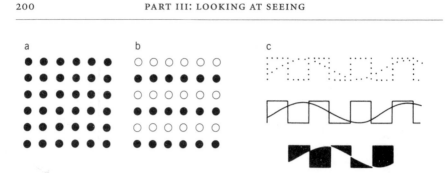

FIGURE 12.7 (a) The *law of proximity*: the dots are seen as arranged in horizontal or vertical lines rather than oblique lines, because they are further apart in the oblique lines. (b) The *law of similarity*: the dots are now seen as horizontal lines because those forming horizontal lines are more similar than those forming vertical lines. (c) The *law of good continuation*: the small dots are seen to form a wavy line superimposed on a profile of battlements rather than as the succession of shapes shown at the bottom of the Figure. The rule is: 'we perceive the organization that interrupts the fewest lines'. ((c) is from R. L. Gregory.)

handling of information in the brain were available, Gestalt psychology as a separate school had withered away – 'dissolved', as David Marr put it, 'into a fog of subjectivism'.

13 Disordered Seeing with Normal Eyes

A neurological approach to vision

In 1888, Louis Verrey, an ophthalmologist from Neuchâtel, described an unfortunate sixty-year-old woman, Mme R., who, while busy in her garden two years earlier, had suffered a violent attack of giddiness, headache and vomiting.[1] Ever since, her sight had been affected in a remarkable way: in the right half of her visual field she had lost all sense of colour – everything appeared in shades of grey; in the left half, colours were entirely normal. Nearly two years after the original attack, Mme R. had a severe stroke and died within a few days. Examination of her brain showed long-standing damage in the lowest part of the left occipital lobe. Complete loss of sight in one half of the visual field following damage to the occipital lobe on the opposite side is, as we have seen (Chapter 11), not uncommon. What was odd about Mme R. was not the restriction of the visual loss to one half of the field of view, but the restriction to just one aspect of vision – colour. Similar clinical observations had been made before, in Germany and in England, but the significant thing about Verry's work was that, for the first time, it linked the loss of colour sense to damage in a particular small area of the occipital lobe.

If localized brain damage could cause loss of the sense of colour alone, could damage in different locations cause selective losses of other aspects of vision – the ability to recognize shape or individual objects or movement or to see things as three-dimensional?

Ten years before Verrey's paper, Hermann Munk had reported some strange observations to the Physiological Society of Berlin. A dog who had had parts of both occipital lobes removed, and had then had several days to recover from the operation, seemed to be entirely normal except for a peculiar disturbance of vision. The dog could certainly see – he roamed freely indoors and in the garden, avoiding obstacles, crawling under a stool, stepping over people's feet or the bodies of other animals – yet he no longer seemed to recognize people or dogs with whom he had been friendly, he no longer responded to a threatening gesture and, most remarkably, even when he was thirsty he ignored his water bucket, and when he was hungry he paid no attention to his food bowl unless the food could be smelt.[2] Munk called the condition *mind-blindness* (*Seelenblindheit* in German), and supposed that it resulted from a loss of visual memory.

Following Munk's experiments, what looked like a similar mind-blindness was noted in several patients, one of whom, described by Hermann Wilbrand, showed a feature that necessitated a slight change in Munk's hypothesis.[3] Wilbrand's patient, a highly intelligent, sixty-four-year-old German woman, showed convincing mind-blindness – though she could see the butcher's cart coming down the street and the objects in her own display cabinet, they seemed unfamiliar and she could not recognize them. But she had not lost her visual memory. She could not recognize a pencil by sight, but when one was put into her hands while her eyes were shut she had no difficulty in visualizing it; the smell of paraffin evoked the visual image of her lamp. It seemed that her visual memory was intact and that her failure to recognize objects was the result of a failure to make associations between current perceptions and the memories of past perceptions.

A year after the publication of Wilbrandt's paper, Heinrich Lissauer,[4] only twenty-seven years-old and working in Wernicke's clinic in Breslau, suggested that mind-blindness could be of two kinds. It could, as in Wilbrandt's patient, be the result of a failure to relate current perceptions to past experience – this Lissauer called *associative mind-blindness* – but it could also, he said, be the result of a failure to form adequate current perceptions. To regard recognition as a two-stage process is certainly an oversimplification, but Lissauer's subdivision proved useful clinically and is still used by neuropsychologists who work with patients.[5]

As a further example of associative mind-blindness, Lissauer described an eighty-year-old salesman who, following a few days' illness, was incapable of finding his way outside his own room, and had great difficulty in choosing the right clothes when dressing or the right cutlery when eating. On one occasion he put the wrong end of the spoon into his soup; on another he put his hand into a cup of coffee. He was found to have lost the right half of the visual field in both eyes, but vision in the left half of the field was adequate, and loss of the right half could not account for his disabilities. He was unable to identify the most common objects by sight, confusing an umbrella with a leafy plant, a coloured apple with a portrait of a lady, an onion with a candle, and a clothes brush with a cat. But he was not generally confused and he identified objects promptly and correctly as soon as he could touch and feel them. He could match strands of wool of different colours accurately, but often named the colour incorrectly. Because he could copy simple drawings of objects, and could draw the actual objects, though laboriously and clumsily, Lissauer concluded that his disability was not in forming visual perceptions but in associating them with previous experience.

In 1891 Freud introduced the term *agnosia* (from the Greek for 'without knowledge') to indicate an inability to recognize objects normally, despite normal working of the sense organs and the nerves to the brain. Agnosia applies not only to vision – people sometimes speak of auditory or olfactory agnosia, or a kind of agnosia in which the patient has normal senses of touch and pressure but cannot determine the shape of objects by feeling them – but it is used mostly in connection with disturbances of the visual system, and it has completely replaced the term mind-blindness.* In the century since Freud introduced the term, a vast literature on visual agnosia has developed, but all I want to do here is to talk about a small number of cases, chosen because in each of them the patient has a particular kind of disability – often more than one – and such disabilities throw light on the way visual information must be handled in the visual parts of our brains. The various kinds of visual agnosia differ in the class of perceptions that cause problems, and also in whether the agnosia is the result of a failure to form adequate perceptions or a failure to recognize what is perceived.

VISUAL FORM AGNOSIA – A FAILURE TO PERCEIVE SHAPE

In 1988, DF, a thirty-four-year-old Scotswoman, holding a private pilot's licence and working in northern Italy as a freelance commercial translator, collapsed and lost consciousness from carbon monoxide poisoning while taking a shower at her home.[6] She was admitted to hospital in deep coma, and when she recovered consciousness was found to be blind. After about ten days her vision gradually returned, though she had a number of residual problems, and six months after the accident she still had a very poor perception of the shape or orientation of objects in front of her.

The peculiar nature of her disability became clear from a series of tests done by Melvyn Goodale, David Milner and their colleagues, about fifteen months after the accident.[7] To test her ability to detect orientation, she was seated in front of a large slot in an upright disc, given a card to hold, and asked to turn it so that its orientation matched that of the slot. She did this very badly, even failing to distinguish between vertical and horizontal. On the other hand, when she was asked to 'post' the card through the slot, she did this as well as normal subjects. What is more, analysis of a video recording showed that, like the normal subjects, she began to orient the card correctly even as it was being raised from the starting position. Posting a card through a slot is a more natural activity than orienting a card to match the orientation of a slot, and you might wonder whether DF simply failed to understand what she was being asked to do in the tests in which she performed so badly. This explanation is made unlikely by a further test in which she was asked to shut her eyes and to orient the card to match the orientation of an *imaginary* slot at 0°, 45° or 90° from the vertical. This she did as accurately as

*So completely that the term has recently been reintroduced (without a hyphen) to refer to the apparent inability of autistic children to appreciate that other people have minds. See Baron-Cohen, S. (1995) *Mindblindness: an Essay on Autism and Theory of Mind*, MIT Press, Cambridge, Mass.

normal subjects. Since, in the orientation experiment with the real slot, she had failed in a task she performed easily with the imaginary slot, it seems that information about orientation obtained by looking at the real slot was *not* available to her for purposes that involved conscious thought – though her ability to 'post' the card through the real slot showed that it *was* available to her for the direct visual control of movement.

To see whether a similar pattern applied to other aspects of visual form, Goodale and his colleagues did a further set of experiments. Normal people are extremely good at discriminating between shapes. A very slight elongation of two opposite sides of a square turns it into an obvious oblong, and a standard test for patients with visual form agnosia – the Efron squares test – is to see how good they are at discriminating between a series of rectangles of equal area, ranging from a square to an oblong with length twice its breadth. Goodale and his colleagues prepared five pairs of plaques based on the Efron shapes and showed these plaques to DF, two at a time, in all possible combinations. In each case, she was asked to say whether the two plaques in front of her were alike or different. In this test she scored no better than chance. She was then shown the plaques one at a time, and asked to use the index finger and thumb of her right hand to indicate the width of the plaque in front of her. There was almost no correlation between the width of the plaque and the separation of her finger and thumb; and repeated tests with the same plaque gave a very wide scatter. Yet when DF was asked to reach out and pick up a plaque, her performance was indistinguishable from that of a normal subject. As she reached out, she adjusted the gap between her finger and thumb to the width of the plaque, and this adjustment was complete well before contact was made. Again, you might wonder whether DF had properly understood what was required of her in the tests in which she performed badly, but again that explanation is unlikely because when she was asked to use her finger and thumb to indicate the width of an *imagined* plaque of some specified width she performed as well as a normal subject. Just as in the orientation experiment with the slot, her failure in the real case in a task she did successfully in the imagined case showed that information about shape obtained by looking at a real situation (the real plaque) was *not* available to her for purposes that involved conscious thought. It must, though, have been available to her for guiding a grasping movement under direct visual control, since she had no difficulty in picking up the real plaque in a normal fashion.

We shall come back later to the cause and significance of this striking feature of DF's disability, but there is one other point that I want to mention here. Figure 13.1 (left and middle) shows the result of asking DF to copy line drawings of an apple and an open book. She was unable to recognize the drawings she was copying, and her copies bear little relation to the originals, though the dots in the copy of the open book presumably represent the text. Her incompetence was not simply the result of lack of skill in drawing, for when she was asked to draw an apple and an open book from memory – Figure 13.1 (right) – the results were easily recognizable, though she herself, when shown them later, had no idea what they represented. It seems clear that DF's disability is in forming a conscious perception of what she is looking at, not in associating that perception with past experience.

ASSOCIATIVE VISUAL AGNOSIA

Clear-cut cases of *associative* visual agnosia – agnosia caused by failure to associate perceptions of what is being looked at with past experience, and therefore failure to understand those perceptions – are sufficiently rare for their existence to have been a matter of controversy until fairly recently.[8] They are also tricky to diagnose; to be sure that a patient who can see objects but not identify them is suffering from an associative visual agnosia, you need to exclude three other possibilities – that the patient is demented, or has a linguistic problem, or is simply incapable of forming adequate visual perceptions. Lissauer's eighty-year-old salesman looks a strong candidate, but Lissauer's assumption that his patient *was* capable of forming adequate perceptions was based on an ability to draw, and that ability was limited. The patient's sketch of a pocket watch showed the round outline, the ring attachment for the chain, and a ridged winding button, but he always drew very slowly, he often started wrongly, and he failed to draw many simple objects such as a beer bottle, a water jug, and a cup. So let us look at a much more recent case described by A. B. Rubens and D. F. Benson in Boston in 1971.[9]

Their patient was a forty-seven-year-old doctor, a heavy drinker and at times suicidal, who had been found unconscious with a very low blood pressure – possibly the result of taking twenty tablets of chloral hydrate that were found to be missing. More than six weeks after this episode he still had unusual behavioural difficulties whenever his behaviour depended on vision, and these difficulties

model   copy   memory

FIGURE 13.1 Drawings by DF, a patient unable to perceive shape. She was unable to copy drawings of an apple or an open book, but was much better at drawing these objects from memory. (From A. D. Milner and M. A. Goodale.)

closely paralleled those described by Lissauer eighty years earlier. Most striking was his inability to recognize familiar objects placed before him. He was unable to name them, or to explain what they were, or to demonstrate their use, but he could do all these things without difficulty if he was allowed to feel them. Though he was a doctor, when asked to identify a stethoscope by sight he replied 'a long cord with a round thing at the end' and suggested it might be a watch. He failed to identify a comb or a toothbrush, and suggested that a key was 'perhaps a file or it could be a tool of some sort'. Like Lissauer's patient, he was blind in the right half of the visual field, but vision in the left half was good. Like Lissauer's patient, too, he had difficulty in reading but not in writing, and he could match coloured wools correctly but not name them correctly. One difference from the earlier case was that he remained unable to recognize members of his family, or the hospital staff, or his own face in the mirror. (Loss of facial recognition had been a very transient symptom in Lissauer's patient.) What makes Rubens and Benson's investigation relevant here, though, is that they provided much stronger evidence that their patient could form accurate perceptions, yet fail to understand them. Figure 13.2 shows the skill with which he copied a drawing of a pig and of a railway engine; but when asked what he had drawn he could identify the pig only as 'Could be a dog or any other animal' and the railway engine as 'A wagon or a car of some kind. The larger vehicle is being pulled by the smaller one.' As well as copying a drawing, he could draw competently from memory, he could manage tests in which he was asked to move the tip of a pencil through a visual maze or to remember several simple geometrical shapes, and he could even pick out a pre-designated geometrical form hidden in a complex illustration. His performance in all these ways makes it very unlikely that his disabilities arose from any

FIGURE 13.2 Drawings by a patient unable to associate visual perceptions with past experience. He made good copies but had only rather vague ideas about what he was copying. (From A. B. Rubens and D. F. Benson.)

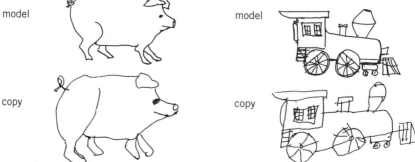

difficulty in forming visual perceptions of objects in front of him rather than a failure to associate those perceptions with previous experience.

Patients suffering from the associative type of visual agnosia are sometimes impaired to different degrees in recognizing different categories of objects. The literature includes a case in which the deficit was said to be limited to people and animals, while recognition of trees, flowers and common objects was described as normal;[10] and also a contrasting case in which it was the recognition of inanimate objects that was most affected.[11] In another case, the patient – a seventy-seven-year-old retired shopkeeper who woke one morning to discover that he was unable to read or to recognize common objects by sight – was found to be very bad at recognizing pictures of objects but surprisingly good at naming 'action' pictures – digging, peeling, swimming.[12]

VISUAL AGNOSIA FOR FACES

One of the distressing features of Rubens and Benson's patient was his inability to recognize faces of people familiar to him, but this is unusual in patients with visual agnosia for objects. Conversely, most patients who have difficulty in recognizing faces, can identify ordinary objects without problems – though there are a few reports of patients who, as well as finding it difficult to identify faces, have difficulty in identifying individual members of other visually similar categories: the bird watcher who lost the ability to recognize birds, the farmer who lost the ability to recognize his cows, the car enthusiast who lost the ability to recognize individual makes of car.[13] The 'double dissociation' between the loss of ability to recognize faces and the loss of ability to recognize common objects – each often occurring without the other – strongly suggests that the visual recognition of familiar faces must involve machinery that is at least partly different from that involved in the visual recognition of ordinary objects.

On reflection, it is not too surprising that recognizing a particular face as familiar should be different from recognizing a key as a key, or a chrysanthemum as a chrysanthemum. As Antonio Damasio and his colleagues have put it, 'Faces are not only *many*, and *different*, and *unique*, but they are also *similar*.'[14] Yet we can usually tell from a glance, not only the sex, race, and approximate age of the owner of the face, and which of a whole gamut of emotions the face expresses, but also something more demanding. We generally know at once whether the face we are looking at is one of the thousand or more faces that we have stored in our memory. 'I never forget a face!' people say, and, though that is certainly untrue, our capacity for recognizing faces is astonishing – and judging by the way we go on learning the faces of new acquaintances and new public figures, almost unlimited.

A few individuals can do even better and can remember faces in sufficient detail to produce accurate drawings of them. In 1844 a London physician, A. L. Wigan described a painter 'who succeeded to a large part of the practice and (as he thought) to more than all of the talent of Sir Joshua Reynolds' and who claimed to have painted 300 portraits in one year.[15] His secret, he said, was to have one brief sitting of about half an hour, during which he made a few slight sketches, and then to paint the portrait entirely from memory. Since this same painter later spent thirty years in an asylum, he may not have been the most reliable of witnesses, but similar skill in drawing faces from memory certainly exists. Some years ago the wood engraver Joan Hassall, who was also an excellent portraitist, visited Cambridge to make a portrait drawing of the economist Maurice Dobb. I took her to Dobb's rooms after breakfast, and by lunchtime she had produced two indifferent attempts and one superb drawing – an excellent likeness, sensitive and full of character. She told me that the failures had been done first, with Dobb sitting in front of her. He had then gone into town to do some shopping, while she made the successful drawing from memory.

Such virtuoso performances are of course rare, but the ability to recognize familiar faces is not simply common but virtually universal. It is an ability that we acquire without conscious training, and we begin to acquire it extraordinarily early. Despite the undeveloped state of their visual systems, newly born babies are said to track moving schematic faces in preference to similar head-shapes without features or with the features in incorrect positions.[16] And there are claims that babies a few days old can reliably discriminate between their mothers' faces and those of strangers.[17] The evolutionary acquisition of our remarkable skill in facial recognition is presumably explained by the advantages it brings to a highly social animal.

Agnosia for faces is so bizarre and so devastating for the patient that one might expect it to be easily recognized, in spite of its rarity. In fact, the first description of a case was by Wigan in 1844,[18] and though a trickle of cases were recorded in the last part of the 19th century and the first part of the 20th, it was not until after the Second World War that the condition was studied as an entity in its own right. This late recognition is probably the result of the remarkable ability of patients suffering from agnosia for faces to use other clues – context, voice, speech, posture, and motion – to identify people. My colleague Rosaleen McCarthy tells me of a patient with facial agnosia who was so good at picking up 'other clues' that she was better at recognizing a neighbour's toddler from a distance than the toddler's own mother. In any event, in 1947 Joachim Bodamer published his observations on three German soldiers who had suffered brain injuries leading to what he called *prosopagnosia* – from the Greek *prosopon*, a face – a term whose acceptance probably contributed more than a little to the general recognition of the condition.[19] Curiously, of Bodamer's three cases, one would not now be classified as prosopagnosia – the patient was able to recognize faces but described them as grossly distorted (the mouth twisted, one eyebrow too high, the nose askew) though the rest of the world looked normal – and another was unusual in that the main difficulty seems to have been in perception: faces appeared 'strangely flat; white with dark eyes, as if in one plane, like white oval plates ... all the same'. Here are some excerpts from the report of a more typical case of prosopagnosia, in a hospital in Modena, taken from a paper by Ennio De Renzi:

> A 73-yr-old, right-handed, public notary suddenly complained of blurred vision and had to be taken back home as he was unable to find his bearings. When first seen, he was found to have a dense left hemianopia [i.e. he had lost the left half of the visual field in both eyes] ... Visual acuity was 10/10 in both eyes ... Some difficulty in finding his way about was apparent ... but the most impressive disorder was his inability to identify familiar faces by sight and to learn to recognize new faces. If his wife, daughter or secretaries approached him without speaking, he gave no evidence of recognition. He spent many hours with his doctor and with a psychologist ... but never succeeded in recognizing them ...
>
> ... Verbal memory and visuo-perceptual tasks, inclusive of those requiring the matching of front-view with lateral face photographs, were correctly performed, while he scored at chance level on unknown-face memory tests and failed to recognize photographs of famous faces.
>
> The patient was checked 12 months later. He was reported by his collaborators to be as efficient as before the disease in dealing with legal questions and in remembering his clients' business, but the inability to recognize familiar persons by sight persisted unchanged. Once he asked his wife 'Are you ... ? I guess you are my wife because there are no other women at home, but I want to be reassured.'[20]

The loss of the left half of the visual field is significant because it suggests damage to the right occipital lobe, a suggestion that was confirmed by a 'CAT scan' (see p. 221). The patient's ability to match full-face photographs with photographs of the same faces in profile is significant because it implies that his failure to recognize familiar faces could not have been the result of failure to form adequate facial perceptions; the defect must therefore have been in relating those perceptions to past experience.

It would be wrong, though, to think of recognition of faces as involving only two stages, each of which can be defective. A defect in recognizing familiar faces may or may not be associated with a defect in recognizing facial expression, and vice versa.[21]

One patient who was unable to recognize familiar faces or to interpret facial expressions – she could not reliably distinguish between a smile and a frown – was shown to be still able to lipread to a normal extent.[22] A poor performance in recognizing familiar faces may or may not be accompanied by a poor performance in matching full-face and profile photographs; De Renzi's seventy-three-year-old public notary, who was unable to recognize the faces of his wife, daughter or secretaries, had no difficulty with tests of this kind.[23] To explain these various 'dissociations' it is necessary to suppose that, after the face has been perceived, analysis of the perception involves different processes devoted to different objectives – recognition of familiarity, naming, recognition of facial expression, lipreading – and that these different processes work in parallel.[24] Selective damage to the machinery involved in the individual processes can account for the great variety of ways in which individual patients may be affected.

Strong support for this *modular* view of 'facial processing' comes from some remarkable experiments by Justine Sergent and her colleagues at the Montreal Neurological Institute.[25] They used two of the modern methods of brain scanning: positron emission tomography (PET scanning), to measure local changes in blood flow; and magnetic resonance imaging (MRI) to obtain an accurate picture of the anatomy of the brain so that areas of increased blood flow could be reliably identified. (The principle of these scanning methods, and also of computer aided tomography – CAT scanning – is explained on pages 221–2.) In the experiments, each of seven normal, right-handed young men was given different visual tasks to perform, and the changes in blood flow in different regions of the brain were monitored, so that the areas of increased flow associated with each of the tasks could be accurately identified.

The most interesting results came from comparing the changes in blood flow during four tasks. One was simply to decide whether gratings projected onto a screen in front of the subject had vertical or horizontal bars. Another involved recognizing photographs of common objects. A third was to say whether photographs of *unfamiliar* faces were male or female. And the last depended on being able to recognize photographs of *familiar* faces. In each case, the subject indicated his decision by pressing one or other of two buttons with his right hand, so that one area in which blood flow was increased in all the experiments was the part of the left motor cortex involved in these movements.

Comparison of the pattern of increased blood flow in each of the four tasks showed, first, that, in all four tasks there was an increased flow in the primary visual cortex and in some of the secondary visual areas of both occipital lobes; and, secondly, that each of the three more complicated tasks – identification of common objects, judging the sex of unfamiliar faces, and recognition of individual familiar faces – led to different patterns of increased blood flow, showing that separate areas are involved in particular tasks.

Encouragingly, the different areas shown to be involved in 'processing' faces in these experiments fit rather well with areas found to be damaged in various patients suffering from prosopagnosia.[26]

Covert Recognition in Facial Agnosia

A very surprising feature of some cases of prosopagnosia is that though the patient presented with a familiar face is not conscious of recognizing it, there is evidence of unconscious or 'covert' recognition. This was first shown in the early 1980s by several investigators who detected changes in the electrical conductance of the skin in patients when they were shown familiar faces, but not when the faces were unfamiliar.[27] A change in skin conductance in response to seeing a familiar face is, of course, analogous to a positive response in a lie-detector test, but there is no suggestion that the patients were lying. The most impressive evidence for covert recognition, though, comes from a series of investigations, by Edward de Haan, Andy Young and Freda Newcombe in Oxford.[28] They studied a nineteen-year-old patient, PH, who had had a motorcycle accident followed by twelve days of coma. When he recovered from the coma he found that he was unable to recognize faces.

They first excluded the possibility that PH, though he claimed to have no ability to recognize faces, simply had an impaired ability and lacked the confidence to make a definite claim that a face was familiar if he was not sure. They did this by a 'forced choice' technique. In repeated trials, PH was shown pairs of faces, one familiar and the other unfamiliar, and asked to point to the familiar one and to guess if he wasn't sure. As he was correct in only 51 per cent of the trials – i.e. insignificantly different from chance – they concluded that he had no overt ability to recognize faces.

They then did a number of tests for covert recognition, of which I shall describe two.

The first was a *matching test*. PH was shown pairs of photographs of faces, one complete and the other (which was always taken from a different angle) masked so that only the eyes, nose and mouth were visible. Each time, he was asked to say as quickly as possible whether the two faces were of the same person. In this test, a normal subject can spot that two photographs are of the same person significantly faster if the faces are familiar. PH was very much slower in making all decisions, but, like the normal subject, he was significantly quicker if the faces were familiar. This implies that he was able to differentiate between familiar and unfamiliar faces, even if he could not consciously recognize them as familiar.

The second test was designed to see whether the information that a face was familiar, though not available to introspection, could be used to *allot the face to a category*. PH was shown a series of faces of famous television stars or famous politicians, each accompanied by a name in a cartoon-like speech bubble. The name in the bubble could be that of the person in the photograph, or of another person in the same occupation, or the name and the face could be quite unrelated. PH was asked to ignore the face and to identify as quickly as possible whether the name in the bubble was that of a politician or of a television star. When normal people are given this test, despite being asked to ignore the face, they are influenced by it, and they take longer to reply if the face does not correspond to the name. PH took longer to make all decisions, but he too was slower if the face did not correspond to the name. Again this implies that covert recognition had occurred.

The conclusion from these and other tests has been nicely put by de Haan and his colleagues: 'PH's problem is not that he does not recognize faces; it is that he is unaware of the recognition that has taken place.'

PH's covert recognition of faces is in some ways reminiscent of DF's ability to use visual information that she was not conscious of for posting a letter through a slot or picking up a plaque. There is, of course, nothing mysterious about using information, obtained through sense organs, that does not reach consciousness; it is something we do all the time. The rate and depth of our breathing is controlled through sense organs that monitor the levels of carbon dioxide, oxygen and acid in our blood, though we are quite unconscious of these levels. What makes PH and DF interesting is that they can make some use of information that a normal subject would be conscious of but which they are not.

There is, though, a significant difference between the two cases. DF was able to make use of information she was unconscious of for the visual control of movement, but not for answering oddball questions about the orientation of slots or the width of plaques. It seems reasonable to suppose that, in the course of evolution, mechanisms for the visual control of movement arose long before mechanisms for thinking about formal geometrical questions, and that the neural pathways involved may therefore be largely separate.* With PH there is no obvious evolutionary dichotomy of this kind. Recognition must involve the comparison of the perception of a face – that is to say, some kind of encoding of the structure of that face – with some stored set of representations of faces. In the normal subject, recognition then unlocks access to a very variable amount of biographical information. It looks as though, in PH, it is this second stage that was largely blocked.[29]

Some support for this view comes from an investigation by Justine Sergent and Michel Poncet of a fifty-six-year-old Frenchwoman, Mme V., who had suffered from prosopagnosia since an attack of encephalitis fifteen years earlier.[30] Although profoundly prosopagnosic, Mme V. showed clear evidence of covert recognition – most strikingly in tests in which she was asked to match thirty photographs of faces of French celebrities familiar to her with photographs of the same people taken thirty years earlier. Matching photographs taken thirty years apart is much easier if the faces are familiar; and, both in speed and in accuracy, Mme V's performance fell in the normal range for 'familiar' faces, and was much

*Corresponding, probably, to the two streams of visual information leaving the primary visual cortex described by Leslie Ungerleider and Mort Mishkin, who believed, however, that the first stream carried information used to decide where an object was, and the second information used to decide what it was. (See Ungerleider, L. G. & Mishkin, M. (1982) In *Analysis of Visual Behavior*, (Eds D. J. Ingle, M. A. Goodale & R. J. W. Mansfield), pp. 549–86, MIT Press, Cambridge, Mass.)

better than the performance of a control group of Canadian subjects for whom the French celebrities were unfamiliar. (The possibility that Mme V. was simply unusually good at matching unfamiliar faces was excluded by repeating the experiment with photographs of Canadian celebrities, when her performance fell within the normal 'unfamiliar' level.) Sergent and Poncet argued that if *overt* recognition by Mme V. was prevented by greatly reduced access to biographical information, she should be helped by being given information which narrowed the field without providing any specific cue. They therefore presented her with photographs of eight familiar faces and asked her to identify them, which she was unable to do. They then told her that the eight faces belonged to individuals sharing the same professional occupation. 'She spent about 10 seconds examining the faces, and then answered that they were politicians. Then, without any prompting, she correctly identified 7 of them, pointing to each face while telling their names. She also recognized the eighth face ... providing information about his political affiliation and the ministry position he once held, though she could not retrieve his name.'

VISUAL AGNOSIA FOR MOVEMENT

Nearly all types of agnosia are rare but agnosia for movement – motion blindness as it is sometimes called – is so rare that only three cases have been described; and it was only after J. Zihl and his colleagues in Münich had described the most recent case that the existence of the condition was generally accepted.[31] Their patient was a forty-five-year-old woman who, nineteen months earlier, had been admitted into hospital in a stupor after three days of headache, vertigo, nausea and vomiting; but at the time they saw her, her chief complaint was that she was unable to detect movement. The following extracts from the case report give a vivid impression of the kind of difficulties she experienced:

> She had difficulty ... pouring tea or coffee into a cup because the fluid appeared to be frozen, like a glacier. In addition, she could not stop pouring at the right time since she was unable to perceive the movement in the cup .. when the fluid rose. [She] complained of difficulties in following a dialogue because she could not see the movements of the face and, especially, the mouth of the speaker. In a room where more than two other people were walking she felt very insecure and unwell ... because 'people were suddenly here or there but I have not seen them moving'. [She] experienced the same problem but to an even more marked extent in crowded streets ... which she therefore avoided as much as possible. She could not cross the street because of her inability to judge the speed of a car, but she could identify the car itself without difficulty. 'When I'm looking at the car first, it seems far away. But then, when I want to cross the road, suddenly the car is very near.' She gradually learned to 'estimate' the distance of moving vehicles by means of the sound becoming louder.

Careful examination of her visual abilities showed that she could not detect movement in depth at all. When she was seated in front of a table, and a wooden cube was moved towards her or away from her in the line of sight, she had no sensation that the cube was moving, although she noticed that both the position and the apparent size of the cube had changed. In the central part of her field of vision she could detect movement in other directions provided it was slow. In the periphery of her field of vision – and for good evolutionary reasons connected with the detection of predators and prey peripheral vision is particularly important in movement detection – she could tell whether objects were stationary or moving, but she had no idea of the direction of movement.

Zihl's patient showed a clear dissociation between an ability to detect shape and colour, which was near normal, and an ability to detect movement, which was very defective. An equally clear dissociation, but in the reverse direction – so that in the affected part of the visual field only moving objects were detected – was described by Captain George Riddoch, of the Royal Army Medical Corps, in several soldiers wounded by bullets or shrapnel in the First World War.[32] Perhaps the most striking was Lieutenant-Colonel T., who was wounded by a bullet in the right occipital region while advancing with his men. Though dazed he carried on, and noticed nothing wrong with his vision at that time. About a quarter of an hour later he lost consciousness and remembered nothing further until he found himself in Rawalpindi General Hospital eleven days later, by which time the bullet had been removed. About a month later, when he was transferred to Bombay, he realized that he could see nothing on the left side. He used to miss pieces of meat on the left side of his plate, but in a good light he could detect movement in the blind area. After his transfer to England, careful examination showed that for stationary objects he was completely blind in the whole of the left half of the visual field, but he could detect movement in the whole of this area. The 'moving things' had no shape, and the nearest approach to colour was a shadowy grey. The residual vision for movement was not helpful. When travelling on a train, Col. T. could read the names of the stations and enjoy the scenery from the windows on his right; but he saw nothing but vague movements from the windows on his left. The 'consciousness of something moving' made him continually want to turn his head, and what he saw – or rather, what he failed to see – when he did, was so off-putting that he hated journeys.

VISUAL AGNOSIA FOR DEPTH

If you look around the room, then shut one eye and look around again, you will – unless you are one of the 2–5 per cent of people who lack stereoscopic vision – have no doubt that binocular vision makes an important contribution to our ability to see things in three dimensions. But because, even with one eye, our surroundings don't look flat, our appreciation of three dimensions cannot depend solely on binocular vision; we must get other visual clues about the distribution of objects in space. The nature of these clues has been well understood by artists since the Renaissance. Things that are nearer tend to block our view of things that are further; they look bigger, and we see more detail. Because nearer things look bigger we see objects in perspective. Shadows tell us about bumps and hollows; and colours change – distant mountains tend to look bluer. Most dramatically, if we move our heads, nearer objects appear to move faster than further objects; so that, sitting in a railway train, we see the landscape apparently rotating around the most distant point.* This apparent relative motion is of no help to painters wanting to represent three-dimensions on a flat surface, but camera movement is widely used in television and the cinema to increase the impression of depth.

Oddly, although the contribution of binocular vision must have been obvious, and the optical principles involved have been understood for centuries, it was not until 1836 that anyone paid much attention to stereoscopic vision. Then Charles Wheatstone realized that because the two eyes are some distance apart they must look at solid objects in their vicinity from slightly different viewpoints. By making drawings from similarly separated viewpoints, and using an arrangement of mirrors to present each eye with the appropriate drawing, he created the illusion of looking at a real scene. When he read about Fox Talbot's photographs, he realized that photography was just what he needed to make the pictures from the two viewpoints. The stereoscopes that were soon to be found in English drawing rooms, along with stereoscopic photographs of the Holy Land, used lenses instead of mirrors but the principle was the same. If, in looking at a real scene, the two eyes are fixated at a not too distant point, then the images of any other point at the same distance from the eyes will fall on corresponding points in the two retinas – see Figure 13.3. Points slightly nearer or slightly further will stimulate points on the retina that do not quite correspond, and the nature of the disparity provides the clue to the depth.

We now know that the disparity is detected by cells in the visual cortex that receive information from not-quite corresponding points on the two retinas (see page 228), and the development of the connections of these cells during early life depends on the correlation of the inputs from the two eyes.[33] If the two eyes are not properly aligned – in other words if the child squints – the connections do not develop as they should; and if the squint is not corrected soon enough the child may end up without stereoscopic vision.[34] Lack of stereoscopic vision may also, though much less frequently, be the result of damage to the posterior part of the right hemisphere.[35]

---

*In fact, the landscape appears to rotate about any point on which we care to fix our gaze. If we choose the most distant point the whole landscape rotates in the same direction.

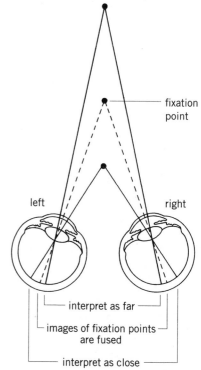

fixation point

left

right

interpret as far

images of fixation points are fused

interpret as close

FIGURE 13.3 The mechanism of depth perception. If we look at an object, that object, and others at the same distance from us, produce images on corresponding parts of the two retinas. Objects that are a little nearer or a little further will produce images slightly removed from corresponding points, and the direction and size of the disparity allow us to estimate the relative distances of the different objects in the scene and therefore to perceive depth. (From E. R. Kandel, J. H. Schwartz and T. M. Jessell.)

Unlike failure of true stereoscopic vision, failure of the monocular mechanisms for depth perception is very rare – fortunately so, for the effects are far more serious. A classical case, from the First World War, was described by George Riddoch.[36]

Riddoch's patient, Captain de W., was wounded at Gallipoli by a piece of shrapnel that entered the left frontal region of his brain and lodged in the back of the right occipital lobe, from where it was removed about a month later. Not surprisingly, given the route of the shrapnel, he was left with several different neurological problems, but I shall mention only those concerned with depth perception. The following extracts are from Riddoch's paper:

> When I [Riddoch] was carrying out some test one day he [Captain de W.] suddenly said: 'Everything seems to be really the same distance away. For example, you appear to be as near to me as my hand' (he was holding his hand 1.5 inches from his face and I was sitting about 5 ft away from him). This loss of appreciation of the relative position of things that he sees quite well has been the means of giving him many a fright. Two vehicles approaching one another in the street always seem to be about to collide. A person who is crossing the street is sure, he thinks, to be run over by a taxi that is really yards away . . . When waiting with his wife to cross the street, the house

on the other side, the policeman and the refuge all seem to be the same distance away and quite near to him.

His most interesting defect is inability to appreciate depth or thickness . . . The most corpulent individual might be a moving cardboard figure, for his body is represented in outline only. He has colour and light and shade, but still to the patient he has no protruding features; everything is perfectly flat. . . .

A stair is a flat inclined plane with no protruding steps, and yet he knows from the light and shade that he ought to see the steps . . . A landscape is like a painted picture or a piece of stage scenery.

Hollow vessels appear to have simply shape and no depth. Looking into a kidney-tray he sees only a flat kidney-shaped white metal plate, part of it in shade and part not.

Very similar symptoms were described in a paper by Gordon Holmes, then consulting neurologist to the British Expeditionary Force, and Captain Gilbert Horrax of the U.S. Army, about a soldier who had been wounded by a machine gun bullet that passed right through the back of his head.[37] Most strikingly:

When shown a cardboard box 18 cm square and 8 cm deep [the patient] described it as a piece of flat cardboard no matter at what angle he saw it, and was surprised when it was placed in his hands to discover it was a box . . . A glass tumbler appeared 'a piece of flat glass' the shape of which varied according as it was presented to him.

As Riddoch says, with masterly understatement, after describing his patient: 'It would appear from this and similar cases . . . that stereoscopic vision [by which he meant merely three-dimensional vision, not necessarily true stereopsis] depends on something more than the possession of binocular vision and the ability to appreciate differences in light and shade.'

THE SIGNIFICANCE OF 'DEFICITS'

Neurologists' jargon tends to run to the Greek and the highly polysyllabic but there is an exception. The selective, but personally devastating, losses of particular visual activities that we have been discussing in this chapter are usually referred to as *deficits* – as if the neurologist were an accountant and the patient a profit and loss account. I shall come back to some of these deficits later, when we look at the neural machinery responsible for visual analysis, but there is an important general point. The selectivity of deficits caused by local lesions is the result of the modular nature of the cerebral cortex, and of the way visual information is processed in parallel by the different modules, and to different ends. It is because the selectivity arises in this way that neurologists put so much more emphasis on *dissociations* between deficits than on *associations*. If John Doe cannot recognize faces but has no difficulty in recognizing colours the straight-forward explanation is that different modules are responsible for the two activities. An alternative explanation is that only a single module is involved and that one of its activities is less sensitive to damage than the other. But if, in addition, Richard Roe has no difficulty with faces but cannot recognize colours – in other words, if there is a *double dissociation* – the alternative explanation is ruled out and different modules must be at least partly responsible for the two activities.[38] In contrast, because the effects of haemorrhage, blood clots, carbon monoxide and bullets are unlikely always to be neatly circumscribed to fit individual modules, an association between deficits does not imply that the same module is involved; modules that are physically close are likely to be damaged together. It is therefore more significant that, say, prosopagnosia and loss of colour sense are sometimes found separately than that they are often found together.[39]

BLINDSIGHT

There is just one other kind of 'disordered seeing with normal eyes' that I want to mention. Some patients with damage to the primary visual cortex claim to be completely blind in the affected part of the visual field yet their behaviour is influenced by what appears there. This paradoxical finding – given the name blind-sight by Larry Weiskrantz – was first noticed by E. Pöppel and his colleagues at the Massachusetts Institute of Technology, who asked brain-damaged patients to move their eyes in the direction of a spot of light that was flashed briefly in the blind part of their visual fields.[40] Although the patients were sure that they saw nothing they tended to move their eyes in the right direction.

The following year a similar, though much more striking, result was reported by Larry Weiskrantz, Elizabeth Warrington and their colleagues in London.[41] They studied a 34-year-old man, DB, who had lost most of his right primary visual cortex in an operation to remove a tumour, and, as a result, was blind in the left visual field except for a crescent of vision in the left upper quadrant. They started by using the procedure of Pöppel and his colleagues, but the results were only just significant. They then asked DB to point to where he guessed the flash occurred, and found him surprisingly accurate. In later experiments, when asked to guess between alternatives, he could distinguish between a vertical and a horizontal line (97 per cent accuracy for an exposure of 100 milliseconds), and between a circle and a cross (90 per cent accuracy).[42] He could tell whether a circular patch of parallel lines was oriented with the lines horizontal or not, provided the deviation from horizontal was at least 10°. It was even possible to measure his visual acuity by seeing whether he could detect the presence of bars in a grating. This extraordinary performance was not accompanied by any visual sensation. If by chance the stimulus extended beyond the blind area, DB immediately reported it, but otherwise he denied having any visual experience. In some parts of the blind field he claimed to have no sensation of any kind; in other parts he said there was no sensation of vision but there were sensations of, for example, 'smoothness' or 'jaggedness' in the discrimination of a circle and a cross. Experiments with other patients with blindsight have shown similar results; some are even able to distinguish between light of different wavelengths, though without any sensation of colour.[43]

Attempts have been made to explain these findings, and others (not quite so impressive) subsequently reported by other investigators, as the result of the scattering of light onto the seeing portion of the retina. Scattering of light certainly occurs but is unlikely to be the explanation because the position of a stimulus falling on the blind spot in the blind half-field could not be 'guessed' by DB, though he could guess the position of the same stimulus in other parts of the blind field. It is significant, too, that he could guess the position of black stimuli on a white ground and could distinguish between a uniform field and a coarse grating of the same average brightness; both abilities are difficult to explain on the basis of scattered light.

Blindsight seems very strange because it is impossible to imagine getting visual information and at the same time being unable to see. A few years ago, though, F. C. Kolb and J. Braun, at the California Institute of Technology, invented a method that allows normal people to experience something rather like blindsight.[44] What they did was to show people visual displays in which a particular area – call it the target area – was defined by some alteration of the background pattern in one quadrant of the display. For example, if the background pattern consisted of short lines tilted slightly to the right, the target area would have similar lines tilted to the left. The viewer was shown the display for a quarter of a second and was then asked to state in which quadrant the target was, and to give an estimate (on an arbitrary 1–10 scale) of his confidence in his statement. With the pattern just described, the position of the target was obvious, but it could be made more difficult to identify by presenting the two eyes with what might be called complementary patterns. For example, if the left eye sees a display with the background lines tilted to the right and the target lines tilted to the left, the right eye sees a similar display but with the tilts reversed. In this situation, with binocular vision and very brief exposures, the viewer sees little crosses, both in the target area and the background, and the target is not consciously distinguishable (presumably because the viewer has no way of knowing from which eye each bit of information is being received). But when asked to guess where the target is, the extraordinary thing is that viewers guess correctly in nearly 80 per cent of the trials, despite claiming not to be able to see any target. Since random guessing would give a success rate of only about 25 per cent, this result is highly significant. It means that the viewer's response to the examiner's question is being strongly influenced by visual information that he or she is quite unconscious of.

(The reason for the brief, quarter second, exposure is that if the viewer looks too long, one eye tends to become dominant, the information provided by the other is suppressed, and the target does become visible.) In other experiments Kolb and Braun used moving dot patterns to define the target, and these gave similar results.

At one time it seemed that blindsight might be the result of subcortical activity, perhaps activity in the roof of the midbrain – the centre for vision in lower vertebrates, and still involved in eye movements in mammals. However, the ability of some patients with blindsight to discriminate colours makes it probable that visual areas in the cortex are also involved. Whatever the explanation, and however counter-intuitive the phenomenon is, we have to accept that it is possible to discriminate using information acquired through vision without being conscious of the visual features providing the information.

Non-invasive methods for scanning the living brain

Computer-aided tomography – CAT-scanning

The trouble with an ordinary X-ray picture of the brain is that it is a two-dimensional representation of a three-dimensional structure, and images of different parts overlap making interpretation ambiguous. The CAT scanner uses a thin pencil of X-rays instead of a broad beam, and an electronic device (a scintillation counter) to detect the X-rays that pass through the brain. The brain is examined a thin 'slice' at a time. For each slice the 'pencil', held in a constant direction, scans in small steps across the slice; the machine is rotated through a small angle round the patient's head and another scan is made. This procedure is repeated until the slice has been scanned in all directions. By taking all the results together the computer can deduce the relative opacity to X-rays of each point in the slice, and hence in the whole brain. The degree of detail that can be seen in the pictures produced by the scanner is only a little less than would be seen if the slices were viewed with the naked eye.

Magnetic resonance imaging – MRI

Magnetic resonance imaging depends on the selective absorption, not of X-rays, but of radio waves of particular wavelengths. When atoms of certain elements, including hydrogen, are in a strong magnetic field, their nuclei behave like tiny bar magnets, aligning themselves with the field. If the atoms are now exposed to a pulse of radio waves of appropriate frequency, some of the energy is absorbed, disturbing the alignment. The disturbed nuclei now behave like a spinning top that has been jolted – they *precess*. The radio waves that are most effective (and therefore absorbed to the greatest extent) are those whose frequency is the same as the natural frequency of the precession. Because this natural frequency depends on the nature of the atom and the atom's environment, seeing which frequencies of radio waves are best absorbed provides a basis for detecting differences in composition. Coupled with suitable scanning methods, the observations can produce high resolution pictures of brain structure comparable with, and often better than, those obtained by CAT-scanning. As with CAT-scanning, however, the production of such pictures is slow.

Because the absorption of radio waves by the nuclei of the iron atoms in haemo-globin varies depending on whether the haemoglobin is bound to oxygen or not, it is also possible to use MRI to measure changes in the oxygenation of haemoglobin in different areas of the brain, and hence to detect areas of increased metabolic activity.

Positron emission tomography – PET scanning

If water containing a radioactive isotope is injected into the bloodstream, the radioac-tivity will accumulate most quickly in parts of the brain with the greatest blood flow. Since activity increases blood flow, if one could measure the amounts of radioactivity in each part of the living brain at successive times one would have a method of seeing which parts are active during any particular task. The difficulty is in determining the distribution of radioactivity in the living brain. PET scanning provides an ingenious solution to this problem, provided that the radioisotope chosen is one that emits positrons – particles like electrons but with positive charges. The point is that very soon after a positron is emitted it will collide with an electron. The charges cancel out, and the very small masses are annihilated, the loss of mass being associated with the release of energy in the form of two gamma rays that leave the collision site in precisely opposite directions. If the brain is surrounded with an array of 'coincidence counters' – detectors that register only if a pair are excited simultaneously – the two gamma rays will excite a pair of detectors, and the computer will 'know' that the two rays must have originated at a point lying on the line joining that pair of detectors. Lines joining such pairs of excited detectors will intersect most densely where the concentration of radioactivity is highest. The computer is programmed to produce pictures of thin slices through the brain with the concentration of radioactivity in each part of each slice indi-cated by different colours.

PET scanning is not only useful for measuring blood flow. By injecting a suitably labelled analogue of glucose that is taken up by cells but not metabolized further, it is possible to use the technique to map areas of high glucose utilization. Using suitably labelled neurotransmitters, it is possible to map the distribution of receptors that bind those transmitters.

14 Opening the Black Box

A neurophysiological approach to vision

Recording from microelectrodes poked into the retina proved to be an effective way of sorting out how visual information was handled in the retina itself; could a similar approach work with the visual parts of the brain? At first sight the prospects did not look good. William Rushton pointed out that to try to discover how the brain works by poking microelectrodes into it is a bit like trying to discover the foreign policy of a central European power by crossing the border on a dark night, entering a border town, and asking random passers-by for their opinion. In fact, like international spying, poking microelectrodes into the brain has led to some dramatic successes; and, as in international spying, success depends on good maps and the patience to build up a picture by putting together a large number of disparate pieces of information.

The basic strategy is straightforward. To see what happens to visual information once it leaves the eyes, we need to be able to follow it along the visual pathway. That pathway, with its main stages, is illustrated in Figure 11.2 (p. 175). The fibres of each optic nerve do not carry their messages straight to the brain. The two nerves first come together and exchange some of their fibres, in such a way that all information about the right half of our visual field, whichever eye it comes from, ends up in the left half of the brain, and all the information from the left half of the field ends up in the right half of the brain. (It says something for the liaison between the two halves of the brain that we are quite unaware of any seam running vertically down the middle of our visual field.) Visual information entering the brain on each side passes first to a part of the thalamus, then to the primary visual cortex, and from there to several adjacent areas of secondary visual cortex. All these sites are richly interconnected by nerve cells, with information being conducted in both directions; and the cortical areas are connected with the rest of the cortex through the association areas. Ideally then, we want to use microelectrodes to follow the activity associated with particular visual stimuli as far as possible along these pathways.

It was only in the late 1950s and early 1960s that experiments of this kind first succeeded. David Hubel and Torsten Wiesel, at Harvard, projected patterns of light onto a screen in front of an anaesthetized cat or rhesus monkey, and recorded from nerve cells in the thalamus and the primary visual cortex.[1] (The anaesthetic is thought not to alter the broad response pattern in these areas of the brain.)

They found that the retina is mapped point by point onto the visual part of the thalamus, and the responses of the thalamic nerve cells are much the same as the responses of the ganglion cells. But in the transfer of information from the retina through the thalamus to the cortex there are two significant segregations. Although all the information about each half of the visual field is sent to the opposite visual thalamus, *the contributions from the two eyes are kept separate, and are sent along different nerve fibres to the visual cortex.* This is important because it is differences between the two retinal images of the same half-field that make stereoscopic vision possible. There is also a segregation, into separate pathways, of the information from different classes of retinal ganglion cells; one pathway carrying information used to detect movement, the other carrying information used to detect colour.[2] Information about shape is carried by both pathways, with the colour pathway handling information about fine detail.

The visual part of the thalamus, though, is not merely a relay and sorting station. As we saw in our Cook's tour, it receives an enormous backflow of information from the visual cortex, one of whose functions may be to allow information currently arriving from the eyes to be compared with information received from the eyes a very short time ago, so highlighting, continuously, changes in the visual scene. It also receives a large inflow from the midbrain, which is thought to be concerned with arousal.

FISHING FOR FEATURES

Hubel and Wiesel's work on the primary visual cortex started unpromisingly, but it took off dramatically as soon as they discovered what kinds of stimuli were appropriate. To start with, like others before them, they found that a few cells responded in the same way as retinal ganglion cells, while most seemed not to respond to light at all. Then one day, when they had spent about five hours recording from a particularly stable but unresponsive cell, they suddenly had the impression that the glass slide they were using to project a dot onto the screen:

> was occasionally producing a response, but the response seemed to have little to do with the dot. Eventually we caught on: it was the sharp but faint shadow cast by the edge of the glass as we slid it into the slot that was doing the trick. We soon convinced ourselves that the edge worked only when [the image of] its shadow swept across one small part of the retina and that the sweeping had to be done with the edge in one particular orientation. Most amazing was the contrast between the machine-gun discharge when the orientation of the stimulus was just right and the utter lack of a response if we changed the orientation or simply shined a bright flashlight into the cat's eyes.[3]

This observation triggered a host of systematic studies of cortical responses to visual stimuli with particular features. In the primary visual cortex of cats and monkeys, Hubel and Wiesel distinguished three new classes of cell on the basis of the character of their receptive fields and of the kind of stimulus that was effective. Unlike the ganglion cells in the retina and the cells in the visual thalamus, the receptive fields of all these cells were not circular but elongated, and the effective stimulus was neither diffuse light nor a spot of light, but a thin bar or line – it might be a bright line on a dark ground, or a dark line on a light ground, or a straight boundary between light and dark areas. And for any one cell, the line had to be in a particular orientation. Some of the cells (*simple cells*) responded best to lines of a particular width centrally placed in the receptive field; others (*complex cells*) were less choosy in those respects but responded vigorously only to lines that moved sideways in a particular direction, (recalling the behaviour of the directionally sensitive ganglion cells that serve as movement detectors in the rabbit retina – see Figure 10.4). Some cells (*end-stopped cells*) responded only if the line or edge ended within the receptive field, or bent so that the orientation was appropriate over only part of the field. In other words, these cells responded to ends of lines or to corners.

Two questions arose from this work. What were all these cells for? And how did they work? Since lines, edges, corners and movement are among the most important features that we use to identify objects in the world about us, the obvious use of the cells was as *feature detectors*. Even curved lines can be thought of as being made up of short segments capable of stimulating simple or complex cells with the appropriate orientation preference. It is probably the absence of end-stopped cells with receptive fields adjacent to the blind spot that accounts for the 'filling in' effect described in Chapter 12. Nor was it difficult to see how the cells might work. Figure 14.1 shows an arrangement suggested by Hubel and Wiesel that could explain the selectivity for lines of particular orientation and width placed centrally in the receptive field. Similarly convergent arrangements were suggested to explain the behaviour of cells with other preferences. The general picture that emerged is that a suitable hierarchy of cells is capable of analysing the form of an image by abstracting features of increasing complexity at successive levels.

LAYERS, COLUMNS AND BLOBS

How are all these cells arranged? The layered structure characteristic of the cerebral cortex as a whole is particularly marked in the primary visual cortex. In general, simple cells are found at levels at which the input fibres from the thalamus end; complex cells tend to be in the more superficial and the deeper layers. This fits in with the hierarchical notion that the more complex cells are further from the input.

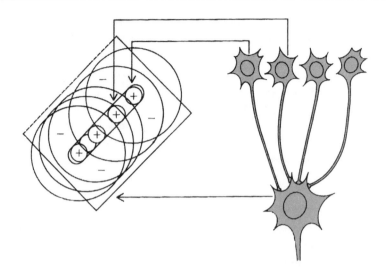

FIGURE 14.1 Hubel and Wiesel's suggestion to explain the existence of cortical cells that are stimulated only by lines of particular orientation falling on the retina. The large nerve cell, bottom-right in the diagram, is assumed to fire only if it is stimulated simultaneously by all four nerve cells feeding information to it. Each of these cells is assumed to have a circular receptive field in the retina, stimulated by light near the centre of the field and inhibited by light falling in the periphery. If the centres of the four receptive fields are aligned, as in the left part of the diagram, the large nerve cell will fire only if a bright line of appropriate orientation falls on that part of the retina. (From D. H. Hubel.)

There are no visible vertical divisions in the primary visual cortex – by convention 'vertical' means perpendicular to the plane of the cortex – but evidence for such divisions was found by Hubel and Wiesel. They noticed that when they pushed their microelectrode vertically into the cortex, all the simple and complex cells encountered by the electrode shared the same character in two respects. In the first place, they were dominated by the same eye – that is to say, the cells responded only to suitable stimuli in that eye, or were, at least, more sensitive to stimuli in that eye. Secondly, all the cells encountered had the same orientation preference. If, instead of pushing the microelectrode vertically, they pushed it very obliquely so that the tip moved nearly parallel to the surface, they found that it tended to encounter, alternately, regions of cells dominated by each eye; and the orientation preferences of the cells encountered tended to change in a regular fashion, though with some breaks in the sequence.

To explain their results, Hubel and Wiesel suggested that the columnar organization of the primary visual cortex is something like the diagram in Figure 14.2. Alternate slabs are concerned predominantly with information from one or other eye (ocular dominance columns), and within each slab there are vertical slices, each slice containing cells with a particular orientation preference (orientation preference columns). Hubel and Wiesel pointed out, though, that because each penetration with the microelectrode gives information only along a single dimension, trying to elucidate a three-dimensional pattern in this way is extremely tedious, and they emphasized that the true structure was likely to be much less rectilinear than an array of ice cubes.

And that prediction was right. More recently several methods have been developed that make it possible to look at the surface of the cortex and see which areas of cells are activated by visual stimuli of particular kinds. The most straightforward method, at any rate in theory, is to treat the surface of the visual cortex of an anaesthetized animal with a suitable voltage-sensitive dye and to see which areas show a change in voltage in response to, say, stimuli with a particular orientation.[4] It turns out that, in each small area of primary visual cortex, the orientation preference columns are arranged, not in parallel-sided slices, but with all the columns coming together at the centre, like a sliced cake.[5] Functionally, the precise geometry is less interesting than the scale of the vertical organization, and what sort of information is segregated. Roughly speaking, each square millimetre of primary visual cortex includes ocular dominance columns from both eyes, each containing a complete set of orientation preference columns.

That the orientation of any line in the retinal image is detected by cortical cells each selective for a narrow range of orientations (roughly $\pm 15°$) explains a striking visual illusion – see Figure 14.3. After looking for half a minute at a grating tilted slightly one way from the vertical, a truly vertical grating appears to be tilted the opposite way – presumably because the cells that responded to the tilted grating are less active than they would have been without the exposure. The fatigued cells (or at least some of them) must be in the brain rather than the retina, and must receive information from both eyes, since the effect is seen even if the tilted grating is viewed with only one eye and the vertical grating with the other.

FIGURE 14.2 Hubel and Wiesel's 'ice-cube model' of the primary visual cortex. For explanation see text. (From D. H. Hubel.)

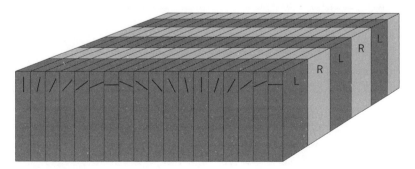

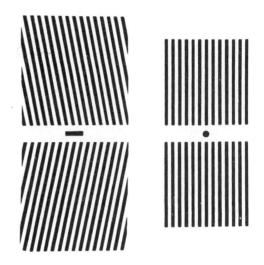

FIGURE 14.3 A visual illusion explained by the existence of orientation-selective cells in the visual cortex. After looking for half a minute at the bar between the left-hand pair of gratings, and then transferring your gaze to the dot between the right-hand pair, you will see that pair appears to be tilted slightly in a direction opposite to the tilt of the left-hand pair. (From C. Blakemore.)

Although most of the cells in the primary visual cortex are dominated by one or other eye, a small proportion have a balanced input from both eyes, and respond much more vigorously when the receptive fields in both eyes are stimulated. Careful examination of these binocular cells by Horace Barlow, Colin Blakemore and Jack Pettigrew, showed that the receptive fields in the two eyes always have the same orientation preference but are not always in precisely corresponding position in the retina.[6] These cells are the *disparity detectors* that make true stereoscopic vision possible – see Figure 13.3 (p. 216).

The existence, in the primary visual cortex, of nerve cells driven mainly by one or other eye, or selectively excited by lines in particular orientations, raises the question: how do these cells get connected so that they work in this way? Is the development of the connections innate – that is to say, wholly under genetic control – or is there some flexibility during development so that what happens depends partly on what the eyes are seeing once they start working? In the 1960s Wiesel and Hubel showed that, though many of the nerve cells in the visual cortex of very young kittens reared in the dark could be influenced by either eye, if the kittens were transferred to a lit environment but allowed to use only one eye, within a few days all the cells had become sensitive only to stimulation of that eye.[7] This change occurred only with kittens between two weeks and four months old, and most readily during the earlier part of this period. Even more remarkably, Colin Blakemore and his colleagues in Cambridge showed that, within this sensitive period, if the closure was reversed, so that a kitten that had been allowed to use only, say, its left eye was now allowed to use only its right eye, the cells in the visual cortex soon switched allegiance, becoming sensitive only to stimulation of the eye that was in habitual use.[8]

An equally striking flexibility was revealed in experiments to look at the development of orientation-selective cells in the visual cortex.[9] Between the ages of two weeks and five months, kittens were reared in an environment in which they saw *either* only vertical *or* only horizontal stripes. Such kittens behaved as if 'they were virtually blind for contours perpendicular to the orientation they had experienced'; for example a kitten that had experienced only vertical stripes would respond to a shaken vertical rod by coming to play with it, but would ignore a shaken horizontal rod; and vice versa for a kitten who had experienced only horizontal stripes. And when the visual cortex of such kittens was examined under anaesthesia, it was found that each lacked nerve cells that responded to lines (in the retinal image) whose orientation was anything like the ignored orientation, and was richest in nerve cells that responded to lines close to the experienced orientation. The conclusion drawn by Blakemore and his colleagues from all these experiments is that during maturation the visual cortex may adjust itself to the nature of the animal's visual experience in a way that facilitates behaviour appropriate to features likely to be experienced.

An odd feature of all the early work by Hubel and Wiesel was that it revealed rather few cells that were sensitive to colour, but this turned out to be because the colour-sensitive cells are collected together in short, round, vertical columns that had been missed by the exploring microelectrodes. These short columns have different staining properties from the surrounding cortex and appear in stained sections cut parallel to the surface as a regular series of blobs forming a polka-dot pattern. Margaret Livingstone and David Hubel showed that they were rich in nerve cells that had circular receptive fields, centre/surround antagonism, and interesting colour discriminating properties.[10]

So far we have looked at the cells in the primary visual cortex, and have ignored the nerve fibres connecting them. Within each column, there are rich connections between the cells in the different layers, and it is presumably the interactions that these connections make possible that account for the different kinds of behaviour of the different types of cell. There are also interconnections between nerve cells in different columns or blobs, sometimes up to about 8 millimetres apart, and the interesting thing about these is that they always seem to connect cells with similar functions: a cell in one column with a particular orientation preference with cells in columns some distance away with the same orientation preference; or a colour-sensitive cell in one blob with colour-sensitive cells in other blobs. The significance of this is that the response of each cortical cell to a stimulus in its receptive field can be modified by relevant events in adjoining fields. There are also much longer connections between nerve cells in the primary visual cortex and nerve cells in adjacent areas of cortex concerned with vision, but these are made by nerve fibres that lie in the white matter underlying the cortex, rather than in the cortex.

Looking at the properties of the retina, of the visual part of the thalamus, and of the primary visual cortex, three things stand out. First, as we have already noted, there is a hierarchy of cells capable of beginning the analysis of the image by successively abstracting features of increasing complexity. Secondly, there is evidence of parallel processing of information of different kinds: information about form, about movement, and about colour, are handled separately, and information from the two eyes is kept sufficiently separate for disparities between the two retinal images to be used to determine depth. Thirdly, the principle of topological mapping is maintained: adjacent points on the retina are mapped to adjacent points in the visual thalamus and the primary visual cortex. What is more, the organization of the primary visual cortex ensures that the different kinds of information that are sent to the brain about any small part of the visual field are all accessible within a very small area of cortex, though there is also provision for sending that information to more distant areas.

FOLLOWING THE TRAIL

Both the evidence for hierarchies of analytical cells, and the evidence for parallel processing of different attributes of the retinal image, fit well with the clinical evidence showing selective deficits in the ability to recognize images of particular kinds – as in prosopagnosia and visual form agnosia – or in the ability to recognize colour or movement or depth. But the site of the trauma or bleeding or clotting that are the usual causes of such deficits is unlikely to be in the primary visual cortex because there the geographical arrangement of areas specialized for different functions is much too fine-grained. With the cortical pattern repeating every square millimetre, no bullet or piece of shrapnel or haemorrhage or clot is going to destroy, selectively, nerve cells concerned with colour or movement or disparity. If selective deficits from such causes are to be explained along these lines, the parts of the brain concerned with vision must contain regions where the painting is with a much broader brush.

In the early 1970s, a number of such regions were demonstrated in the brain of the macaque monkey by Semir Zeki, at University College, London.[11] In an anaesthetized monkey, he recorded from microelectrodes inserted into various areas in front of the primary visual cortex and studied their responses when stimuli of various kinds were flashed onto a screen in front of the animal. Most striking were two areas, one in which virtually all the cells were sensitive to motion – each cell being maximally sensitive to motion in a particular direction[12] – and another in which the great majority of cells were sensitive to differences in colour. In the motion-sensitive area (often referred to as Zeki's area V5) most of the cells responded to moving spots; others preferred moving oriented lines; none showed any colour discrimination. In the colour-sensitive area (Zeki's area V4) some of the cells were also sensitive to linear orientation. In a third region Zeki found that most of the cells responded to stationary lines of particular orientation, and these cells too were insensitive to the colour of the line or of the background. The general picture that emerged from these experiments was that, in both sides of the monkey's brain, outside the primary visual cortex, there are separate areas of visual cortex that are particularly concerned with motion or with colour or with form. This, of course, is just the picture suggested by the clinical studies discussed in the last chapter.

A curious feature of the specialized areas is the way the retinal surface is mapped onto the cortical surface. In the thalamus, in the primary visual cortex, and to a lesser extent in the immediately adjacent area, the mapping is precise, though with much more cortical space being devoted to the central part of the retina concerned with detailed vision. In the motion and colour areas (and in other visual areas not discussed) the receptive fields of the cortical cells tend to be much larger, and the mapping looks chaotic, though this appearance presumably reflects our ignorance of the principles controlling the connections rather than a failure of control during development.

The methods used by Zeki to demonstrate the existence of motion and colour areas in the monkey could not, of course, be used in humans, but in 1991 he and Richard Frackowiak and their colleagues used PET scanning to monitor the changes in blood flow in different parts of the cortex when subjects looked at scenes designed to cause increased activity in such areas.[13]

When they compared the blood flow in the brain of a person looking at a random pattern of *moving* black and white squares, with the blood flow in the brain of the same person looking at a similar pattern in which the squares were stationary, they found that when the squares were moving there was evidence of increased activity on both sides of the brain at the junction of the temporal and occipital lobes. What is more, these areas lay in just those parts of the brain that a CAT scan had shown to be damaged on both sides in the unfortunate patient with agnosia for motion described by Zihl and his colleagues (page 213).

To identify colour areas, Zeki and his colleagues used abstract collages of rectangular coloured papers of various sizes – nicknamed 'colour Mondrians' – of a kind that had been introduced by Edwin Land in his studies of colour vision. They compared the PET scan when a normal person looked at such a colour Mondrian with the scan when the same person looked at a similar collage in shades of grey chosen to give equivalent brightnesses. With the colour Mondrian, there was an extra area of heightened activity on each side of the brain in the lowest part of the occipital lobe near the midline – precisely the region where, a hundred and three years earlier, Verrey had detected damage on the left side in Mme R., the woman who had suddenly lost the sense of colour in right half of the visual field.

Clearly, areas corresponding to the motion and colour areas that Zeki found in the monkey exist in our own brains.

PARALLEL PATHWAYS

The existence of parallel pathways for analysing form, motion and colour makes it convenient to look at the three topics in turn, though ultimately we shall have to face what neuroscientists like to call *the binding problem.*

If we are looking at a passing London bus, the machinery in our brains that is concerned with form will register something shaped like a bus; the machinery concerned with colour will register something that is red, and the machinery concerned with motion will register something that is passing. But what machinery will determine that the three 'somethings' are the same thing – and, for that matter, that that thing is also the source of the noise and the smell of fumes that we are experiencing? For the moment, let us put such problems aside.

THE ANALYSIS OF SHAPE

The idea that, starting with Hubel and Wiesel's simple and complex cells it is possible to construct a hierarchy of cells capable of selecting features of a more and more sophisticated kind is both attractive and tantalizing. It is attractive because – in facial recognition, for example, or in recognizing handwriting – we are incredibly good at differentiating between similar shapes; and an arrangement of the kind suggested provides, in the barest outline, a possible explanation of the sort of machinery that could achieve such differentiation. It is tantalizing because, in the absence of more detailed information, it is not clear how the hierarchical analysis would work or where the process would stop or quite what would be achieved at any given stopping point.

Identifying an object in the retinal image is not quite like identifying a water dropwort using an old-fashioned unillustrated flower-book, where you simply narrow down the possibilities by applying more and more selective criteria until only water dropwort fits the bill. In analyzing the retinal image, many criteria used in the early stages may be irrelevant later – for example, position, size, orientation, contrast. Nor is it clear what sort of object the cell at the apex of an analytical hierarchy might most usefully represent. The traditional extreme example – not quite a *reductio ad absurdum* – suggested by Jerry Lettvin, is the grandmother cell, a nerve cell that fires if, and only if, you are looking at your grandmother.* Since local lesions in the brain do not tend to destroy specific

---

*The story of the origin of the grandmother cell is told in a letter from Jerome Lettvin to Horace Barlow, reproduced as an appendix to Barlow's chapter on 'The neuron doctrine in perception', in *The Cognitive Neurosciences* (1995) (ed. Gazzaniga, M. S.) MIT Press, Cambridge, Mass.

memories, if such cells exist they must exist in multiple copies. But even if security is assured in this way, it is not clear how such cells would be used. What would they project to? The 'bug-detector' cells that Barlow detected in the frog retina make sense because they initiate feeding responses that are virtually automatic. There is no comparable automatic activity appropriate to seeing your grandmother or seeing any other complex stimulus of the kind we are considering. To respond to such a stimulus you need more detailed information about it, and that can't be stored in a single cell. If this information is represented as a pattern of synaptic strengths in an ensemble of cells, it makes sense for the stimulus itself to be represented in the same way. On the other hand, the existence of nerve cells responsive only to very specialized and important stimuli does make it possible to represent a scene containing such stimuli with the minimum number of active neurons. Barlow has pointed out that cells of this kind could 'be active in combinations and thus have something of the descriptive power of words'.[14]

Ignoring these theoretical considerations, let us just consider experimental evidence for the existence of cells activated by shapes more complex than those studied by Hubel and Wiesel in their work on the primary visual cortex. In macaque monkeys, the lowest part of the temporal cortex seems to be concerned exclusively with the analysis of visual information.[15] Damage in this region (known as the inferior temporal cortex) interferes with the monkey's ability to recognize, learn or remember visual stimuli, though it does not prevent the animal from seeing, or even lessen its ability to see detail. In the conscious animal, recording from microelectrodes previously inserted into the inferior temporal cortex shows that many of the cells are selective for some aspect of shape or texture or colour. The receptive fields of these cells – that is, the area of retina in which a suitable stimulus causes a response in an individual cortical cell – tend to be very large, usually including the central area of the retina used for detailed vision, and those cortical cells selective for shape respond to the appropriate shape irrespective of its size or contrast or colour.

The selectivity of some of the shape-selective cells is remarkable. In 1969, Charles Gross and his colleagues at Harvard and Rio de Janeiro described a cell that seemed to be selective for the shape of a monkey's hand, and three years later they described cells that were particularly excited by photographs of faces.[16] Face-selective cells have since been studied in detail in laboratories in Oxford and St Andrews and, more recently, in several laboratories in Japan.[17] They occur not only in the inferior temporal cortex but also in other cortical areas[18] and in the amygdala. And they seem to vary in function. Most respond only to views from particular angles, for example, full-face or right or left profile; a small fraction respond to views from all directions, perhaps by combining the outputs of the more view-selective cells. Most of the cells that respond to faces do not distinguish between the faces of different individuals, but about 10 per cent respond more vigorously if the face is familiar. Some of the cells that do not distinguish between different faces do distinguish between different facial expressions.

David Perrett and his colleagues at St Andrew's wondered whether an important function of the cells selective for particular views of faces might be to provide information about where the viewed monkey was directing its attention. Since attention is often better indicated by gaze direction than by the position of the head, they argued that if such information is important – as it might well be in a social animal – one would expect to find cells whose responses were linked more closely to gaze direction than to head position. And they did find such cells.[19]

The cells with different selectivities tend to be segregated so that, for example, cells sensitive to face identity are separated from cells sensitive to gaze direction or facial expression.[20] Similar segregation in humans could explain the double dissociation that has been observed in human patients between loss of face recognition and loss of the ability to distinguish the direction of gaze or to recognize different facial expressions.[21] It is difficult, though, to translate from the monkey to the human because different areas of cortex seem to be involved in facial recognition tasks.[22]

Recognition of faces is obviously of special biological significance, but it is possible, at least in monkeys, that the machinery involved is not specific for faces. Logothetis and his colleagues trained macaque monkeys to discriminate between very similar computer-generated wire objects, or very similar spheroidal objects, of a kind that would be quite novel to the monkey and would also have no biological significance.[23] When the monkey was expert, they recorded from neurons in the inferior temporal cortex while the (anaesthetized) monkey was shown either the object it had learnt to discriminate or very similar 'distractors'. They were able to find neurons that responded selectively to the learned object, and often only if the object was viewed more or less from a preferred direction. The similarity to the behaviour with faces suggests that training may play an important part in facial recognition, though the preference shown by newly born babies for patterns resembling faces (p. 208) implies that, in humans, some capacity for facial recognition is either built in or can be acquired very rapidly.

A CLOSER LOOK AT MOTION

Macaque monkeys are very good at discriminating the direction of motion of a set of dots that are moving not quite randomly, and they can be trained to indicate whether the direction of motion is, say, to the right or the left. Local damage to cells in the motion area of the cortex renders the corresponding part of the visual field insensitive to motion, though other aspects of vision – acuity, colour sense – are unaffected.[24] Even more remarkable are the results of what was, in a sense, the reverse experiment. The arrangement of cells in the motion area is columnar, with all the cells in a column being sensitive to movement in the same direction, and with the preferred direction changing progressively from column to column. By injecting current through an electrode inserted at one point, William Newsome and his colleagues were able to stimulate a cluster of cells with nearly the same directional preference.[25] Such stimulation was found to influence the directions of motion indicated by trained animals who were presented with arrays of moving dots, biasing the responses in the directions preferred by the stimulated cells. The implication is that activity of cells in the motion area is not merely correlated with the behavioural response but is in the causal chain.

The existence, in the cortex, of cells sensitive to motion in particular directions explains one of the most striking of all optical illusions – the so-called waterfall effect. If you stare at a waterfall for a minute or two and then transfer your gaze to the rocks alongside, those rocks appear to move upwards – yet, paradoxically, without changing their position. In 1925, the Czech physiologist Johannes Purkinje observed the same effect after watching a cavalry parade. (If your environment is so impoverished that it lacks both waterfalls and cavalry parades, you can see the illusory reverse movement just as well by walking across a field rough enough to oblige you to keep your eyes on the ground ahead, and then looking up at the sky.) The cause of the apparent movement is, presumably, fatigue (or adaptation) in the motion-sensing mechanism, resulting in an imbalance between nerve cells signalling movement in opposite directions. The apparent fixed position of the rocks despite the sense of movement reflects the independence of the mechanisms determining movement and position.

The illusion of movement in the waterfall effect can be understood, at least in principle. Much more puzzling are certain static patterns that induce a sensation of movement – a feature cleverly exploited by 'op-artists'. In the Museum of Modern Art in New York there is a black and white painting called 'Current' by Bridget Riley which appears to be in continual shimmering motion.* Isia Leviant has designed a pattern in which concentric circular blue bands are superimposed on a background of close-set alternate black and white spokes radiating from a yellow centre. Most people gazing at his pattern see the bands continually rotating, sometimes clockwise and sometimes anticlockwise, and not all the bands in the same direction. It is not at all clear how the perception is produced, but Zeki and his colleagues have shown that a PET scan of a subject looking at the pattern reveals activity in the classical motion area and its immediate vicinity.[26]

More detailed examination of the responses of nerve cells in the motion area have shown them to be surprisingly varied and subtle.[27] Some are selective for

---

*The shimmering can be seen well in a reproduction of the painting in Bridget Riley's *Dialogues on Art* (Zwemmer, 1995). As if to show that the magic lies in the artistry rather than the ingredients, the picture is described as 'synthetic polymer paint on composition board'.

speed of motion as well as direction. For others, it seems to be movement relative to background that matters – a cell that responds vigorously to a small group of dots moving in a particular direction, may show a much smaller response if those dots are part of a much larger group all moving in the same direction. Cells in an area just in front of the classical motion area have been shown to be sensitive to other kinds of relative motion within their receptive fields. Different cells respond to clockwise and anticlockwise rotation, others to expansion and yet others to contraction. The circuitry that makes such sensitivities possible is not known, but the value of being able to detect the expansion and shrinking of the visual scene as objects approach or recede is obvious. The sudden expansion of the image of the visual scene in the eye of a housefly as it approaches a surface is known to make the fly stop flying and prepare to land. Landing an aeroplane is more complicated, but it seems likely that the cortical nerve cells that respond to expansion of the image on the pilot's retinas have a role comparable with that of their counterparts in the fly.

COLOUR IN THE CORTEX

Thomas Young's trichromatic theory, developed by Helmholtz and confirmed by the later discovery of three types of cone with three different pigments, accounts for so many observations about colour vision that it is easy to assume that it can account for all of them. In fact, as both Young and Helmholtz were well aware, the perception of colours is not straightforward.

Objects seem to have much the same colour even if viewed under very different illumination. This is, of course, true only within limits. When we buy clothes in a shop lit with fluorescent light we ask to see them in daylight, or at least under an ordinary tungsten light. But we don't notice much difference between the way colours look under a cloudy sky or a blue sky, in the morning or at sunset, in electric light or in candle light. So much do we take this for granted that we are dismayed when our photographic film, designed for use in daylight and actually used in electric light, makes the bridal dress look pink. Had we used indoor film and photographed the bride in daylight, her dress would have looked blue.

This ability to see things as having much the same colour under different illumination is known as *colour constancy*. It is obviously useful, since we live in a world in which lighting changes and it must be easier to recognize any object under different lighting conditions if its appearance remains unaltered. The wavelength composition of light reflected from a lemon into our eyes will depend both on the relative extent to which the lemon's surface reflects light of different wavelengths, and on the composition of the light falling on the lemon. The first of these factors is an intrinsic property of the lemon; the second is not. To achieve colour constancy, we have to deduce the reflective properties of the surface of the lemon by making allowances for the composition of the light.

Helmholtz – who was familiar with the idea of colour constancy – believed that it was the result of involuntary and unconscious inference. In a sense it is, but that does not take us very far. Anything inferred, whether consciously or unconsciously, must be inferred from something. From what could we infer anything about the composition of the light falling on our lemon?

The answer to that question was provided most clearly by Edwin Land, the inventor of polaroid film, who was of course familiar with the inferiority of the camera to the visual system when it came to distinguishing colours. In a series of remarkable experiments, starting in the 1950s, he showed that it is only in rather special circumstances that the colour seen by an observer looking at a particular small area is determined solely by the relative intensities of light of different wavelengths coming from that area. If the area is part of a whole scene, the apparent colour depends not only on the light from that area but also on the light coming from other parts of the scene. This effect of the surroundings had been noted before – indeed as early as 1789[28] – and Helmholtz spends many pages of his *Physiological Optics* discussing what he calls 'simultaneous contrast'; but Land took the matter much further. Two of his experiments are very surprising because the results are disturbingly counter-intuitive.

In the first experiment, using a specially designed camera, he took two simultaneous *black and white* photographs of a young woman, one taken through a red filter so that the film recorded only long-wavelength light, the other taken through a green filter so the film recorded only middle- and short-wavelength light.[29] The photographs were positive transparencies so they could be projected onto a screen. When projected separately, using white light, the two black and white images showed the expected differences – for example, the woman's lips and red dress appear light grey in the long-wavelength photograph and black in the shorter-wavelength photograph. When the long-wavelength photograph was projected alone through a red filter, the image on the screen was, of course, in various shades of red grading into black. The surprise came when the image of the long-wavelength photograph (projected through a red filter), and the image of the shorter-wavelength photograph (projected *without* a colour filter), were exactly superimposed on the screen. On classical colour theory, the result should have been a picture in shades ranging from red through pink to white – the result of mixing red light and white light. In fact what appeared on the screen looked like a full-colour portrait – 'blonde hair, pale blue eyes, red coat, blue-green collar, and strikingly natural flesh tones'. Clearly, the classical rules weren't working. As if that were not surprising enough, Land also found that marked changes in the relative strengths of the two projectors did not alter the colours seen.

In the second experiment, Land studied the appearance of a 'colour Mondrian' illuminated by three projectors, one using light of long wavelengths (reddish light), one light of middle wavelengths (greenish light) and one light of short wavelengths (bluish light).[30] The intensity of light from each projector could be individually controlled, and the intensity of the light reflected from any particular rectangle could be measured with a photometer. With the three projectors adjusted to shine lights of similar intensity, the colours of the assembled rectangles looked normal. The surprise was what happened when Land varied the intensities of the light from the individual receptors, always ensuring that there was a contribution from all three projectors, and keeping the sum of the three intensities constant – for example by increasing the amount of red and reducing the blue and green. Over a wide range of relative strengths, the perceived colours of the assembled rectangles did not change: nothing happened.

By adjusting the relative strengths of the three projectors, and using the photometer to measure the amount of light of long, middle and short wavelengths reflected from particular rectangles, it was possible to arrange that the ratios of the amounts of light in the three wavebands reflected from a yellow rectangle with one setting of the projectors was the same as the ratios of the amounts reflected from a green rectangle with a different setting of the projectors; *yet the yellow rectangle looked yellow and the green rectangle looked green with both settings.* This is unequivocal evidence that the ratios of the different wavelengths reflected from a rectangle cannot alone determine the colour seen by an observer. But Land pointed out that behaviour of this kind is seen only if the rectangles are viewed as an assembly. If a single rectangle (or any other single coloured area) is viewed on its own against a dark background, its apparent colour *is* determined solely by the ratios of the lights of the different wavelengths reflected from it. (This is, of course, why the classical colour-mixing experiments gave the results they did.) It seems, then, that when we view the whole scene (the assembly of coloured shapes), we must alter our interpretation of the information coming from any particular area by making use of information from outside that area.

How might this be done? Land's suggestion was that, based on the three cone pigments, there are three independent mechanisms that deal respectively with light of long, middle and short wavelengths. And he supposed that the colour perceived in each small area in a scene is determined not (as the classical Thomas Young theory states) by the ratios of the *absolute* intensities of light in each waveband reflected from that area, but by the ratios of those intensities *after each has been adjusted to take account of the properties of the rest of the scene.* As to how this adjustment was to be achieved, Land had two hypotheses. The first was that the absolute intensity of light in a given waveband reflected from a small area was divided by the *maximal* intensity of light in that waveband reflected from anywhere in the scene. The other was that the absolute intensity of light in a given waveband reflected from a small area was divided by the *average* intensity of light in that waveband reflected from the whole scene.[31] Either hypothesis could account for colour constancy in the face of changes in the composition of the incident light. For example, doubling the strength of long wavelengths in the incident light would double the intensity of long wavelength light reflected from red areas of the scene but it would also double both the maximal and the average intensity of long-wavelength light reflected from the scene. Use of the average intensity rather than the maximal intensity is physiologically more attractive since it simply requires centre/surround antagonism with a small centre and a very large surround.

The sort of adjustment Land suggested explains why changing the relative strengths of the two projectors in his first experiment had no effect on the colours in the image of the young woman. But why did those colours appear so magically in the first place? How did a black-and-white picture and a black-and-red picture combine to produce blue eyes and a blue-green collar? That too follows from Edwin Land's theory though the explanation is less obvious.

Consider the image of the blue-green collar. The light falling on that part of the screen is mostly white light with a trace of red light; on classical theory the image of the collar should look a pale pink.

But now remember that we are working according to the gospel of St Edwin not the gospel of St Thomas. What matters is not the ratios of the absolute intensities of long-, middle- and short-wavelength light reflected from an area but the ratios of those intensities *each expressed as a fraction of the average intensity of light of similar wavelength reflected from the whole scene.* When the white projector alone is turned on, the image of the collar looks pale grey. When the red projector is turned on as well, very little red light reaches the image of the collar, but a good deal reaches other parts of the image; so the *average intensity of long-wavelength light reflected from the whole image* is much increased. The intensity of the long-wavelength light reflected from the image of the collar *expressed as a fraction of that average intensity* is therefore much reduced. Since what determines the perceived colour is not the ratios of the absolute intensities of light in the three wavebands but the ratios of those intensities expressed as fractions of the average intensities, the collar will look blue-green. (It is as though shining red light on parts of the screen subtracts red light from the other parts. Removing red light from white light leaves blue-green light.)

Land showed that computations of the sort required by his theory, and based on the amounts of light reflected from many areas in a scene, predicted rather well the colour perceived in any particular area. What we want to know, then, is whether the brain really judges colour in this way, and, if it does, how the neural machinery carries out the necessary processing. The answer to the first question is that Land's observations are as convincing as they are counter-intuitive, and something like Land's theory is almost certainly right. How the neural machinery works is more of a problem. The information the brain starts with cannot be simply the responses of the three types of cone in each region of the retina since those responses are not directly available to the cortex because of the processing that has already occurred in the retina. It is time we remembered the title of this chapter and looked inside the black box.

As we have seen (page 160) the outputs of the different types of cone converge at the retinal ganglion cells, so that each ganglion cell concerned with colour provides information about the relative amounts of light of different wavelengths falling on its receptive field. Information about wavelength seems to be organized in much the same way in the thalamus, but in the primary visual cortex there is more variety. Margaret Livingstone and David Hubel found some cells with centre/surround antagonism that behaved much like retinal ganglion cells, but they also found other cells that showed 'colour opponency' in both centre and surround.[32] For example, if, in the centre, light of long wavelengths stimulated, and of middle wavelengths inhibited, then in the surround the effects would be reversed. Or there might be a similar arrangement, but with short wavelengths opposed to long and middle wavelengths.

But a feature of all the nerve cells in the primary visual cortex whose responses have been recorded, is that *in each case the response is wholly determined by the pattern of illumination in a very small area of the retina.* Since Land's work shows that the colour perceived in one area in a whole scene is not defined by the wavelengths of light coming from that area alone, it follows that none of the cells in the primary visual cortex have responses that correlate with the colour perceived in the corresponding area of the the visual field. Yet it is perceived colours rather than patterns of wavelengths that we recognize; so there ought to be some nerve cells somewhere whose responses do correlate strictly with the colour perceived in a particular part of visual field, *irrespective of the particular pattern of light on the retina that gives rise to that colour.*

And it seems that there are such cells in the part of the visual cortex that is specially concerned with colour (area V4). Zeki recorded from cells in this area in an anaesthetized monkey, while the monkey viewed a colour Mondrian.[33] The Mondrian could be moved so that the receptive field of the cell under investigation received light from only one rectangle at a time, and the proportion of light of different wavelengths falling on that rectangle could be adjusted to any desired value. He found that the cells fell into two classes. For those in the first class, the only thing that mattered was the proportion of light of different wavelengths reflected from the rectangle. For those in the second, firing was correlated with perceived colour – perceived, that is, by Professor Zeki or his colleagues, though it seems reasonable to suppose that the monkey would have made similar discriminations had it not been anaesthetized.

When Land first proposed his theory, in the 1950s, he called it the *retinex* theory because he did not know whether the relevant machinery was in the retina or the cortex. Later, work on the responses of the retinal ganglion cells, and of the cells handling colour information in the thalamus, suggested that the machinery must be in the cortex. It is, of course, possible that cells whose activity correlates with perceived colour may yet be found in the primary visual cortex, but Zeki's results, as they stand, suggest that the machinery that ensures – as Land puts it – that 'colour is always a consequence, never a cause' must lie beyond the primary visual cortex and in or before the part of the visual cortex specially concerned with colour (area V4). Just how that machinery works remains obscure.

15 Natural Computers and Artificial Brains
A computational approach to vision

But something is missing
—Refrain of Jimmy Mahoney, in Brecht and Weill's opera *Mahagonny*

What worried David Marr in the 1970s was that work on understanding the brain did not seem to be fulfilling the promise of the very exciting work of the previous two decades – the pioneering studies of Hubel and Wiesel; the experiments showing the contribution of individual nerve cells to stereopsis and to the perception of colour; the discovery of 'hand-detectors' in the temporal cortex of monkeys; the kind of work, in short, that had led Horace Barlow to write:

> A description of that activity of a single nerve cell which is transmitted to and influences other nerve cells, and of a nerve cell's response to such influences from other cells, is a complete enough description for functional understanding of the nervous system. There is nothing else 'looking at' or controlling this activity, which must therefore provide a basis for understanding how the brain controls behaviour.[1]

What was missing, Marr felt, was something 'that was not present in either of the disciplines of neurophysiology or psychophysics' – the one concerned with the behaviour of nerve cells; the other with the behaviour of human subjects. 'The study of vision', he felt, 'must . . . include not only the study of how to extract from images the various aspects of the world that are useful to us, but also an enquiry into the nature of the internal representations by which we capture this information and thus make it available as a basis for decisions about our thoughts and actions.' There must exist, he thought – and a similar conclusion was reached independently by Thomaso Poggio in Tübingen – a 'level of understanding at which the character of the information-processing tasks carried out during perception are analysed and understood in a way that is independent of the particular mechanisms and structures that implement them in our heads'.

To understand fully any information processing device, Marr argued, it is necessary to be able to answer questions at three levels. At the top level, what is the goal of the computation, and why is it appropriate? (For vision, the goal is to produce from the 'images of the external world a description that is useful to the viewer and not cluttered with irrelevant information'.) At the intermediate level,

FIGURE 15.1 The Salem girl/witch –
an ambiguous woodcut.

how are the input and output represented, and what is the algorithm – the
sequence of elementary computational steps – that is used to achieve the transformation from input to output? And, at the lowest level, how can the representations and the algorithm be realized physically? 'Trying to understand
perception by studying only neurons [the lowest level]', Marr said, 'is like trying
to understand bird flight by studying only feathers. It just cannot be done.'

He emphasized that, though the three levels are 'logically and causally related',
the relation is a loose one and it is important to consider the three separately.
Some phenomena (after-images for example) can be fully explained at the physical level (the properties of the receptor cells); others (the ambiguity of Figure
15.1, for example) require explanations at one or both of the higher levels. But the
looseness of the coupling between the levels does not imply that the neural
machinery available is equally suited to all possible algorithms.[2] In seeking to
discover which algorithms the brain actually uses, it is therefore wise to look
particularly at those for which the neural machinery seems most apt.

What description, derived from the retinal images of the external world, is
useful to the viewer depends very much on the viewer. The visual system of a fly,
it seems, is unlikely to provide it with any explicit representation of the external
world, but it is extremely effective in providing information which, though very
limited, is precisely what the fly needs. If the images in its eyes loom sufficiently
fast the fly prepares to land. A speck in front of a textured ground and moving

relative to it leads to information about the direction of the speck and its angular velocity being fed to the fly's motor system.[3] The 'bug detectors' cells in the frog retina (see p. 159) would seem to play a similarly limited role. In contrast, our own visual system does give us an explicit picture of the external world, including information about the shapes, positions, colours, brightnesses, visual textures, distances and movements of the objects viewed, and Marr's approach was to look at the computational issues involved.

Of the various kinds of information, Marr felt that shape and position were fundamental. He tells us that he was particularly impressed by an account by Elizabeth Warrington of the different abilities of patients with right-sided and left-sided lesions of the brain in recognizing common objects from unusual views – an account which convinced him that vision alone could deliver an internal description of the shape of a viewed object, even when the object was not recognized. The crucial problem, then, was to derive the actual shape of an object from the retinal images, and he felt that this could only be done in steps.

Starting with the pattern of intensity values at each point in the two-dimensional retinal image, the first step was to make explicit important features of the geometrical distribution of the intensity differences – lines, blobs, edges, terminations and discontinuities, groups, curves. By adding to this 'primal sketch', clues about distance from the viewer (from discrepancies between the two retinal images, visual discontinuities, shading, occlusion of one object by another) it would be possible to derive what he called a '2½-D-sketch'. This would make explicit the orientation and rough depth of visible surfaces, and contours of discontinuities in these quantities, *but all in a 'viewer-centred frame'*. Most of the time, though, we are not interested in the particular and unique view we happen to have of an object; we want to know what shape it is and where it is. The final step is therefore to transform the viewer-centred description of surfaces into a representation of the three-dimensional shape of the object, and its position in the scene, that is independent of the vantage point. That process necessarily involves inferences based on information not available from the scene itself. We see a ball as a ball, but it might be a hemisphere with the flat side turned away from us.

I have started by talking about David Marr because his influence on the way neuroscientists look at vision has been so great, but he was not the first to take a computational approach to the brain, and his approach to vision was atypical in tending to focus primarily on a theoretical analysis of the tasks facing the visual system rather than on simulating the brain's activity on a computer. In 1943, the physiologist W. S. McCulloch and the mathematician W. H. Pitts pointed out the resemblance of the logic circuits in digital computers to networks of real nerve cells, and the parallelism between the binary states of the units in a computer and the all-or-nothing behaviour of nerve cells. After the War this led to the development of computer models of learning and pattern recognition, driven partly by the huge commercial potential of possible practical applications and partly for the light it might throw on psychological mechanisms.

SERIAL V. PARALLEL PROCESSING

There are two different styles of computer. The vast majority of commercial computers conduct their operations one step at a time, but very fast – perhaps ten million steps per second. Because they work in this stepwise fashion they are called *serial* computers. Roughly speaking, they consist of: a *central processor* (i.e. the machinery for controlling everything that happens inside the computer); a large store full of pigeon holes each with its own address and each capable of holding a number; a set of interchangeable programs – the computer *software* – that tell the central processor what to do and in what order to do it; and a printer or screen which prints out or displays whatever the programme, acting through the central processor, orders it to. A tiny bit of the program (translated from 'machine code' into English) might read something like this: 'Read the number in pigeon hole so-and-so and subtract it from the number in pigeon hole such-and-such. If the answer is less than zero, go to instruction 123; if not, proceed to the next instruction on the program.' It is characteristic of computers of this kind that the central processor forms only a small part of the computer, and is physically separate from the storage sites. It is also characteristic that a very small amount of damage or an error in the program is apt to be catastrophic. In the early days of computing, when computers depended on unreliable thermionic valves rather than very reliable transistors, this tiresome characteristic limited the size of a computer, since if it had too many valves the chance of their all being in working order was small.

If we now compare the characteristics of a computer of this kind with the characteristics of the human brain, the differences are striking. In the first place, although our thoughts are sequential – hence William James's famous 'stream of thought' – we know that the brain does not take one step at a time. The optic nerve, for example, conducts a flood of impulses to the brain along about a million parallel paths, and a similar abundance of parallel paths is found wherever we look in the brain. One consequence of this almost ubiquitous parallelism is that the brain can manage with nerve cells that fire at relatively low rates – of the order of 100 per second. Another is the brain's ability to survive small amounts of damage without a disastrous loss of performance. In the jargon of the trade, the brain is said to *degrade gracefully*. Then, there is no obvious separation into processing and storage areas. Information is thought to be stored as modifications of synapses, but the nerve cells whose synapses are modified are not special 'storage' cells but the nerve cells engaged in the operations for which the information is being used. The terms *hardware* and *software* do not quite fit entities in the brain. In a conventional serial computer, hardware seems an appropriate term to refer to the actual machine, because the properties of the machine remain unchanged except for the contents of the store. In the brain, as we have seen, synapses may alter their properties, and with time there may also be changes in the detailed circuitry; so hardware is not quite the right term. In a conventional serial computer, software refers to instructions fed into the machine to modify its behaviour in some desired way or to 'run a program'. There is no obvious equivalent in the brain; and although it is sometimes said that the mind is to the brain as a computer's software is to its hardware, extracting the kernel of truth in that statement from the misleading shell encasing it requires some effort (see Chapter 22).

Given all these differences, it would not be surprising if the performance of the brain, thinking of it solely as a computer, differed markedly from the performance of conventional serial computers. And it does. Serial computers are much better, that is both faster and more reliable, at long or complex arithmetical problems and logical tasks such as playing chess – tasks that we find difficult. They are very much worse at tasks that we find almost effortless, such as recognizing visual or verbal patterns, or recognizing shape from shading, or deriving three-dimensional structures from two dimensional images. Of course, the brain differs from conventional computers in all sorts of ways, and you may wonder why the difference in aptitude for tasks such as playing chess or recognizing shape from shading should be linked to the differences I have listed rather than to any of the other differences – for example, that the brain and conventional computers are made of different kinds of stuff.

There are two reasons. In the first place, the tasks that we are particularly good at tend to be those that require consideration of many pieces of information, each potentially critical in determining the outcome. In a *parallel system* these 'constraints' can be considered simultaneously. If they had to be considered sequentially, the relatively slow rate of neural events would make the whole process impossibly slow. The second reason is that, particularly over the last two decades, work on model computing systems of a highly parallel kind shows that they have unexpected and interesting properties – interesting both because they can be used commercially* and because they have features resembling those we have noted as characteristic of the brain. Because such model systems consist of many interconnected units, they are known as 'neural networks', usually – invariably in their commercial context – without even the fig leaf of quotation marks.

Figure 15.2 is a diagram of one variety of neural network introduced by David

---

*Neural networks are particularly useful for making decisions based on experience, where there is no specific theory or model to apply. For example, credit-card companies know something of the personal details and a great deal about the shopping habits of their card-holders. With information of this kind a neural network can produce lists of names of people who live or work in a district in which it is proposed to open, say, a restaurant, and whose characteristics suggest that they would be likely to use the restaurant and would therefore be worth targeting.

output patterns

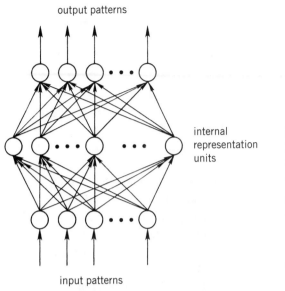

internal
representation
units

input patterns

FIGURE 15.2 A three-layered network. Conduction is only upwards, and each unit in a given layer is connected to every unit in the layer immediately above. (From D. E. Rumelhart, J. L. McLelland and the PDP Research Group.)

Rumelhart and James McClelland.[4] It is simple in conception yet can be 'trained' to carry out surprisingly sophisticated tasks. The network consists of three layers of interconnected units which resemble very simplified neurons. Note that the connections conduct in only one direction, so the flow of information is from the bottom or *input* layer, through the intermediate or *hidden* layer, to the top or *output* layer. Each unit is connected to every unit in the layer above. Activities in units in the input layer are determined by the environment. Activity in each unit in the hidden layer can be increased or decreased by activity in any unit in the input layer. Similarly, activity in each unit in the output layer is controlled by all the units in the hidden layer. The strength of each of the connections is assumed to be adjustable, so the influence of one unit on another in the layer above will depend both on the level of activity of the unit doing the influencing and the strength of the connection – just as the influence of one nerve cell on another with which its axon forms a synapse will depend both on the firing rate of the first cell and the properties of the synapse. Indeed, the strength of the connection is sometimes referred to as the 'synaptic strength'. Besides varying in magnitude, the connection strength can also vary in sign. (A negative connection strength implies that an increase in activity in the influencing cell will *decrease* activity in the cell being influenced.) Because the role of the connections is so central to systems of this kind, the whole approach is often known as *connectionist*.

It turns out that networks no more complicated than this, and containing only a few thousand units, have rather remarkable properties. I want to mention two examples, both impressive in what the network achieves, and both showing features in their behaviour intriguingly reminiscent of real nervous systems.

A CONNECTIONIST DEVICE FOR RECOGNIZING FAMILIAR FACES

Fine nets and stratagems to catch us in
—George Herbert

The first example is a network studied by Garrison Cottrell and his colleagues at the University of California at San Diego, which they trained to recognize familiar faces presented as black and white photographs.[5] Like the network introduced by Rumelhart & McClelland, their network had three layers, with information flowing from an input layer through a hidden layer to an output layer. The input layer consisted of a 64×64 array of units, each of which could discriminate over two hundred levels of brightness. The intermediate or hidden layer contained eighty units, each receiving connections from every one of the units in the input layer. The top layer contained only eight units, each receiving connections from every unit in the hidden layer.

The bottom layer can be thought of as a kind of retina: when it is presented with a photograph, the activity of each of its units will reflect the degree of lightness or darkness of the very small area of photograph monitored by that unit. The primary aim of the experiment was to see whether it was possible to arrange the connection strengths of the very large number of connections in the network in such a way that, when the network was presented with a photograph of one of a collection of familiar faces (each of specified sex), the activity of units in the top layer indicated which face it was, and whether it was male or female.

There are two questions we need to consider. Why should anyone think that any arrangement of connection strengths should enable the network to recognize familiar faces? And, even if such an arrangement exists, how can it be discovered?

To start with, suppose that all the connection strengths are set at the same value. The activity of the different units in the input layer will reflect the local variations in the photograph, but because *each* unit in the hidden layer receives a connection from *every* unit in the input layer, the units in the hidden layer will all be activated to the same degree. Almost all of the information in the photograph will have been lost. But now suppose instead that the connection strengths are varied, so that individual units in the hidden layer differ in the extent to which they respond to activation of different units in the input layer. You can think of a unit in the hidden layer as paying much more attention to some tiny areas of the photograph than to other tiny areas – and which areas are attended to will be different for the different hidden units. What is more, the relative activation of these different units will vary with the choice of photograph, and this variation can therefore be used to distinguish between the different photographs.

But the units in the hidden layer are inaccessible, so how can you know how much each is activated? The answer is to use the same trick and to adjust the strengths of the connections between the units in the hidden layer and those in the output layer so that different patterns of activation in the eighty units in the hidden layer are reflected as different patterns of activation in the eight units in the output layer. This kind of argument does not prove that adjustment of connection strengths will enable the network to perform the task expected of it, but it suggests that it might.

So much for the first question. What about the second? How do you know how to adjust the strengths of the vast number of connections – 328,320 of them – so that the network does what is required? The answer is that you don't know, but David Rumelhart, Geoffrey Hinton and R. J. Williams developed a remarkable technique – the 'back-propagation' learning algorithm – for training networks of this kind, proceeding by trial and error.[6]

For each input pattern (for each photograph of a face) you know what output pattern is desired; it will be a string of eight numbers (i.e. the levels of activation of the eight output cells) which encode the name and sex of the particular face in the photograph. Suppose you start by setting the values of all the connection strengths at random and not far from zero, and then look at the output. It is unlikely to be at all close to the desired output but never mind. Subtracting each number in the output string from the corresponding number in the desired string gives you eight numbers representing discrepancies between what you have and what you want to have. Squaring these numbers and taking the average gives you a single number that represents the discrepancy between what you have and what you want to have. (The point of squaring is that if you take the actual numbers discrepancies of opposite sign will tend to cancel out.) Call this single number the 'error'. What you have to do is to change the connection strengths to reduce this error as much as possible.

The procedure is, starting with the connections to the units in the output layer, to make small trial increases or decreases in the strength of each connection in turn, preserving those changes that reduce the error, and repeating the entire procedure for each photograph again and again until the error stops shrinking. This sounds a herculean task of insufferable tediousness, but it is precisely the sort of repetitive task that conventional computers are extremely good at. So what people working on 'neural nets' usually do – and there is a marvellous irony in this – is not to work on actual artificial 'neural nets' but to simulate their behaviour on conventional sequential digital computers. The conventional computer may take days or weeks finding the right pattern of connection strengths to make the simulated 'neural net' work, but it then works very effectively.

Note that the training technique is quite unbiological and no one supposes that brains learn in this fashion. The interest of networks trained in this way, whether the ultimate aim is to design a device or to understand how the brain works, lies, not in the wholly artificial training process, but in the *emergent* properties of the trained network – that is to say, properties that could not readily have been predicted from the properties of the individual units and their arrangement.

Cottrell's network was trained by being shown sixty-four photographs of eleven different faces, along with a few photographs that were not faces. When training was complete, the network was 100 per cent accurate in recognizing any of the 64 photographs of faces that had been used in the training. That is not as impressive as it sounds, because each of these photographs is unique in innumerable ways, and the network may merely have been using quite trivial features in the photograph to make the identification. A more stringent test was to ask the network to identify unfamiliar photographs of the familiar faces, and in this test the network was 98 per cent accurate. A still more stringent test was to show the network photographs of *unfamiliar faces* and to see whether it could decide, first, if the photographs were of faces, and then, if they were, whether they were of male or female faces. The network was 100 per cent accurate in the first of these tasks and 81 per cent accurate in the second. That a network of about 4000 units, trained on only eleven faces, can be 81 per cent correct in identifying the sex of unfamiliar faces is remarkable. It is not safe to equate the units in a 'neural network' with neurons, but 4000 neurons would occupy about a twentieth of a square millimetre of human cerebral cortex.

One characteristic of our own ability to recognize familiar faces is that we can do almost as well when the face is partly obscured. So it is interesting that Cottrell and his colleagues found that obscuring a fifth of the photograph of a familiar face by placing a horizontal bar across it did not much impair the performance of their network unless the bar obscured the forehead, when accuracy fell to 71 per cent. (It is thought to be the hairline that gives the forehead its importance in facial recognition.)

How does the network manage to be so effective? Although each unit in the hidden layer receives connections from every unit in the input layer, if the connection strengths were made very unequal it would be possible for particular units in the hidden layer to concentrate on crucial areas of the photograph. You might then expect many units in the hidden layer to concentrate on areas that provide information about such things as the length of the nose or the distance between the eyes, or the distance between the nose and upper lip, dimensions that seem to be useful in describing the differences between different faces. But that is not how the network works.

The basis for this negative statement is an examination, in the trained network, of the strengths of the connections that link units in the input layer with units in the hidden layer. One of the advantages of simulating the 'neural network' on a conventional computer is that the computer can not only carry out the tedious back-propagation process, but it can also be programmed to reveal the strength of all the connections, once training is completed. For each of the eighty units in the hidden layer, you can therefore discover the strengths of the connections it receives from every unit in the input layer. You can then display the pattern of the strengths of these connections by taking a 64×64 grid, representing the 64×64 units in the input layer, and making each grey square in the grid whiter or blacker according to the sign and magnitude of the connection strength. The face-like pictures obtained in this way for many hidden units showed that each hidden unit received useful information from the whole of the input layer. There was no evidence of concentration on particular parts of the picture.

A DEVICE FOR DEDUCING SHAPE FROM SHADING

Turns them to shapes, and gives to airy nothing
A local habitation and a name.
—Shakespeare, *A Midsummer Night's Dream*

My second example of the power of a three-layered 'neural network' is concerned with the deduction of shape from shading. Sidney Lehky & Terrence Sejnowski were interested in the way the brain is able to deduce information about shape from the continuous gradations of light and dark found on the shaded surface of three-dimensional objects, and this without any independent knowledge of the direction of the light source.[7] Using the back-propagation learning algorithm, they showed that a three-layered network with less than 200 units could be trained to estimate the maximum and minimum curvature at points on the smooth matt surface of simple curved solids – they were shaped roughly like tea cosies – illuminated by light from one unspecified direction. The experiment also produced an unexpected and important finding. When, in the trained network, they looked at the connections linking the input units to the hidden units, it became clear that some of the hidden units were receiving *excitation* from input units forming a band across the array, and *inhibition* from units flanking that band. For other hidden units the pattern was reversed. These are, of course, precisely the patterns found in the 'simple cells' in the primary visual cortex, cells that have always been thought of as 'bar' or 'edge' detectors – see Figure 14.1. This does not prove that the simple cells in the primary visual cortex are concerned with the determination of shape from shading, but it does show that detecting contours is not their only possible function. More generally, Lehky and Sejnowski point out that it is unsafe to deduce the function of a cell simply from the character of its receptive field; it is also important to consider the pattern of connections it makes with subsequent stages. That is, of course, more easily said than done.

GENERAL FEATURES OF NEURAL NETWORKS

A characteristic of neural networks, illustrated by the examples we have just looked at, is that each entity is represented by a pattern of activity distributed over many computing units, and each computing unit is involved in representing many different entities. It may seem surprising that a new item can be stored in a neural network without disturbing the many items already stored. What makes it possible is what Hinton, McLelland and Rumelhart call a 'conspiracy effect'.[8] If a very large number of connection strengths are varied slightly, and all the variations are in the direction that helps the new pattern, the total help for the new pattern will be the sum of the many small effects. For unrelated patterns, the modifications may help or hinder, so they will tend to cancel out.

The 'distributed representation' of entities in neural networks turns out to be an efficient way of using computing units, and it has other advantages that seem to be relevant to computation in the brain. The ability of neural networks to complete patterns of activity that are incomplete means that any part or parts of the pattern can help to elicit the complete pattern. An ability of this kind may well provide a general explanation of our striking skill at recalling scenes or events from very partial (and even partly incorrect) descriptions of them.

At first sight, the idea of distributed representation may appear to conflict with the overwhelming evidence for localization of function in the brain. In fact there is no conflict because the two patterns are relevant at different scales. There are certainly separate modules responsible for different kinds of activity, but the distributed representations occur within those modules. Hinton and his colleagues put it epigrammatically: 'the representations are local at a global scale but global at a local scale'.

The sort of simple *feedforward* networks we have been discussing are, of course, peculiar in that there were no lateral connections between units in any layer, and no recurrent connections – that is, connections transferring information from a higher to a lower layer. Both kinds of connection are common in real nervous systems, and they have been studied in neural networks.

The usefulness of lateral connections is that if they are excitatory they provide a very simple method for arranging mutual excitation between units representing entities that are mutually consistent; if they are inhibitory they provide an equally simple method for arranging mutual inhibition between units representing entities that are mutually incompatible. Marr and Poggio used both kinds of connection in a network that was able to detect depth in random dot stereograms[9] – pairs of areas of random dots that, viewed with a stereoscope give the appearance of depth although there is no obvious structure to give a clue to disparity.*

*Invented by the Hungarian Bela Julesz in 1960. He took two pages covered with an identical pattern of random dots and displaced all the dots in a small square area on one of the pages slightly sideways. When the two pages were viewed with a stereoscope the square appeared to lie in a plane in front of or behind the rest of the page.

The introduction of recurrent connections into a network creates two important possibilities. If a downstream unit can feed information to a unit further upstream, there is a possibility of present activity being influenced by activity in the very recent past. This is, in effect, very short term memory, and it can be useful in controlling the flow of information into the more downstream parts of the network. It is different from the memory that is embedded in the pattern of connection strengths in the network because that pattern has no record of the individual events that produced it or of their timing. Recurrent connections and short term memory are almost certainly involved in attention. The second possibility is of cyclical activity, which can be stable or oscillatory or chaotic. Stable or oscillatory cyclical activity can be used to control activity in particular neurones or to anticipate it. As networks become more complicated, of course, the terms upstream and downstream become meaningless, but that does not lessen the efficacy of recurrent connections. The very extensive reciprocal connections between the different visual areas in the cerebral cortex are an indication of their importance in the brain.

HOW DO NETWORKS LEARN?

It is clear that neural networks can do rather sophisticated things if they have the right patterns of connection strengths, but it is far from clear how such patterns are to be achieved. In the network used by Marr and Poggio for handling random-dot stereograms, part of the pattern was incorporated into the design, and this was sufficient for the network to do what was required.[10] In the networks for facial recognition and for deriving shape from shading that we discussed earlier the process of adjusting the connection strengths so that the network could do something useful was not only extremely tedious but also required a teacher. If regions of the brain work in anything like this fashion there must be some way in which connection strengths (synaptic strengths) are adjusted by the activity of the neural network itself, which implies that the information necessary to adjust the strength of each connection must be available locally. How might this be done?

There is, as yet, no general answer to this question, but I want to mention two suggestions. The first was made by the Canadian psychologist Donald Hebb in 1949. The idea that synapses might become more effective through use had been around for a long time, but Hebb realized that, for learning, something more specific was required. He suggested that increases in synaptic strength occurred if, and only if, the presynaptic and postsynaptic cells fired simultaneously. McLelland, Rumelhart and Hinton have described a simple network, based on this principle, that can learn to associate two patterns repeatedly presented together; and they have suggested that a large scale network of this kind presented with suitable inputs could learn to associate, say, the appearance of a rose with the smell of a rose.[11] So far, direct evidence that learning in the mammalian brain involves synaptic modification of the kind envisaged by Hebb has been found only in the hippocampus and the cerebellum.

The second suggestion is an ingenious algorithm – the 'wake-sleep algorithm for unsupervised neural networks' – invented by Hinton and his colleagues.[12] Mixing homage and whimsy, its designers also call it a Helmholtz machine because of Helmholtz's emphasis on the role of inference in the interpretation of sensory input. The machine consists of a multilayered neural network, like those we have already met except that there are also recurrent connections, so that each cell receives an input from every cell in the layer above as well as every cell in the layer below; information can flow downwards as well as upwards. The bottom up connections ('recognition' connections) convert the input into representations in successive hidden layers. The top-down connections (referred to as 'generative' connections) reconstruct the representation in one layer from the representation in the layer above. The system works by alternating between what they call 'wake' and 'sleep' phases. In the 'wake' phase, neurons are driven by information coming upwards through the recognition connections, and the synaptic strengths of the *generative* connections are adjusted to increase the probability that they would reconstruct the correct activity pattern in the layer below. In the 'sleep' phase, neurons are driven by the generative connections, and the synaptic strengths of the *recognition* connections are adjusted to increase the probability that they would produce the correct activity pattern in the layer above. In this strange reciprocating fashion – the two systems pulling themselves up by each other's bootlaces – the network is able to learn to recognize patterns. There is no supervision and the adjustment of the strength of each synapse depends only on information available locally.

Whether any part of the nervous system works like this is, of course, not known. Such a mechanism could, though, provide a possible explanation of the richness of reciprocal connections.

*   *   *

We have now looked at four different ways of trying to understand vision: studying illusions by comparing what we seem to see with what we are looking at; seeing how damage to different parts of the neural machinery by injury or disease affects our visual perceptions; exploring that machinery directly so far as it is accessible to us; and examining the behaviour of model systems consisting of units something like simplified neurons interconnected in ways something like the ways in which real neurons are interconnected. The picture that emerges is coherent, but it is far from complete. Some important topics – the synaptic changes associated with memory, the binding problem, the curious phenomena of 'neglect' in certain brain-damaged patients, and the question of visual attention, will be considered later. Perhaps the most worrying gap is our almost complete ignorance of the neural correlates of consciousness – the events in our nervous systems that we assume must occur if we are to have particular conscious experiences, and which we are tempted to believe might somehow account for those experiences. In their experiments on anaesthetized monkeys, Semir Zeki and his colleagues succeeded in finding cortical neurons whose activity correlated with the *perceived* colour of a small area in a colour Mondrian, rather than the mixture of wavelengths reflected from that small area (see p. 241). Activity of such neurons is probably a necessary condition of monkeys perceiving colours, but unless *anaesthetized* monkeys can perceive colours it cannot be a sufficient condition. It is, perhaps, symptomatic of our problems that in these experiments the active neurons were those of a monkey but the perception was by the investigators.

IV TALKING ABOUT TALKING

Words are the tokens current and accepted for conceits, as moneys are for values
—Francis Bacon (London, 1605), *Advancement of Learning*

the race is not to the swift but to the verbal
—Steven Pinker (Princeton, 1994), *The Language Instinct*

Once we have learned to talk, the ease and precision with which we can influence thoughts in each other's minds by talking is as impressive as the ease and precision with which we can know 'what is where by looking'. And it is our monopoly of speech, with its more recent by-products, reading and writing, that almost certainly accounts for our success as a species – a success which may conceivably now threaten our survival. In this section, I want to look at language from three viewpoints.

Just as the study of disordered vision in patients with normal eyes revealed much about the ways the brain handles visual information, so the study of abnormal speech in patients with normal hearing and normal vocal tracts reveals much about the way the brain handles language. I want to discuss the various kinds of aphasia, and also what has gone wrong in patients who have lost the ability to read (alexia) or the ability to write (agraphia).

Secondly, I want to say something about the structure of language, and the acquisition of the ability to talk by very young children – topics that have been revolutionized in the last thirty years by the work of Noam Chomsky.

And thirdly, I want to look briefly at the evolution of language, and to examine a little more closely what it is that we humans have a monopoly of.

16   In the Steps of the 'Diagram-Makers'

The picture we have of the way language is organized in the human brain was broadly drawn by a few outstandingly able physicians studying a small number of individual patients with severe speech disorders in the second half of the 19th century. For a mixture of reasons – not all bad – that picture was largely discarded by the more holistically inclined physicians of the first half of the 20th century, who dismissed their predecessors as 'diagram makers', but failed to produce any theories as useful as the diagrams they discarded. The second half of that century saw, first, the resurrection of something like the earlier picture, and then its refinement and elaboration through the work of neurologists, of psychologists, and of linguists. Familiarity with the flow diagrams describing patterns of information-processing in computers has made the diagram-makers' approach seem no longer mechanical and simplistic but the natural way to think about problems which, whatever else they are, are problems in information processing. To adapt Sir William Harcourt's famous phrase about socialists: 'We are all diagram-makers now.'

One of the first of the diagram makers was Carl Wernicke, who, as we saw earlier (p. 172–3), described a kind of aphasia distinct from that described by Broca both in the site of the lesion and the nature of the disorder. In Broca's aphasia, the lesion was in the left frontal lobe, just in front of the motor area concerned with the mouth and tongue (Figure 16.1), and the predominant feature was difficulty in finding and speaking words (giving a sparse telegram-like style), yet near normal comprehension. Broca's aphasia was therefore attributed to the malfunctioning of the 'centre for motor images [i.e. representations] of words'. In contrast, in patients with Wernicke's aphasia the lesion was in the left temporal lobe adjacent to the area concerned with hearing, and the main features were loss of comprehension, and a fluent speech sounding superficially normal but lacking in content and with many wrong words, malformed words and invented words. Wernicke's aphasia was therefore attributed to the malfunctioning of the 'centre for auditory images of words'. In his book *The Shattered Mind*,[1] Howard Gardner gives modern American examples of the speech in each of these kinds of aphasia in English speaking patients; here are two brief extracts.

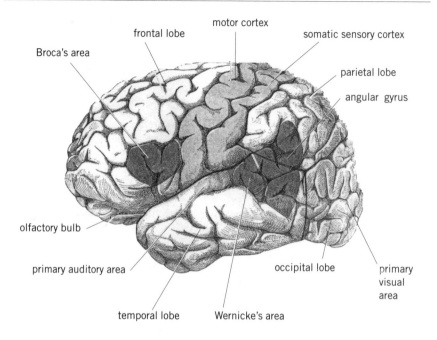

FIGURE 16.1 The left cerebral cortex showing particularly Broca's area, Wernicke's area and the angular gyrus. (From N. Geschwind.)

Broca's aphasia following a stroke in a thirty-nine-year-old former Coast Guard radio operator, DF:

HG: Were you in the Coast Guard?

DF: *No, er, yes, yes, ... ship ... Massachu ... chusetts ... Coastguard ... years. (He raised his hands twice, indicating the number 'nineteen'.)*

HG: Oh, you were in the Coast Guard for nineteen years.

DF: *Oh ... boy ... right ... right.*

HG: Why are you in the hospital, Mr F—?

DF: *(Pointing to his paralyzed arm) Arm no good, (then to his mouth) Speech ... can't say ... talk, you see.*

HG: What happened to make you lose you speech?

DF: *Head, fall, Jesus Christ, me no good, str, str, ... oh Jesus ... stroke.*

HG: I see. Could you tell me, Mr F—, what you've been doing in the hospital?

DF: *Yes, sure. Me go er, uh, P.T. nine o'cot, speech ... two times ... read ... wr ... ripe, er, rike, er, write ... practice ... get-ting better.*

Wernicke's aphasia in a seventy-two-year-old retired butcher, P.G.

HG: What brings you to the hospital?

PG: *Boy, I'm sweating, I'm awful nervous, you know, once in a while I get caught up, I can't mention the tarripoi, a month ago, quite a little, I've done a lot well, I impose a lot, while, on the other hand, you know what I mean, I have to run around, look it over, trebbin and all that sort of stuff.*

(HG attempted several times to break in, but was unable to do so. Eventually he rested his hand on the patient's shoulder.)

HG: Thank you Mr G—. I want to ask you a few—

PG: *Oh sure, go ahead, any old thing you want. If I could I would. Oh I'm taking the word the wrong way to say, all of the barbers here whenever they stop you it's going around and around, if you know what I mean, that is tying and tying for repucer, repuceration, well we were trying the best that we could while another time it was the over there the same thing . . .*

Wernicke assumed (correctly) that the two centres were linked by a nerve tract, and he pointed out that damage to this tract would disconnect the centres and give rise to a third kind of aphasia. Because the 'centre for auditory images of words' was intact, comprehension would be normal; but because that centre could no longer feed information to the 'centre for motor images of words' the speech produced by the patient would be similar to that in Wernicke's aphasia. Such a disconnection syndrome, now always referred to as *conduction aphasia*, was described by L. Lichtheim,[2] in Berne, eleven years later, and included a striking feature which, curiously, Wernicke had failed to predict. Because the connection between the two centres is interrupted, the patient finds great difficulty when asked to repeat words dictated to him.

Interest in difficulties with repetition led Lichtheim to discover several further kinds of aphasia and to elaborate Wernicke's scheme, introducing 'concept centres' that were supposed to be concerned with the meaning of language. The details of Lichtheim's schemes and the interpretations based on them have not stood the test of time, but the general notion that the various aphasias can be explained *either* in terms of damage to local areas with specific functions, *or* in terms of disconnection of those areas by the interruption of the pathways between them, remains valid.

SUDDEN LOSS OF THE ABILITY TO READ

We learn to read and to write by building on our ability to speak; so it is not surprising that all the aphasias we have been discussing tend to be associated with difficulties in reading and writing. But loss of the ability to read (or to read and to write) sometimes occurs with little or no loss of speech. In 1891 and 1892, Jules Dejerine – working at the Bicêtre, where thirty years earlier Broca's famous patient had died – described two cases which have become classics in the medical literature.[3]

The first was a sixty-three-year-old navvy, Monsieur S., who found one morning that he was quite unable to read his newspaper. He could recognize neither letters nor words, though he could still recognize his own name, and, surprisingly, could also read figures. He was unable to write, either spontaneously or to dictation. His comprehension of spoken language was normal, though in speaking he often used wrong or malformed words. He had no difficulty in recognizing objects or people. Nine months after he lost the ability to read and to write, Monsieur S. died, and a post-mortem examination of his brain showed a single lesion in the left hemisphere, in the region of the *angular gyrus* (Figure 16.1) – the gyrus that lies just behind Wernicke's area, and links the parts of the cortex concerned with hearing and with vision. Damage in this area, Dejerine concluded, prevents the conversion of representations of the appearance of words and letters into representations of their sounds, and also the conversion of representations of their sounds into representations of their appearance. The first, he assumed, results in the loss of the ability to read – *alexia* – and the second in the loss of the ability to write – *agraphia*.

Dejerine's second case was a case of alexia without agraphia, and it showed, among other things, the importance of the transfer of visual information between the two cerebral hemispheres. The patient, Monsieur C., was a highly intelligent and musical sixty-eight-year-old Parisian fabric designer and merchant. In October 1887 he had a number of transient attacks of numbness and slight weakness in the right arm and leg, with very slight difficulty in speech. These attacks continued for several days, but were not severe enough to prevent him from taking long walks, during which he would read the shop signs and posters. On the sixth day he was shocked to find that he was unable to read a single word; yet his speech, his writing, and his recognition of objects and people seemed entirely normal. His vision had previously been particularly good – and needed to be, since designing fabrics involved counting threads and working out patterns on paper marked in millimetre squares. Thinking that there must be something wrong with his eyes, and that his problem might be solved by getting appropriate glasses, he consulted the distinguished ophthalmologist E. Landolt, who referred him to Dejerine.

Both Dejerine and Landolt noted that, as well as suffering from alexia, Monsieur C. was blind in most of the *right* half of the visual field of each eye, and had lost any sense of colour in the remainder of that half of the field. (This clearly points to severe damage to the left occipital lobe, including the area critical for colour sense that Verrey was to discover in Mme R. in Neuchâtel in the following year – p. 201.) Visual acuity (with glasses) in the *left* half of the visual field was normal in both eyes.

Though he was wholly unable to recognize words or letters, Monsieur C. retained some ability to read figures. He could also recognize *Le Matin* – his regular newspaper – by its heading, but he could not recognize other newspapers. Despite his almost total alexia, he could write, either spontaneously or to dictation, almost as well as before the attack (Figure 16.2a). But though he could write, he was quite unable to read what he had written, unless the writing was so large that he could follow the shapes of the individual letters with his finger. He also had great difficulty in copying either print or cursive writing, needing to have what he was copying constantly in front of him, and proceeding slowly and laboriously to copy the forms of the letters, almost as if they were unfamiliar designs (Figure 16.2b); he could not reproduce print as cursive writing or vice versa. His ability to read music was completely lost, though he could still sing operatic arias accurately, and with his wife's help could learn new ones. He could do arithmetic adequately, if slightly untidily, and on behalf of a friend he negotiated an annuity so skilfully that the insurance office assumed he was a professional agent and offered him 700 francs commission. His ability to recognize the heading of *Le Matin* suggests that he could recognize patterns without difficulty, and this fits with his continued ability to play cards.

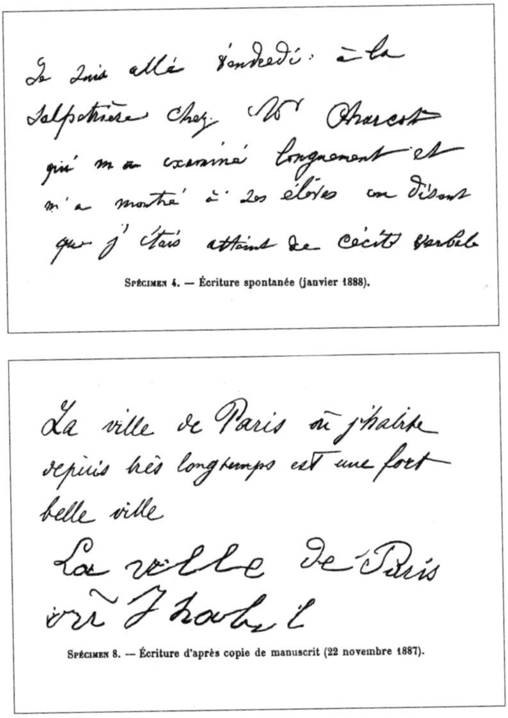

Spécimen 4. — Écriture spontanée (janvier 1888).

Spécimen 8. — Écriture d'après copie de manuscrit (22 novembre 1887).

FIGURE 16.2 The handwriting of Dejerine's patient, Monsieur C., who had lost the ability to read without losing the ability to write: (a) shows his spontaneous writing, which was almost the same as before his illness; (b) shows the great difficulty he had in copying script put in front of him. (From J. Dejerine.)

For more than four years, Monsieur C.'s condition remained much the same. Occasionally he became agitated and despairing, once even thinking of throwing himself off the column in the Place de Vendôme, but generally, with the help of his wife, he led a more or less normal existence. Then, in January 1892, during an evening card party, he had a slight stroke. There was no loss of consciousness, and only transient weakness on the right side of the body, but his speech became incomprehensible, with wrong words and non-words. By the following day the weakness had gone but his speech was worse, and he was reduced to using signs because he found that attempts to write produced only meaningless scrawls. His intelligence and ability to understand speech was unaffected, and continued unaffected until he passed into a coma and died, ten days later.

At autopsy, his brain showed two distinct kind of lesion: old, yellow lesions that presumably accounted for the features that had persisted for over four years, and fresh lesions that presumably accounted for the features that followed the final attack. The old lesions had totally destroyed the visual cortex of the left hemisphere, and they had also damaged the part of the white matter in the left hemisphere that contains fibres from the hindermost part of the corpus callosum – the part responsible for exchanging visual information between the two hemispheres. During the four years before the final attack, because the left visual cortex had been destroyed, letters and words could have been seen only through signals reaching the right visual cortex. For these letters and words to have been understood, information from the right visual cortex would have had to be transferred to speech areas in the left hemisphere, but that was made impossible by the damage to the white matter containing fibres from the posterior part of the corpus callosum. This, Dejerine assumed, explained the alexia. In the final attack, the new lesions destroyed the angular gyrus and adjoining parts of both the temporal and parietal lobes. This explained the subsequent agraphia and the incomprehensible speech.

If Dejerine's hypothesis to account for the alexia was correct, a similar interruption of callosal fibres carrying visual information from the right hemisphere to the left *in a patient in whom the visual cortex was intact on both sides* ought to cause alexia in the left half of the visual field but not on the right. This is precisely the picture that John Trescher and Frank Ford, at Johns Hopkins Hospital, found nearly half a century later, in a young woman whose corpus callosum had been partly cut through so that a small tumour could be removed.[4]

THE ECLIPSE OF THE DIAGRAM-MAKERS

Despite the major advances made by Wernicke, Lichtheim, Dejerine, and others following the same approach, their conclusions were subjected to increasing criticism from the early years of the 20th century.[5] The criticisms came from several directions.

First, the claims for precise localization of different functions were questioned, and the anatomical evidence was sometimes found to be shaky – though not always as shaky as the critics claimed. In 1906, Pierre Marie, the most famous of Charcot's pupils, and an inveterate opponent of Dejerine,* claimed that the loss of speech in Broca's famous first patient, Leborgne, had nothing to do with the lesion in the left frontal lobe, but was the result of the extension of that lesion backwards so that it affected Wernicke's area.[6] Marie supported this claim by a drawing of the left side of Leborgne's brain, which was still uncut in its glass jar in the Dupuytren Museum. For nearly seventy years Marie's view was accepted, but in 1979 the brain – which since the collapse of the walls of the Dupuytren Museum during the Second World War had languished in the cellars of the Paris Medical School – was removed from its preserving alcohol and CAT-scanned.[7] The result was clear; it showed that Broca was right and Marie wrong.

A weakness of quite a different kind was that the interpretation of Broca's aphasia as simply the result of the loss of 'motor images', and of Wernicke's aphasia as simply the result of the loss of 'auditory images', failed to account for some of the characteristic features of each. It failed to account, for example, for one of the most conspicuous features of Broca's aphasia, the almost complete absence of any normal syntax in the utterances. In a typical patient with Broca's aphasia, sense is expressed largely by a succession of nouns; verbs are few, and exist only in very simple forms with no indication of tense or person or number; there are few adjectives, and very few conjunctions or prepositions or articles or pronouns – what linguists call 'function words'. And – although this was realized only in the 1970s – this failure to cope with syntax does not only affect the patient's speech. It also affects comprehension.[8]

A third criticism was that, although the diagram-makers subdivided speech disorders into categories that were nicely explained by the disruption of specific centres or of specific pathways in their models, the majority of actual cases did not fit into these categories. This was awkward but could be explained. It was only if the disease process disrupted a single centre or a single critical pathway that the observed deficit would slot neatly into a category. Given the messy nature of the disease processes mostly involved – trauma, haemorrhage, blood clots, tumours, abscesses – such 'pure cases' would be uncommon. This was a reasonable defence, but it did raise questions about the choice of cases on which models were to be based. It was difficult to be confident about models based on the characteristics of infrequent cases.

Finally, the whole 'localizationist' approach – the preoccupation with

---

*He once accused Dejerine of doing his science as some others played roulette, hedging his bets by gambling on two outcomes. When challenged to a duel, he published a letter saying that neither the personal honour nor the scientific integrity of Dejerine were at issue. (Lhermitte, F. & Signoret, J.-L. (1982) *Revue Neurologique*, 138, 893–919.)

pinpointing damaged centres or pathways, and deducing the consequences of such damage – became unfashionable. In the years just before the First World War, and between the Wars, first Gestalt theory (see p. 198), and later Lashley's mistaken view that the association cortex acted as a single unit with all parts working in the same way (see p. 186), seemed to open broader horizons. For the whole of that period, the tendency was to emphasize the similarities between different disorders of language and to try to explain the different disorders by a common cause. Language disorders were held to represent a failure of 'intellectual function', though in Broca's aphasia a motor deficit might be superimposed. If Lashley was right, the attempt to correlate particular kinds of disruption of language with damage to particular areas of the cortex was misguided. The prevailing view was what we should now call 'holistic' – a term coined by General Smuts about that time to refer to the tendency in nature for elements or parts to act as a whole.[9] (Since then the term has shed some of Smuts mysticism – the 'holistic universe' – but retains his confusing spelling.)

EMERGING FROM ECLIPSE

By the middle of the century, the Gestalt school had withered and Lashley's view of the cortex was recognized as mistaken. Neurologists, including some who often expressed anti-localizationist views, produced clear accounts of aphasic syndromes associated with specific anatomical lesions.[10] 'Deficits' began to be assessed more carefully, by the use of standardized tests that were both objective and quantifiable. In the 1960s the work of Norman Geschwind and his colleagues in Boston, and in particular two massive articles on 'Disconnexion Syndromes' by Geschwind in the journal *Brain*, swung the pendulum back to the localiza-tionist view.[11] It is worth looking a little more closely at two of these disconnex-ion syndromes, first because of their bearing on the localisationist v. holistic controversy, at any rate in connection with speech, and secondly because they show that curious – and sometimes even apparently inconsistent – collections of symptoms can be explained by relatively simple models.

In 1962, Geschwind and Edith Kaplan published a paper about a forty-one-year-old policeman, PJK, who had had a tumour removed from deep in his left frontal lobe.[12] Some months after the operation he still had some motor weakness on the right side but he could write fairly well with his right hand. There had never been any motor weakness on the left side, yet he had great difficulty in writing with his left hand. Since he was right-handed, clumsiness in writing with his left hand was to be expected, but his difficulty was not just clumsiness. He surprised himself by making strange mistakes – writing 'yonti' for 'yesterday' for example – and even when he typed with his left hand the mistakes continued. There were other linguistic problems too. When common objects were put in his left hand without his being able to see them, he could not say what they were. He identified a ring as an eraser, a watch as a balloon, a padlock as a book of matches. Failure to recognize objects by manipulation is a well-recognized condition, but closer examination showed that a failure of recognition was not the cause of his difficulty. In the first place, he examined objects as if he were recognizing their salient features. He pushed his finger into the hole in a thimble, ran his thumb over the teeth of a comb, rubbed the bristles of a toothbrush. And the salient features must indeed have been recognized, for if the object was taken away (without being seen) PJK was able to draw it with his left hand, and would point to it when he saw it. Even more remarkably, when he was asked to demonstrate the uses of unseen objects placed in his left hand, he would manipulate the object correctly, but usually give an incorrect verbal account. Given a hammer, he made hammering movements but said 'I would use this to comb my hair.' Given a key, he pretended to insert it in a lock and turn it, saying that he was 'erasing a blackboard with a chalk eraser'. Given a pair of scissors, he held them correctly and made cutting movements but said that he would use them 'to light a cigarette with'. Yet PJK had no difficulty in speaking. So, as Geschwind put it: If he could speak normally and knew what he was holding in his left hand, why couldn't he say what it was?

And there was a parallel puzzle. When PJK was asked to do things with his right hand he had no difficulty. He could draw a square, wave goodbye, and pretend to brush his teeth or comb his hair. There was clearly no problem with comprehension. Yet, although there was no weakness on the left side, when he was asked to do similar things *with his left hand*, he made many errors. Asked to draw a square he drew a circle; asked to show how he would brush his teeth he would go through the motions of lathering his face or combing his hair. So, if there was good comprehension and no weakness in his left hand, why couldn't he do what was asked?

The answer to both these puzzles, Geschwind and Kaplan suggested, was that damage to the corpus callosum led to a disconnexion between parts of the brain which needed to act in consort for the appropriate response to occur. PJK could not say what he held in his left hand, because the part of the brain that recognized what he held was disconnected from the part needed for speaking. (The former would be in the right hemisphere; the latter in the left.) He could not carry out commands with his left hand because the part of the brain needed for comprehending the commands was disconnected from the part of the brain needed to carry them out. (The former would be in the left hemisphere; the latter in the right.) Fifteen months after the operation to remove his tumour, PJK died, and the findings at autopsy strongly supported this hypothesis.

The second disconnection syndrome of special interest here is sometimes known as 'isolation of the speech centre'. A particularly remarkable and tragic case was described by Norman Geschwind, Fred Quadfasel and José Segarra in 1968.[13] The patient, in the care of Quadfasel, was a twenty-one-year-old woman who had been found in her kitchen unconscious and not breathing, with the unlit gas jet of the water heater turned on. When she regained consciousness, twenty-nine days later, she was almost completely paralysed in all four limbs, incontinent and confused; and for nearly ten years she remained in that state. She could recognize people and indicate satisfaction or dissatisfaction by repetitive movements of the hands or by smacking the lips. She never spoke spontaneously except for the phrases 'Hi Daddy', 'So can Daddy', 'Mother' and 'Dirty Bastard', and she did not seem to understand what was said to her. When questions or phrases were put to her, she repeated the questions or phrases in a normal voice and with good articulation. This *echolalia* – as the neurologists call it – was very marked, but occasionally, when the phrase was a stereotyped one, instead of repeating it she would complete it. Thus 'Ask me no questions' might elicit 'Tell me no lies'. Told 'Close you eyes', she might say 'Go to sleep'. Sometimes her reply would be a conventional phrase triggered by a word used by the examiner. When asked 'Is this a rose?' she might reply 'Roses are red, violets are blue, sugar is sweet, and so are you'.

The combination of lack of spontaneous speech, lack of comprehension, echolalia, and a tendency to complete stereotyped phrases, had been described before,[14] but Quadfasel's patient showed a new and striking feature. Very early in her illness it was noticed that she tended to sing along with songs coming over the radio, or to recite prayers along with the priest during a religious broadcast. If the radio was turned off during a familiar song, she would continue to sing correctly both words and music for a few lines – or longer if the examiner hummed the tune. Remarkably, she could even learn new songs, both tunes and words.

Over nine years of careful observation she was never heard to make either a statement or a request; and, except in the peculiar responses described above, she never gave any indication that she understood what was said to her. It was as if the machinery for hearing speech, for producing speech and for remembering speech was still working, but working in isolation. Nearly ten years after the initial disaster, Quadfasel's patient died. A detailed study of her brain showed that Broca's area, Wernicke's area, and the pathway connecting them, had survived the carbon monoxide poisoning, and so had the auditory pathways, the parts of the motor cortex concerned with speech, and most of the hippocampal regions on both sides – regions which are known to be necessary for forming memories. But these areas were surrounded by areas of destruction, both of cortex and the underlying white matter, that could well have disconnected the speech areas from the rest of the cortex, while leaving a connection with the hippocampal regions. The symptoms and the pathological lesions were therefore consistent.

COGNITIVE NEUROPSYCHOLOGY

Geschwind was so successful in restoring respectability to the classical localizationist approach that modern textbooks often refer to the *Wernicke–*

Geschwind model though Wernicke died more than twenty years before Geschwind was born. Despite the usefulness of that model, work over the last thirty years, particularly by the 'cognitive neuropsychologists', has shown that it is neither adequate nor wholly accurate.

'Cognitive' simply means 'concerning knowledge'. What cognitive neuro-psychologists do is to use the patterns of behaviour shown by subjects with brain damage as a basis for making inferences about the machinery underlying cognitive processes in normal subjects. Such work has, of course been going on for a century and a half, but about thirty years ago those engaged in it – like the Molière character who discovered that he had been speaking prose for more than forty years without knowing it – began to think of themselves as cognitive neuro-psychologists. Unlike most of their predecessors, in their clinical work they tend to use this approach exclusively; they work with a much finer brush; and they are ready, if necessary, to draw inferences about the nature of normal cognitive processes even where they are unable to tie their hypotheses to actual neural structures or to computationally explicit models. This third characteristic has led to their being accused of sitting on a one-legged stool,[15] but in fact they are very ready to use support from other legs when it is available; and even when it is not, their detailed observations have often proved extremely useful in suggesting or constraining theories of normal cognitive function.[16]

Here we are concerned only with language. The most striking contribution of the cognitive neuropsychologists has been to show that the machinery handling language often behaves as if it consists of interacting modules – modules on a scale much finer than anyone had envisaged.[17]

Take patients who have difficulty in understanding words. Each of us must have a store of representations of the meanings of words – a *lexicon* – and difficulties could arise either from interference with that lexicon or with the access to it. And it would not be altogether surprising if it were easier to lose the meaning of abstract words than of concrete words, or of infrequent words than of common words. But some patients are much better at understanding abstract words; Elizabeth Warrington described a patient, a former senior civil servant, who failed to understand *hay, needle, acorn* or *geese*, but had no difficulty with *supplication, arbiter, hint* or *vocation*. And, more generally, it turns out that the words that patients cannot understand are often restricted to quite narrow categories – colours, proper names, body parts, foods, living things, objects, names of actions (standing, sitting, walking).[18] A patient studied by Elizabeth Warrington and Tim Shallice failed to understand *camel, wasp* or *buttercup*, but produced excellent definitions of *torch, thermometer* and *helicopter*.

Patients who have difficulty in finding words when they are speaking show a similar picture. Categories selectively lost include colours, proper names, shapes, letters, numbers, objects and actions.[19] J. Hart, R. S. Berndt and A. Caramazza described one patient who could not remember the names of common fruits and vegetables.[20] He could not name a peach or an orange but had no difficulty with an abacus or a sphinx. Whether an object can be named or not sometimes depends on how it is presented. François Lhermitte and Marie-France Beauvois described a patient who could not name objects shown to him, but could name them if he was allowed to handle them.[21] This, of course, is reminiscent of cases of visual agnosia, but that could not have been the explanation because he could mime the uses of the objects he was shown; he simply couldn't think of their names. Patients have even been described who can retrieve the names of objects shown to them only if they can write the names down: 'I can't say it but I can stick it down.'[22]

Significantly, difficulty in comprehending words in a particular category is not necessarily associated with difficulty in retrieving words in the same category. We shall come back to this point.

THE ALEXIAS AND THE AGRAPHIAS

> there are more problems to acquired reading disorders than are dreamt of in our current psychology.
> — John Marshall & Freda Newcombe, 1973

Some of the most fascinating observations of the cognitive neuropsychologists have been made on patients in whom lesions of the brain have caused disorders in reading or writing. In the UK, disorders of reading tend to be referred to as *dyslexias* – from the Greek for 'bad speech' – rather than *alexias* – from the Greek for 'without speech' – on the grounds that the patient is usually not wholly without speech. Although this is reasonable, I shall follow the American custom and use *alexia*, so as to avoid confusion between the alexias that are the result of brain damage in adults, which are what I want to talk about, and the better known (though less well understood) *developmental dyslexia* that distresses so many children and their parents.

When we first learn to read a language such as English, with its idiosyncratic spelling, we rather obviously use two different methods. To start with, using very simple 'print-to-sound' rules, we convert individual letters or very small groups of letters into sounds which are then combined to form a word. As we progress, we learn more sophisticated rules, but the process is essentially the same: we are following the *phonological* route. Very soon, though, we also build up a lexicon of familiar words whose spelling and pronunciation are stored in our brain, and which we can read at a glance, without any 'sounding out' or guidance from rules. In reading these words we use what infant-school teachers call 'look-and-say' and what cognitive neuropsychologists call the *lexical* or *whole-word recognition* route. In English there are many words – *yacht, colonel, women, island, gnat, choir* – whose spelling is such that only the use of the lexical route can guarantee their correct pronunciation, but fluent readers will also use this route for words with regular spelling. For pronounceable *non*-words – *plam, fleb* or *toth*, for example – only the phonological route is generally available, though occasionally pronunciation will be influenced by analogy with similar real words.

If there are these two separate routes, they might be expected to fail independently, and with predictable results. Failure of the phonological route from print to speech should render the patient incapable of reading pronounceable non-words. Failure of the lexical route should make the patient read irregular words of the *yacht, colonel, women* variety incorrectly, since they would tend to be 'regularized' by application of the 'print-to-sound' rules – *island*, for example, would be read as *izland*. The more irregular the word the less likely it is to be read correctly by the phonological route.

From various case reports in the 1970s, it became clear that patients who become alexic following some kind of damage to the brain (and whose problems are not caused by difficulty in recognizing the letters on the page) often make errors that fit into one or other of these two patterns.[23] Those whose errors suggest that there is a block in the phonological route are said to be suffering from *phonological alexia* (also called *deep alexia*); those whose errors suggest that there is a block in the lexical route are sometimes said to be suffering from *lexical alexia*, though for historical reasons (and perhaps partly to avoid the oxymoron) they are more often said to be suffering from *surface alexia**. As there is a *double dissociation* (see pp. 207 and 218) between the two patterns of error, it follows that the machinery involved in the two routes from letters to speech must be at least partly separate.

What about understanding the words you are reading? After all, there is little point in being able to read unless the words are understood. Because speech evolved many millennia before writing, it is tempting to assume that the comprehension of written words must be preceded by the translation of visual patterns into sound patterns – of squiggles on the page into the sounds that combine to form words. And that must be true for reading by the phonological route; the child beginning to read 'sounds out' the word a letter or a syllable at a time, and then puts the bits together to make a recognizable word. But for reading by the lexical route the position is more controversial. The words that we learn to read by the schoolteacher's 'look-and-say' are, almost invariably words that we already understand. So in learning to read them we are establishing a connection between the appearance of the printed or written word and both its sound and its

---

* The phrase *surface dyslexia* was originally introduced by J. C. Marshall & F. Newcombe to refer to alexia associated with over reliance on simple print-to-speech sound rules, leading to 'regularization' errors.

meaning; and there is no *a priori* reason why the connection between the appearance and the meaning should be routed through the sound.

In learning to use a particular word we must establish three kinds of connection: a direct connection between the sound of the word and its meaning; a connection, direct or indirect, between the sound of the word and its appearance (or, more precisely, that aspect of its appearance that is defined by the spelling); and a connection, direct or indirect, between the appearance of the word and its meaning. A possible pattern of connections is shown in Figure 16.3. We must have, in our brains, representations of the sound pattern, the meaning and the appearance, and although we are ignorant of the form of these representations, we can deduce something about the pattern of connections both from normal linguistic behaviour and, more strikingly, from abnormal behaviour when the machinery is damaged. (This will be true even if, as many 'connectionist' investigators believe,[24] representations of the sound patterns, appearance and meaning of individual words are encoded as patterns of synaptic strengths in a network of processing units, so that any given synapse may be involved in the representations of many words.)

Since we can talk and understand before we can read or write it is obvious that the connection between the sound pattern and the meanings of a word can be formed without any involvement of the representation of its appearance (path A in Figure 16.3). But now we get to the controversial bit. Is the connection between the appearance of a word on the page and its sound pattern a direct one (path B), or does it involve the representation of the word's meaning (paths C and A)? Is the connection between the appearance of a word and its meaning a direct one (path C), or does it necessarily involve the representation of the word's sound pattern (paths B and A)? Or are there both direct and indirect pathways?

There are three kinds of language disorder whose characteristics together provide tentative answers to these questions.

In 1979 Myrna Schwartz and her colleagues published a careful study of a woman in her early sixties who was suffering from a progressive presenile dementia.[25] Her spontaneous speech showed a profound *anomia* – an inability to find names for things she wished to refer to – and tests in which she was asked to match spoken words to pictures showed her to have a vocabulary of word meanings comparable with that of a child of six. Despite this she could read aloud easily, and in her reading she had no problems with words that she could not understand. When the names of forty animals were presented to her, she read all of them aloud almost perfectly, yet she could match only six of the names to pictures of the animals. Indeed with nearly half of the names she failed to recognize that they were animals. She was able to read pronounceable non-words, so her phonological reading was intact, and it must have been in use since she made occasional regularization errors – making *pint* rhyme with *hint* and pronouncing the *w* in sword. But phonological reading alone could not account for her

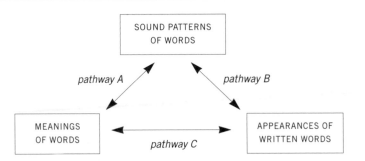

FIGURE 16.3 Possible connections between the sound patterns of words, the meanings of words, and the appearances of written words.

correct pronunciation of words such as *hyena, leopard, tortoise, liquor, blood, police, pursuit, sew, shoe* or *oasis*. She often commented on the discrepancy between her ability to read a word and to understand it: 'Hyena…hyena…what in the heck is that?' Given her competent reading and lack of comprehension, it is difficult to avoid the conclusion that the lexical route must include a path from the printed word to the word sound that does not depend on comprehension. It looks as though, in Figure 16.3, pathway B must exist.

The second kind of disorder relevant here is exceedingly rare. It is called *word-meaning deafness* because the sufferer can hear spoken words and can repeat them correctly but cannot understand them. The clearest description was published by an Edinburgh physician, Byron Bramwell, in 1897, and was largely forgotten until its significance was pointed out in 1981.[26] Bramwell's patient was a young woman who had had some kind of cerebral attack eleven days after the birth of her third child. For three or four days she had been totally deaf, then her hearing had gradually recovered so that she could hear the ticking of a clock though she remained unable to understand any words spoken to her. In contrast, she could understand written language. By the time she was seen by Bramwell (seven weeks after the birth of her baby) she could, with difficulty, understand short, simple, distinctly spoken sentences, particularly if one word was said at a time. She spoke spontaneously and seemed to be able to say what she wanted to say, but often had to ask for the answer to be written down before she understood it. The striking thing was her ability to repeat and to write down spoken sentences that she could not understand, *and then to understand what she had written*. Her inability to understand a sentence after hearing it, memorizing it and repeating it aloud implies that the route from sound pattern to meaning (path A in Figure 16.3) was impaired. That she could also understand the same sentence after writing it down and reading it implies that there is a route from appearance to meaning that does not involve the sound pattern. In other words, path C in the Figure must exist.

The third kind of disorder that throws some light on the pattern of connections between representations of appearance, of sound pattern and of meaning
is the condition of phonological (or deep) alexia.[27] Patients with this condition
cannot read pronounceable non-words, because the phonological route is
blocked, but their most remarkable characteristic is that in reading aloud (presumably by the lexical route) they tend to make frequent 'semantic errors'. Instead
of reading the word that is written they substitute a word that is related in
meaning though not related in either spelling or sound. The substitute word may
be a synonym (shut for close), an opposite (small for large), an associated word
(knife for fork) or what linguists call a superordinate word (animal for cat). Not
all the errors are semantic, but the semantic errors are significant because they
may be explained by supposing that, in addition to a direct connection between
representations of appearance and of sound pattern (path B) there is a direct connection between representations of appearance and of meaning (path C). If the
path (path B) between representations of appearance and of sound pattern is
impaired, the patient can select the appropriate sound pattern only by deciding
the meaning (via path C) and then selecting the sound pattern appropriate to that
meaning (via path A). If the decision about meaning is not quite on target, the
sound pattern chosen will not be the right word but a word related in meaning
to the right word – a semantic error. You can almost see this happening in the
response of a patient with phonological alexia reported by the Russian
neuropsychologist Alexander Luria.[28] Asked to read the word *Holland*, the
patient answered: '*It's a country, not Europe...no...no...not Germany...it's
small...it was captured ...Belgium? ...That's it, Belgium.*' In a normal subject the
direct connections between representations of appearances and representations
of sound patterns would prevent such errors.

Turning now from reading to writing (and leaving aside difficulties in actually
forming the letters), you might expect to find a similar dichotomy between disorders caused by damage to the phonological route and disorders caused by damage
to the lexical route. And that is just what is found.[29] There are patients who can
write regularly-spelt words and pronounceable non-words, but who tend to write
words with irregular spelling as if they were regular. Because they seem to rely on
sound-spelling correspondences rather than using their lexicons, they are said to
have *lexical agraphia*. And there are other patients who can write both regularly-
and irregularly-spelt words but cannot write non-words. Because they seem to be
able to use their lexicons, but are at a loss when all they have to guide them is the
sound of a spoken non-word, they are said to have *phonological agraphia*.

KANJI AND KANA

Elegant evidence for the existence of both phonological and lexical routes in reading and writing comes from the study of Japanese patients. The characters in Japanese writing are of two kinds. *Kanji* or ideograms, were originally taken from Chinese, and like Chinese ideograms tend to represent whole words. More recently – about a thousand years ago – the Japanese introduced *Kana* or phonograms, each representing the sound of a vowel, or a consonant plus a vowel. A well-educated Japanese will have memorized several thousand different Kanji, but there are only seventy-one Kana. Most writing nowadays will be in a mixture of Kanji and Kana; it is possible to write entirely in Kana though only young children would normally do so. Because each Kanji character corresponds to a whole word, and there is little logical connection between the character and the sound of the word, Kanji characters should correspond to irregularly spelt words in English, and reading or writing using Kanji should be by the lexical route. In contrast Kana writing follows strict phonological rules, so words written wholly in Kana should behave like regular English words in that they can be read by the phonological route; but because (unlike regular English words) they are not usually written that way, the lexical route may not be available. If this interpretation is right, Japanese patients with alexia or agraphia should show the same kind of selective deficits as similar patients speaking alphabetic languages; and they do. In 1900, K. Miura described a patient (with Broca's aphasia) who could neither read nor write in Kana yet could do both in Kanji.[30] Since then cases have been reported with the reverse picture – normal reading and writing of Kana, but alexia and agraphia for Kanji. And many other patterns of dissociation have been seen.[31]

PARADOXICAL PATIENTS: ARE THERE SEPARATE INPUT AND OUTPUT LEXICONS?

All this seems to fit together, but there is an awkward extra piece to the puzzle. The first clearcut case of phonological *alexia* to be described was RG, a French sales manager for agricultural machinery who, at the age of sixty-four had a tumour removed from the junction of the parietal and occipital lobes of his left hemisphere. He was studied by Marie-France Beauvois and Jacqueline Dérouesné, at the Salpêtrière Hospital in Paris, who found that his reading of *words*, whether their spelling was regular or irregular, was almost normal, but he could read only 10 per cent of pronounceable *non-words*.[32] Clearly his lexical route for reading was intact and his phonological route was damaged. But they also found that he was suffering from agraphia. And now the rabbit jumps out of the hat. Writing *non-words* from dictation, even long and complex non-words, his performance was faultless. Writing *words* from dictation, he always chose the spelling that most straightforwardly represented the sound, so his spellings were phonologically satisfactory but often wrong. The conclusion was as inescapable as it was surprising: the patient who had a clearcut *phonological* alexia also had a clearcut *lexical* agraphia. As Shallice has put it, '...the one kind of material RG cannot read is the only kind he can write correctly, and the type he cannot write is what he can read'.[33]

The combination of a phonological alexia and a lexical agraphia in RG is only one of a number of paradoxes. Alan Allport and Elaine Funnell described a patient who had fluent spontaneous speech with a severe lack of nouns but normal use of grammatical function words.[34] His reading aloud had precisely the reverse pattern. And we saw earlier that patients with difficulty in understanding words in a particular category do not necessarily have any problem in finding words in the same category. An obvious explanation of all paradoxes of this kind is to suppose that separate lexicons are used by the input and output processes, so that particular entries may be absent or damaged in one but not the other. (Here, lexicon is used to mean simply a collection of representations of word-forms divorced from their meanings.) Whether that sort of explanation is correct, though, is uncertain and highly controversial. In general, the asymmetry could also arise from a difference in the characteristics of the input and output processes using the same lexical entries – possibly even using the same patterns of synaptic strengths involving the same synapses in the neural network, if the 'connectionist' view is correct.

LEFT BRAIN V. RIGHT BRAIN

> *Something I owe to the soil that grew—*
> *More to the life that fed—*
> *But most to Allah Who gave me two*
> *Separate sides to my head.*
> — Rudyard Kipling, in *The Two-Sided Man*

Ever since Broca said 'We speak with the left hemisphere', neurologists and others have been fascinated by the unexpected asymmetry. For nearly a century, there was thought to be a tight link between right handedness and the role of the left hemisphere in speech, and since it is the left hemisphere that controls the right hand it was generally thought of simply as the dominant hemisphere. The difficulty in testing the tightness of the link is that it is easy to tell whether someone is right- or left-handed but difficult to know which hemisphere is controlling speech. In 1960 Juin Wada introduced the technique of injecting a small quantity of sodium amytal – a fast-acting barbiturate – into either the right or the left carotid artery of a talking patient, and seeing which injection makes the talking stop. (Because the technique carries a slight risk, it is used only when it is clinically justified.) Using this technique, Brenda Milner and her colleagues, in

Montreal, showed that although in 96 per cent of right-handed patients speech was controlled by the left hemisphere (as expected), in 4 per cent it was controlled by the right hemisphere.[35] With left-handed patients the outcome was even more surprising. In a substantial majority of left-handed patients (70 per cent) speech was controlled by the *left* hemisphere; in 15 per cent it was controlled by the right hemisphere; and in 15 per cent both hemispheres were involved.

It is clearly wrong, then, to think simply of a dominant hemisphere; Milner's results show that a hemisphere may be dominant for language and not for handedness, or vice versa. Back in 1876, Hughlings Jackson, on the basis of observations on a patient with a right sided cerebral tumour, suggested that the right hemisphere might be critical for 'visuoperceptual' tasks, and there is now plenty of evidence for a special role for the right hemisphere in object recognition, and spatial analysis, and also in the recognition of musical pitch, timbre and harmony.[36]

Even with language, lateralization is more subtle than appears at first sight. A spoken sentence is characterized not only by its vocabulary and grammar but also by its *prosody* or *intonation* – the modulation of pitch, tempo, and stress that gives colour and melody, that makes the sentence sound happy or sad, angry or resigned or indifferent. Noticing that a severely aphasic patient who was unable to use words in the ordinary way to convey meaning could nevertheless use them in moments of stress to express feeling, Hughlings Jackson concluded that the neural machinery involved was different, and he wondered whether the emotional aspects of speech were the responsibility of the right hemisphere.[37] It now looks as though he may have been right. Kenneth Heilman and his colleagues, in Florida, described a series of patients with lesions in the right temporo-parietal region who had difficulty in distinguishing between the same sentences read in four different moods, though they could understand the sentences without difficulty.[38] And Ross and his colleagues, in Texas, described a series of patients with lesions in the right frontal and parietal lobes who were unable to produce speech with normal prosody.[39] One patient in this series, a schoolteacher, who spoke with a feeble, monotonous, colourless voice found she could no longer maintain discipline in the classroom or easily control her own children. She attempted to compensate by adding phrases such as 'God damn it, I mean it' but these phrases, too, were uttered without emotion. It was not that she did not feel the emotion; she simply could not convey it in her speech. Yet she had no difficulty in finding the right words to convey her meaning, and she could both understand the speech of others and appreciate its emotional content. Another patient, a sixty-three year old man with right sided lesions involving both frontal and parietal lobes, spoke in a flat, monotonous way even when he was describing his experiences in liberating prisoners from Nazi concentration camps, or the trauma of his son being shot to death the previous year.

Abnormal prosody can also be present in patients with lesions restricted to the left frontal lobe. An interesting example was described by G. H. Monrad-Krohn just after the Second World War.[40] During a German air raid on Oslo in 1941, a young Norwegian woman suffered a severe injury to the left frontal region of her skull. When she recovered consciousness, nearly four days later, she was completely paralysed on her right side and totally speechless, but both movement and speech soon began to improve. Her speech, when it first returned, was marked by the lack of grammar characteristic of Broca's aphasia, but two years later the only abnormality was in the prosody. Her allocation of stress and pitch to the different syllables of words in a sentence was so abnormal that she was generally assumed to be German, and shopkeepers and strangers were deliberately unhelpful to her. And yet she had never been out of Norway, nor had she consorted with foreigners. Significantly, she had no difficulty in singing in tune – presumably because the right hemisphere was intact.

Ross and his colleagues noticed that the speech of both of his patients with abnormal prosody, and of others like them, was not accompanied by appropriate emotional gestures of the face, limbs or body; and they suggested that emotional gestures, like prosody, are controlled by the right hemisphere. This fits in with an observation, first made by Hughlings Jackson, that (emotional) gesturing is often preserved in patients with aphasia. On the other hand, studies of the effects of strokes on deaf patients who are fluent in American Sign Language – a language that relies entirely on gestures – show that it is lesions in the *left* hemisphere that make the patient incapable of 'signing'.[41] So it looks as if gestures used to convey meaning are controlled by the left hemisphere, whereas gestures used to convey emotion are controlled by the right hemisphere. In other words, gesture shows the same dichotomy as speech.

The distinction between sounds conveying meaning and sounds conveying emotion in speech has special features in speakers of 'tonal' languages such as Mandarin Chinese and Taiwanese. In both of these languages, changes in pitch in the course of a syllable – falling or rising tones – can alter the meaning of a word, and so are an essential part of it. But, as in other languages, changes in pitch are also used as one of the ways of adding emotional colour to a sentence. What you might expect, then, is that damage to appropriate areas in the left hemisphere would interfere with the use of tones to define meaning, whereas damage to areas in the right hemisphere would reduce the changes in pitch that provide the emotional colour, while leaving those that are necessary to convey meaning. And that is what has been found.[42]

THE SPLIT BRAIN

The human corpus callosum contains more than 200 million nerve fibres. Since that is at least 200 times as many as the optic nerve, there must be a huge interchange of information between the left and right cerebral cortices. What would be the effect of stopping that interchange by disconnection of the two hemispheres?

We have already met three patients in whom damage to a part of the corpus callosum contributed to the clinical picture. In the first of them – Dejerine's Parisian merchant who suddenly found himself unable to read – it was the combination of damage to nerve fibres from the corpus callosum and damage to the visual cortex in the left occipital lobe that caused the symptoms. In the second – the young woman at Johns Hopkins Hospital who had alexia in the left visual field, following damage to her corpus callosum during removal of a small tumour – the symptoms are clearly a direct consequence of the disconnection. In the third – the policeman who had had a left frontal tumour removed, studied by Geschwind and Kaplan – the very peculiar symptoms were almost certainly a consequence of the disconnection, although there was also considerable damage to the left frontal lobe. In all these cases, the disconnection was only partial. What would be the effect of complete disconnection?

In the 1950s, Roger Sperry and his collaborators at the Californian Institute of Technology (Caltech) published a series of papers showing dramatic effects of cutting the connections between the two hemispheres in cats and monkeys.[43] Their most striking experiments involved cutting not only the corpus callosum (and the very much smaller anterior commissure), but also the optic chiasma, so ensuring that *information from each eye was transmitted only to the cerebral cortex on the same side.* When animals treated in this way were taught to get a reward by responding to some visual stimulus, they found that if the lesson had been learnt using (say) the right eye, the animal responded only if the stimulus was presented to the right eye. There seemed to be no transfer of information between the hemispheres; it was as if the animal had two separate brains.

The clarity of Sperry's work led to a renewed interest in the possible thera-peutic value of cutting the human corpus callosum in patients with severe epilepsy unresponsive to drugs. In the 1960s, Joseph Bogen, who had known Sperry at Caltech, operated on a number of patients who were carefully investi-gated, both before and after surgery, by Michael Gazzaniga, then a student of Sperry's. Later, Gazzaniga investigated another series of similar patients at Dartmouth Medical School. In all these patients, the optic chiasma was, of course, not touched, so it was not possible to direct information to one or other hemisphere by presenting it to one or other eye. There is , though, a simple way round this difficulty. If you fix your eyes on a point straight in front of you, infor-mation about everything visible to the right of that point will be sent to the left visual cortex, and information about everything visible to the left of the point will be sent to the right visual cortex (Figure 11.2; p. 175). With a screen and a suitable arrangement of projectors, it is easy to flash suitable stimuli – pictures, diagrams, printed words – to the left or right of the screen, to test the patient's responses.

By making the flashes brief, the risk of the patient's gaze wandering from the fixation point during the exposure to the stimulus can be made negligible.

From the patient's point of view, the most important result of operations of this kind was that, not only did they prevent the epileptic seizures from spreading from one hemisphere to the other, but (for reasons that are not clear) the seizures also became much less frequent. In many cases, cutting the corpus callosum had no obvious effects on the patient's ordinary behaviour or personality or emotional state. Tests in which stimuli were presented to one or other visual field gave results in line with what could be predicted from Sperry's animal experiments. When printed words or pictures of objects were flashed onto the right side of the screen, and therefore transmitted to the left (speaking) hemisphere the patient said what they were. When they were flashed onto the left side of the screen, and therefore transmitted to the right (mute) hemisphere, the patient said nothing, and if questioned, denied seeing anything. There was, though, no doubt that the right hemisphere was working. Using the left hand, the patient was able to draw an object flashed onto the left half of the screen (and therefore seen using the right hemisphere). In fact the right hemisphere proved to be better at drawing three-dimensional objects than the left.

This was the picture in the majority of 'split-brain' patients, but in a few of them, and most conspicuously in a fifteen-year-old boy, PS, there was evidence for linguistic ability in both hemispheres. (The explanation is probably that, in these patients, damage to critical areas of the left hemisphere at an early age led to some language development in the right hemisphere – PS had neurological problems when he was only two.) For the first two years after his corpus callosum had been cut, the linguistic ability in PS's right hemisphere was limited to understanding. In one experiment by Michael Gazzaniga and Joseph LeDoux, he was presented with a picture of a chicken claw on the right side of the screen (left hemisphere) and a picture of a snow scene with a snowman and a car stuck in the snow on the left side of the screen (right hemisphere), and he was asked to choose from a series of picture cards the card most obviously linked to the scene presented on the screen.[44] The correct answer was a picture of a chicken for the chicken claw, and a picture of a shovel for the snow scene. PS pointed to the chicken with his right hand and to the shovel with his left. When asked what he had seen, he answered (presumably from his left hemisphere), 'I saw a claw and I picked the chicken, and you have to clean out the chicken shed with a shovel.' Gazzaniga and LeDoux's interpretation of this 'creative fabrication' is that PS's speaking left hemisphere had to explain why PS's left hand was pointing to the shovel. It was unaware of the snow scene, because of the cut corpus callosum, and had to supply the best theory it could given the limited information it had to work with.

About two years after his corpus callosum had been cut, PS began to be able to use his right hemisphere for speech as well as for understanding, though the speech tended to be brief descriptions. The question, then was: how would PS's left hemisphere cope with speech from his right hemisphere? To answer this question, Gazzaniga and his colleagues did an ingenious experiment.[45] PS fixed his gaze on a point on a screen on which five slides were projected briefly in turn. Each slide contained two words, one falling to the left of the fixation point and one to the right. The slides read:

MARY	ANN
MAY	COME
VISIT	INTO
THE	TOWN
SHIP	TODAY

The left hemisphere therefore received the message, 'Ann come into town today', and the right hemisphere the message, 'Mary may visit the ship'. As soon as the last slide had been projected, PS was asked to recall the story. His response and the ensuing conversation with his examiner went like this:

PS: Ann come into town today. (i.e. the message received by the left hemisphere.)

E: *Anything else?*

PS: On a ship.

E: *Who?*

PS: Ma.

E: *What else?*

PS: To visit.

E: *What else?*

PS: To see Mary Ann.

E: *Now repeat the whole story.*

PS: Ma ought to come into town today to visit Mary Ann on the boat.

It looks as though the linguistically dominant left hemisphere gets its message out first, but then hears the piecemeal and garbled message originating from the right hemisphere, and tries to reconcile the two.

LOCALIZING LEXICONS (AND OTHER MODULES)

Deciding whether particular processes occur in the left or in the right hemisphere, or in both, is only the first step in trying to link particular kinds of linguistic behaviour to particular bits of the brain's machinery. During the last two decades, modern scanning methods (see pp. 221–2) have provided new ways of making that link. CAT scanning and MRI scanning make it possible to localize damaged areas in the brains of patients with speech disorders, so that the characteristics of the disorder can be correlated with the pattern of damage to the brain. That, of course, is still the traditional approach of the diagram-makers, though achieved non-invasively in the living patient, rather than at autopsy. But MRI scanning can also be used to measure changes in local bloodflow and oxygen content, and PET scanning can be used to measure changes in local bloodflow, in glucose consumption and in neurotransmitter uptake. Both of these methods can therefore reveal which areas of brain become active when a normal subject, or a patient with a neurological disease, engages in a particular linguistic task. Any such task is likely to involve more than one process, but by comparing the patterns observed in two or more closely related tasks it is, in principle, possible to link the differences in pattern with a chosen process.

There are, though, many snags. Neither the spatial resolution nor the temporal resolution of PET scanning or MRI scanning is really up to the job. Changes in blood flow of less than about 5 per cent are not detectable. The neurons involved in a particular process may be widely distributed over the network. Important changes in what a particular bit of brain is doing may not be reflected in changes in blood flow or in the consumption of glucose or oxygen. And there is another hazard, which stems from the nature of language. In one of his celebrated sermons, the Revd F. A. Simpson – once described (fortunately wrongly) as 'Cambridge's last eccentric' – speaks of the 'mind of man' ranging 'unimaginably fast and far, while riding to the anchor of a liturgy.'[46] The subject in the laboratory chair is not very different from the worshipper in the pew. It is easy to set subjects tasks that depend on sorting out the sound or the appearance or the meaning of words; but it is not easy to restrict the range of their thought while doing the set task. The word whose spelling is being considered will have a meaning. The pronounceable non-word will arouse echoes of words. Even unpronounceable strings of letters will set off hares in people who live in a world of acronyms.

Despite all these hazards, the localizers have had a good deal of success. Here are three examples.

In 1992, David Howard and his colleagues described experiments in which they used PET scanning to locate the lexicon used for recognizing spoken words and the lexicon used for recognizing written words.[47] Subjects were presented with spoken words (all familiar nouns) from a tape recorder, at the rate of forty words per minute, and were required to repeat each word as it was heard. As a control, the tape recording was run backwards, producing a succession of pieces of gibberish with the same auditory complexity and mixture of pitches, and the subject was asked to say a word (always the same word) after each piece. For the part of the experiment concerned with recognizing written words, the subject faced a screen on which the words were projected in turn, and was asked to read each word aloud. As a control, the subject was presented with short strings of letter-like squiggles, and asked to say a word (always the same word) after each string.

Comparison of the auditory and visual control runs showed, as expected, that presentation of auditory stimuli that were complex *but without words* led to increased blood flow in the primary auditory cortex and adjacent association cortex in both hemispheres. Similarly, presentation of visual stimuli that were complex *but without words* led to increased blood flow in the primary visual cortex and adjacent association cortex in both hemispheres. More interesting results came when the changes in blood flow seen during the control runs were subtracted from the changes seen during the experimental runs. For spoken words, there was a clear increase in blood flow close to and slightly in front of Wernicke's area. For written words, there was no increase in Wernicke's area but a clear increase about sixteen millimetres further back, close to, though not quite coincident with, the angular gyrus. It seems that Wernicke and Dejerine were not far wrong in placing the 'centre for auditory images of words' and the 'centre for visual images of words' where they did. And the absence of any increase in blood flow in Wernicke's area following the presentation of *written* words supports the view that recognition of written words does not require them to be first recoded into phonological form and then recognized by the (input) lexicon for spoken words.

My second example has to do with locating the machinery that makes it possible for us to store a short sequence of numbers or letters in our short-term memory. Anyone who has ever tried to remember an unfamiliar telephone number for long enough to make a call knows that you can hold the sequence almost indefinitely if you continually repeat it to yourself. And you don't need to say it aloud, or even to mutter it under your breath. Just rehearsing it mentally is good enough. This is as true of short strings of letters or of unconnected words as of strings of numbers. Even if the sequence is presented visually – as in a telephone directory – we tend to store it in phonological rather than visual form; we rehearse it as a succession of sounds 'in our head'; we don't see it as a string of figures 'in our mind's eye'.

Alan Baddeley pointed out that this behaviour could be explained if there were a small store containing unstable representations of the information being remembered, and if, through some kind of feedback loop, passage of this information through machinery whose activity normally precedes articulation led to renewal of the representations in the store.[48] The whole arrangement, which he called *the articulatory loop*, can therefore be thought of as consisting of two components: the *phonological store* and the *subvocal rehearsal system* – 'subvocal' because actual articulation of the words is not necessary. The question is, where in the brain are these two components?

In 1993, E. Paulesu and his colleagues described experiments that answered this question.[49] They used PET scanning to compare the changes in local blood flow in the brain of normal (right-handed) subjects when they were given tasks of two kinds. The first kind of task was to remember the order of six consonants presented, one per second, on a television screen. The subject was told to rehearse the sequence mentally but not actually speak. Two seconds after the completion of the sequence a single letter was projected on the screen and the subject had to indicate – by pointing to *yes* or *no* symbols – whether the displayed letter had been in the sequence. This task would involve both components of the articulatory loop. (Similar control experiments in which Korean letters were presented – letters that the subjects could not translate into sounds – allowed the experimenters to discount changes in blood flow associated with the presentation and remembering of patterns without linguistic significance.) The second kind of task was to decide whether two letters rhymed. Consonants were presented on the screen at one second intervals, and the subject was told to point to the *yes* symbol if the presented letter rhymed with the letter B, permanently displayed on the screen. This task was thought to involve the subvocal rehearsal system but not the phonological store. (Again control experiments were done with Korean letters, the subjects being asked to say whether the presented letter looked similar to a letter permanently on display.) Comparison of the blood-flow patterns in the different conditions showed that use of the phonological store led to increased blood flow in the *left supramarginal gyrus*, i.e. part of the left parietal lobe just above the fissure of Sylvius. This is just the site in which, more than twenty years earlier, Elizabeth Warrington and her colleagues had found lesions in three patients who had great difficulty in remembering short, spoken lists of words.[50] Comparisons of the blood-flow patterns in each of the experimental conditions with the appropriate control suggested that the subvocal rehearsal system was associated with Broca's area. Broca, Wernicke and Geschwind would all have been delighted.

These two examples show the use of modern scanning methods to localize bits of linguistic machinery in normal subjects. My last example is concerned with the discovery of a totally unexpected abnormality in subjects with developmental dyslexia. Children may have difficulty in learning to read for all sorts of reasons, and not all children who are labelled dyslexic are labelled correctly. But there is a well-recognized (if less well-defined) kind of reading difficulty that tends to run in families, that is commoner in boys than in girls, and in which, as well as difficulty in reading, there is also some difficulty in repeating aloud words that have just been heard. This syndrome is developmental dyslexia, and the difficulty with word repetition is thought to be the result of a difficulty in analysing words into their basic units of sound (phonemes). Recently, Guinevere Eden and her colleagues, using MRI scanning to measure changes in oxygen content, made the startling observation that, when adults with developmental dyslexia looked at patterns of moving dots, they failed to show the expected increase in activity in the part of the visual cortex specially concerned with movement (Zeki's area V5).[51]

What has dyslexia to do with moving dots? A possible explanation of this curious finding is that developmental dyslexia is familial, and that the gene (or genes) responsible for the dyslexia happens also to be responsible for the lack of response to moving dots of the area of the visual cortex concerned with motion. In other words, there is no direct connection. A more interesting suggestion is that the inability to respond to the moving dots reflects a defect in timing, and that this same defect is responsible for the difficulty thought to be experienced by developmental dyslexics in breaking spoken words into their phonemes – a difficulty that could account for reading problems since reading involves linking phonemes and letters or groups of letters. Which of these explanations is correct (if either) remains to be discovered.

These examples of the use of modern scanning methods are chosen from a large literature, and I chose them because they provide clear-cut and interesting answers. But they illustrate the weakness as well as the strengths of these extraordinarily impressive techniques. Scanning can say where the action is, not what it is. So far, there is nothing in the field of the neural machinery of language comparable with the work of Hubel and Wiesel, and what followed from it, in the field of vision.

17   Chomsky and After

Colourless green ideas sleep furiously.
— Noam Chomsky, *Syntactic Structures*[1]

Language, Darwin pointed out in *The Descent of Man*, is 'certainly not a true instinct, for every language has to be learnt. It differs, however, widely from all ordinary arts, for man has an instinctive tendency to speak, as we see in the babble of our young children...'.[2] In the same paragraph, discussing the song of birds that sing instinctively yet learn the actual song from their parents, Darwin refers to the claim by the eighteenth-century lawyer and naturalist, the Hon. Daines Barrington – the man reputed to have persuaded Gilbert White to write his *Natural History of Selborne* – that these sounds 'are no more innate than language is in man'. But how innate is language in man? The unexpected answer, provided by Noam Chomsky nearly two centuries later, is: very much more than anyone had supposed.

Before Chomsky, the conventional view was that, despite our instinctive tendency to speak, learning a language was much like acquiring any other art. Everything about the language being learnt had to be learnt by experience – which the behaviourists thought of as 'conditioning' – or by 'generalization' from experience. Not everyone took the conventional view – as Chomsky himself points out, it did not become the conventional view until the middle of the nineteenth century[3] – but it seemed reasonable; it explained the existence of an extraordinary variety of languages, often with very different grammatical structures; it was compatible with the ability of children to learn any language to which they are exposed at an early age; and it seemed to be confirmed by the total lack of language in those rare cases in which, through peculiar circumstances, children have grown up without contact with any language – the 'Wild Boy of Aveyron', the Indian 'wolf-children' Kamala and Amala (real Mowglis), and more recently and less exotically Anna in Illinois, Isabelle in Ohio and Genie in California.[4] Why, then, did Chomsky reject it?

He was impressed by two facts. First, although each language has a finite number of words, speakers of the language are able to combine selections of these words to produce a virtually infinite number of sentences – sentences that are recognized by other speakers as grammatically well-formed, and many of which will never have been heard or written before. (The sentence you are now reading almost certainly satisfied both these criteria at the time I wrote it.) To generate sentences with these characteristics, the speaker must be applying – almost invariably unconsciously, of course – a set of rules which specify how words can be combined to generate 'all and only' the sentences of the language. (The inclusion of 'and only' is necessary to exclude the vast number of word combinations that do not form grammatical sentences.) Chomsky calls such a set of rules a *generative grammar*, and he thinks of it as 'a device of some sort for producing the sentences of the language under analysis'.[5]

The second fact that impressed Chomsky was the ease with which young children acquire the ability to produce and understand sentences that they have never met before. This ability, which is acquired rapidly, without any formal instruction, and almost irrespective of the child's intelligence, implies a knowledge of the language's generative grammar – a grammar that must be extremely complex and abstract in character to account for the subtleties of syntax that the child can cope with both in speaking and understanding. It is inconceivable, in Chomsky's view, that the knowledge could be derived solely from the child's own experience of language; what he calls 'the poverty of the input' is too great. Something must therefore be innate. But what is innate cannot be closely related to the generative grammar of the language being learnt, since in appropriate circumstances the child can learn any language, and languages differ very widely. It must therefore be something more fundamental – a *universal grammar* that sets limits on the 'range of possible grammars for possible human languages', and so provides the child with a basis for acquiring the generative grammar of whatever language it is exposed to. Armed with the principles of this universal grammar, and a predisposition to apply them to heard utterances, the child can overcome 'the poverty of the input' and learn to talk the language it hears.

In the last four decades this hypothesis, in various forms, has revolutionized linguistics. But is it right? To answer that question properly would require a much more detailed discussion than is appropriate to this book or than I am competent to handle. A splendidly lucid, accessible and entertaining account of the problem has recently been given by Steven Pinker in his *The Language Instinct*.[6] All I want to do here is to make a few general points.

The young child learning a language has two different tasks: acquiring a vocabulary of words that stand for concepts, and acquiring the rules for arranging sequences of words. Apart from a small number of onomatopoeic words – *buzz, hiss, thud, quack* – and a larger number of compound words or words obviously derived from other words – *teapot, toothbrush, football, painter, runner* – the relation between a word and the concept it refers to is arbitrary, (ignoring, of course, the word's etymology, which the child can be assumed not to know). Acquiring a vocabulary is therefore very largely a matter of rote learning, which is tedious but straightforward. It is also something that young children are fantastically good at; Pinker estimates that an average six-year-old has learnt to understand about 13,000 words.[7]

The more awkward task is acquiring the grammatical rules, and here the crucial question is whether the constraints provided by an innate universal grammar are likely to be sufficient to compensate for 'the poverty of the input' and enable the child to derive a generative grammar from what it hears. There is no doubt that young children can learn some grammatical habits by hearing correct examples. Before they are three, many English-speaking children will add an *s* to the third person singular of verbs in the present tense. And they must also be able to generalize, because they will add an *s* (logically but incorrectly) in situations in which they cannot possibly have heard it – 'I want to be a piggy if Sarah be's a piggy'. But adding an *s* to the third person singular of English verbs in the present tense is just one very simple rule. Languages have such complicated rules and the rules are so different in different languages, that Chomsky may seem to be asking too much of universal grammar.

His answer to such doubts is that, although there are indeed a very large number of rules that are different in different languages, the rules do not vary independently. For example, in English the subject normally precedes the verb, the verb precedes its object, the preposition (by definition) precedes the noun it governs, and the adjective precedes the noun it describes. In Japanese, the reverse is true in each case. (Prepositions become postpositions.) Because similarly consistent patterns are found in many languages, for those languages it is possible to subsume many individual rules under far fewer 'super-rules'. If the super-rule about consistency of word order that we have just been considering forms part of the child's innate universal grammar, all that is required for that rule to be implemented in a particular language is for the child's brain to be programmed in such a way that, noting which order is used in any one situation, it switches its machinery to stick to that order. This is, of course, only one example, but it suggests that what is expected of universal grammar may, after all, be reasonable.

The use of the word 'machinery' here is deliberate. The existence of an innate universal grammar implies the existence of some genetically determined neural machinery capable of providing the restrictive framework within which the acquisition of an actual language can occur. Ultimately, we may hope to understand the nature of the neural machinery and its behaviour, but meanwhile there is no reason to doubt its existence. The principles of stereoscopic vision were understood long before 'disparity detectors' (see p.228) were discovered in the visual cortex.

There is some independent evidence that the child's acquisition of particular grammatical rules is under genetic control. A few children, who are normal in all other ways, seem to have peculiar difficulty in generalizing some grammatical rules. In 1990, Myra Gopnik, at McGill, described a large family of whom sixteen members, spread over three generations, suffered from this rare condition (*familial dysphasia*).[8] The affected members had difficulties with (among others) the rules for changing tense ('Every day he kisses his nanny; yesterday he – his nanny.'), and the rules for forming plurals ('Here is a picture of an imaginary animal called a wug; in this picture there are several –'). They also failed to detect some types of grammatical error. The difficulties appeared in both speech and writing, though their discovery required careful testing, since failure to infer general rules from heard examples was often compensated for by the learning of each word as a lexical item (one book; several books). The distribution of affected members in the family suggested that the condition was caused by a single dominant gene.

Before we leave Chomsky, there is one other point. Although the usual purpose of a sentence is to convey meaning, the grammatical rules that constrain sentence building are not much concerned with the meanings of the individual words, though they are very fussy about just what part of speech each word is. A consequence of this unconcern is that sentences may be grammatically impeccable yet semantically maddening. The strange sentence of Chomsky's at the head of this chapter makes the point with panache.

BICKERTON AND THE HAWAIIAN PHENOMENON

If, in learning to speak, children normally use their innate universal grammar to compensate for 'the poverty of the input', the contribution of the universal grammar should be particularly conspicuous when the linguistic input is particularly poor. Derek Bickerton and his colleagues believe that their studies of language in Hawaii, in the early 1970s, provide evidence for an effect of this kind.[9]

In 1875 a Reciprocity Treaty established almost free trade between Hawaii and the USA. Together with the introduction of irrigation a few years later, this gave an enormous boost to the production of cane sugar, which increased nearly tenfold by 1890, and doubled again by the turn of the century. The labour that made such an increase possible was the indentured labour of poor immigrants from China, the Philippines, Japan, Korea, Portugal and Puerto Rico, a 'mixed multitude' that by 1900 formed two thirds of the total population and a much higher proportion of the population on the plantations. Lacking a common language, the immigrants were forced to speak to one another, and to their masters, using a scanty makeshift language or *pidgin*. Pidgins routinely arise in such situations but they are very limited. They vary from speaker to speaker, are poor in vocabulary and even poorer in syntax. They tend to lack articles, conjunctions, prepositions, tenses and auxiliary verbs; their sentences are very short, without subordinate clauses and sometimes even without verbs, and the word order is irregular, often reflecting the word order customary in the speaker's mother tongue. The first pidgin to arise in Hawaii was based on the Hawaiian language, but, following the annexation of Hawaii by the USA in 1898, this gradually gave way to a pidgin based on English.

Here are two examples from Bickerton's 1984 review. (In these and in later examples Bickerton's spelling has been modified to make the words more recognizable.) The first is from an immigrant who came to Hawaii from the Philippines in 1913:

> Good, dis one. kaukau [=chow-chow = food] any-kin dis one. pilipin islan no good. no mo money.

Or in conventional English:

> It's better here than in the Philippines. Here you can get all kinds of food; over there there isn't any money [to buy food with].

In this sentence 'dis one' means 'here', but it can be used as a substitute for various pronouns. Another immigrant from the Philippines, wanting to convey the sense of the English sentence 'And then he got married' said:

> En den marry dis one,

'dis one' meaning 'he'. It is placed after 'marry' because this immigrant's native language always puts the verb before the subject.

Pidgins are created by adults. When young children grow up in a polyglot community in which the only common language is a pidgin, they develop a much more sophisticated common language called a creole. (If a child's parents share a common language – and the child spends enough time with its parents – the child may, of course, also learn that language.) Creoles, unlike pidgins, are consistent from speaker to speaker, have a wide vocabulary, a regular word order, and a syntax that is only slightly less complex than those of ordinary languages. Here, from Bickerton's review, is a description of ball lightning:

> One time when we go home inna night dis ting stay fly up.

Or in conventional English:

> Once when we went home at night this thing was flying about.

Note the presence of the subordinate clause: 'when we go home inna night'. The use of 'stay' is a standard Hawaiian creole way of showing that the action of the verb continues. Here 'stay fly' replaces the English 'was flying'. Another of Bickerton's examples of a creole sentence with English words and un-English grammar is:

> The guy gon lay the vinyl been quote me price.
> (The guy [who was going to lay the vinyl] had quoted me a price.)

Unlike English, Hawaiian creole does not require a relative pronoun (who) to introduce the relative clause in square brackets.

The interesting question is, where do the new features found in a creole syntax come from? Are they inventions on the part of the first generation of children who have a pidgin as their linguistic input, or are they features transmitted from pre-existing languages? It is this question that Bickerton and his colleagues attempted to answer by studying the pidgin speakers and creole speakers of Hawaii.

Of course, the kind of history that gave rise to the linguistic pattern found in Hawaii is not peculiar to Hawaii; there are many examples of polyglot communities of immigrant labourers being created on tropical islands or littorals to exploit some local resource. What makes Hawaii special is that the events are so recent. Bickerton and his colleagues could talk to people who had come to Hawaii, or been born in Hawaii, during the time the Hawaiian creole was developing.

They studied particularly the speech of two groups of working-class people aged between seventy-five and ninety-five – mostly in the lower half of that range. Those in one group had arrived in Hawaii as immigrants and would not have been exposed to any form of English, even a pidgin, in their childhood. Those in the other group had been born in Hawaii and so would have been exposed to whatever was the common language in the plantations around the turn of the century. It turned out that those who had been born in Hawaii far surpassed those who had immigrated, in the range of syntactic structures that they used in their speech. The immigrants all spoke some form of pidgin, those who had arrived earliest tending to speak a more primitive variety, perhaps because the later arrivals were exposed to creole as well as pidgin. Those who had been born in Hawaii spoke a creole, and the speech of those born later than 1905 already showed most of the features present in current (1970s) Hawaiian creole. The results suggested that *creolization* occurred in a single generation. A similar conclusion was reached more recently from a study of hundreds of verbatim records of statements by Hawaiian residents of many ethnic groups in the period 1880–1920.[10]

Bickerton argues that the source of creole grammar must be mainly the innate linguistic abilities of children because there are too many differences between the structures of Hawaiian creole and the structures of any possible donor languages – English or the children's ancestral languages – for derivation from another language to be likely. If innate linguistic abilities are generally the main source of creole grammar, creoles from different regions of the world should be similar in their grammatical structures. Recent studies show that they are – in fact, Bickerton claims, they are more similar to one another than they are to the structures of any other language.[11] This similarity, he concludes, *'derives from the structure of a species-specific program for language, genetically coded and expressed...in the structure and modes of operation of the human brain'.*

Bickerton's 'species-specific program for language' is, of course, very close to Chomsky's universal grammar, though there is some difference in emphasis. Chomsky's universal grammar can be thought of as providing a mechanism that allows information in the linguistic input to flip switches that engage sets of rules appropriate to the language being learnt, and that disengage sets of rules that are inappropriate. In thinking about the development of creoles Bickerton is particularly concerned with the situation where there is very little information in the linguistic input. Here the rules that are operative will be partly those defined by the immutable part of the program and partly those defined by the switches when they are in their 'default positions' – that is to say, their position before there is any linguistic input. The grammatical structures common to creoles are therefore simply the structures defined by the rules before any switches are flipped.

If this hypothesis is right, why don't all children grow up speaking a creole language? Bickerton's answer is that this is precisely what they try to do, but because those around them persist in speaking English or French or whatever, the child has to modify the basic creole grammar until it fits the linguistic input. He points out that, if this interpretation of what children do is correct, where the structure of the language being learnt deviates from the pattern found in creoles, children should make errors in the direction of the creole pattern. And they do. For example, young children tend to follow the creole pattern and use a change in intonation rather than a change in word order to convert a statement into a question. And it is only children and creole speakers, according to Bickerton, who use a negative subject with a negative verb. He compares the child's:

'Nobody don't like me.'

with the Guyanese creole:

'Non dag na bite non kyat.'
(No dog didn't bite no cat.)

Bickerton's claim must, by the way, be a slight overstatement since the Guyanese creole construction is nicely paralleled in Alfred the Great's 9th-century translation into Anglo-Saxon of Boethius's description of the effects of Orpheus's harp. In English this reads:

No hart shunned not no lion, nor no hare no hound, nor no beast knew-not no hatred nor no fear of others, for the sweetness of the sound.[12]

But why did natural languages come to deviate from the creole pattern in the first place? Bickerton's answer is that 'all biological programs seem to allow for this space for variability and would probably be maladaptive if they didn't.'[13] And that sounds right; after all, why evolve switches that flip unless there is sometimes an advantage in flipping them? Bickerton has an appropriate illustration. In some species of bird, the innate version of the song is produced only by individuals that have been deafened or isolated at birth.[14] Birds reared under normal conditions learn and use the variant of the song that is current locally. Which is where this chapter started.

18   Monkey Puzzles

'It's good to talk!' we are told in countless British Telecom advertisements; and the ability to talk has such obvious survival value that we automatically assume it evolved by natural selection. Yet if we are asked to point to the steps in that evolutionary pathway we are in some difficulty. The trouble is that the evolution of linguistic ability must have involved subtle changes in the vocal tract, and even more subtle (and less understood) changes in the brain, and neither vocal tracts nor brains leave much in the way of fossils. We are therefore forced to depend on indirect evidence.

Thirty years ago, Norman Geschwind and Walter Levitsky showed that the functional asymmetry of the human brain was accompanied by a small but measurable anatomical asymmetry in the region of Wernicke's area – the area, in the left temporal lobe, in which damage tends to cause a 'fluent' aphasia with difficulty in comprehension (see pp. 173 and 261).[1] A few years later, other investigators showed that a similar asymmetry was present in the endocranial cast of the Neanderthal skull found at La Chapelle-aux-Saints, suggesting that special language areas were perhaps present in the Neanderthal brain.[2] More recently still, it was found that endocranial casts of *Homo habilis* showed strong development in the regions corresponding to Wernicke's and Broca's areas (though without any evidence of asymmetry) and this led to the suggestion that the evolution of cortical language areas might have begun early in the history of the genus *Homo*, nearly two million years ago, or even earlier.[3] Against such interpretations, it has been argued that studies of the topography, the fine structure, and the connections, of the cortex in the frontal and temporal lobes in monkey brains suggest that these regions contain areas homologous to the language areas of the human brain, so the existence of such areas is no proof of significant linguistic ability.[4] A problem with such arguments is that the great development of the different regions of the cortex between our primate ancestors and ourselves makes claims about homologies themselves open to question.

Arguments about the date our ancestors began to talk, based on the development of the vocal tract, are only a little more helpful.

It is well known that our ability to produce a wide range of speech sounds depends on the characteristics of the vocal tract above the larynx.[5] The anatomy of this region in the human adult is peculiar. All other living terrestrial mammals, and also new-born human babies, have a flattish tongue and a high larynx that meets the soft palate at the back of the throat; because of this, fluid and food pass on either side of the larynx, and the animal or baby can drink and breathe at the same time. The adult human larynx is much lower, and the back of the tongue curves downwards, so that above the larynx and behind the back of the tongue there is a sound-chamber whose shape can be adjusted and which can, at will, be shut off from the nasal passages. It is these features that give us our great range of vowels and our ability to make non-nasal sounds – but at the cost of choking much more easily; which is presumably why the adult plan is not present from birth.

The crucial question is: what is the shape of the vocal tract in the various hominids? That question can only be answered by arguing from the anatomy of the bony structures that have been fossilized, and in particular from the base of the cranium, which is flattish in living monkeys and apes and in the new-born human, but arched in the adult human. Such arguments cannot prove what the soft tissue was like, but it is interesting that the australopithecines have flat bases to their crania, that arched bases appear first in *Homo erectus* fossils in Kenya nearly two million years ago, and that the earliest fully arched bases are found in some of the 'archaic *Homo sapiens*' fossils 300,000 to 400,000 years ago.[6] Curiously, Neanderthal fossils have cranial bases that are even less arched than those of *Homo erectus*. Whether this implies that their vocal tract, and therefore their speech, was primitive is controversial; one suggestion is that the flat base of Neanderthal skulls is related to the protruding midface – thought to be related to the need for large air passages to warm cold air – and was not associated with an ape-like vocal tract.[7] In any case, as Pinker has pointed out, 'e lengeege weth e smell number ef vewels cen remeen quete expresseve'.[8]

Evidence from our extinct ancestors and their relations proving so limited, can we learn more about the evolution of speech by looking at our living relations? There have been two different approaches. One is to look at the communication between monkeys in the wild. The other is to see to what extent apes can be taught to understand human speech and to talk – or, given their very limited vocal abilities, to communicate in other ways.

VOCAL VERVETS

In 1967, Tom Struhsaker reported that vervet monkeys in the Amboseli National Park in Kenya gave different alarm calls in response to at least three different kinds of predator.[9] Seeing a leopard, or one of the other vervet-hunting cats, male vervets give a loud series of barks and females a single high-pitched chirp; other vervets respond to either of these signals by running into the trees, where their greater agility and small size – about the same as a domestic cat – make them relatively safe. Seeing either of the two kinds of eagle that habitually prey on vervets, individuals of both sexes give a short double-syllable cough; other vervets hearing this, whether they are in trees or on the ground, look into the air and may run into bushes – an appropriate response, since both kinds of eagle use a long, fast swoop to seize vervets both in trees and on the ground. The third kind of alarm call is used when a vervet sees a snake. It is described as a *chutter*, and the response of other vervets is to stand erect on their hind legs and peer into the grass around them; they may then mob the snake from a safe distance. There are also other, minor, alarm calls induced by baboons, by unfamiliar humans, and by carnivores that rarely prey on vervets (lions, hyenas, jackals).

How should we interpret these alarm calls? It is tempting to be thoroughly anthropomorphic and interpret them as if they had been made by us in similar situations. They would then be the equivalent of learnt words, used voluntarily with the intention of warning other members of the immediate community of dangers so that they could take appropriate precautions. But are they learnt? Are they voluntary? Does the monkey have an intention of warning others, or, indeed, any intentions at all? Do the individual calls refer to specific dangers, or – though it seems less likely – does the apparent 'naming' of the type of predator merely reflect the degree of fear induced by the different types? In other words, ought we to think of the call as the equivalent of a word or phrase or as more akin to a cry of surprise or a gasp of pain? Not all of these questions can be answered, but some can. The basis for answers comes from various studies, but particularly from the work of Dorothy Cheney and her husband Robert Seyfarth, who have spent the last twenty years looking at vervet monkeys in the wild.[10]

To study the monkeys' vocal behaviour, Cheney and Seyfarth used tape recordings in two different ways. By analysing their recordings electronically they could display each signal as a sound spectrogram showing the amount of energy present at each frequency at any time. This helped them distinguish vocal signals that sounded the same to the human ear, though, judging by the behaviour induced, not the same to the monkeys. The other use was to play back recordings through a hidden loud-speaker while the monkeys were being filmed, so allowing the investigators to see how monkeys in different situations responded to the various signals. Experiments of this kind showed that the characteristic responses to the different kinds of alarm call could be obtained in the absence of any predator – the alarm call alone producing the appropriate defensive response.

Whether the individual alarm calls are learnt or innate is uncertain. Cheney and Seyfarth never heard infant vervets make alarm calls in the first month, and recorded only a few in the first six months; but those few were acoustically normal, suggesting that the ability to produce the different types of call is innate. What young vervets did have to learn, was to restrict the calls of each type to a narrow range of stimuli. For example, infants under a year old would give eagle alarm calls not only to the two main predatory eagles, but also to other birds of prey and to vultures, bee-eaters and even a falling leaf. This tendency for the young to respond to too wide a range of stimuli is, of course, found in humans. When I was staying with friends in Connecticut many years ago, each morning during breakfast the mail would be pushed through the front door and land on the floor. And each time, my friends' two-year-old son would look up from his bowl and say brightly, 'Mailman's come! Then one afternoon, as the little boy's mother took a book out of a high shelf, several loose leaves fluttered to the floor. 'Mailman's come!' said the little boy.

Young vervets need also to learn the appropriate responses to alarm calls. Very young vervets respond to a leopard alarm call by running to their mothers; slightly older ones often make quite inappropriate responses such as running into the bushes; it is only after 6–7 months that most individuals show the adult response of running up a tree. Examination of films of young vervets, just after a leopard alarm call, showed that those that responded only after looking at an adult were more likely to make the correct response, suggesting that the young vervets learn by example.

The alarm calls appear to be voluntary since they are not made unless there are other vervets around to hear them, and parent vervets are more likely to make them when their young are close by. Whether there is a conscious intention to communicate on the part of the monkey cannot be known; nor, if there is such an intention, is it clear precisely what the monkey is trying to convey.

Judging from the circumstances in which it is used and the response it generates, you might suppose that in giving the leopard alarm call the monkey is saying, in effect, 'Look out! One of those dangerous large cats is around and up to no good! You'd better get into the trees pronto.' And 'in effect' that is what it is saying – in the limited sense that, if vervets could speak English, the effect on other vervets of uttering those sentences would be similar to the effect of giving a leopard alarm call – though the verbosity might give the leopard a slight advantage. But what we want to know is not what might produce the same effect in English-speaking vervets but what information is actually contained in the leopard alarm call.

But why fuss about this, anyway? The answer is that it bears on an important general point. One body (person, animal, organization, machine) can use language or quasi-language to control complex behaviour in another body in two ways, ways which place very different demands on the communication channel. One way is to transmit explicit detailed instructions. This requires the transfer of a good deal of information, which makes severe demands on the communication channel, and is expensive in resources and time; but if the communication channel can cope it is very flexible. The other way is to transmit much simpler signals from a limited stock. The demands on the communication channel are modest, so transmission is quick and cheap, and, provided the receiver of the messages is programmed (genetically or by experience) to understand them, the behaviour they control can be very complex. Which method is appropriate depends on a balance of factors. Just before the First World War, when telegraphy was slow and expensive, Reuters news agency found it worthwhile to have a code-book containing a million combinations of five letters, each representing a particular message. Provided the message the sender wanted to send was in the codebook, it could be sent very quickly and very cheaply. Given the small number and relatively simple character of vervet alarm calls, and the restricted uses to which they are put, it seems likely that, in their information content, they are more comparable with the codes in a telegraphic codebook than with English sentences.

Though we cannot know for certain what the monkey making the alarm call is thinking or feeling, we suspect that it is feeling afraid and is conscious of the cause of the fear, and we cannot rule out the possibility that it is anxious to warn and even to give advice. Yet we hesitate to attribute conscious states to the return-ing worker bee whose dance – informing its sisters of the direction and approx-imate distance of the source of the pollen – carries a message every bit as complex as the message carried by the monkey's alarm call.[11] So why are we comfortable in attributing conscious states to the calling monkey?

The general reason, of course, is that monkeys resemble us much more than bees do; but there is also a more specific reason. Both in generating messages, and in responding to them, the monkey shows much more flexibility. We have just seen one example of this in the narrowing of the range of stimuli producing alarm calls as young vervets learn. Another is the ability to remember a warning and produce an appropriate response later. Cheney and Seyfarth describe an occasion in which playing a snake alarm call to three vervets led two to respond immediately, while the third did nothing immediately but behaved in the char-acteristic manner when re-entering the area some time later. A further example may be the alleged ability of vervets to use the alarm calls deceitfully – giving a leopard alarm call during an intergroup encounter that is going badly, and so making all the participants scatter into the trees. (A devil's advocate, though, would point out that we cannot be sure that there is an intention to deceive, and that insects can also deceive. Female scorpion flies expect their mates to present them with a dead insect. Some males catch their own, but others pretend to be females and steal the insects given them by would-be husbands.[12])

Perhaps the most striking example of flexibility in the use of alarm calls is the ability of vervet monkeys to learn to take advantage of the alarm calls made by the East African 'superb starling', a brightly coloured songbird that lives in the same areas in Amboseli. These birds warn of terrestrial predators by making a harsh chatter, and they warn of birds of prey by a rising or falling whistle. Hearing the chatter, vervets often run into the trees; hearing the whistle, they tend to look up into the sky. In fact, the vervets seem to treat the starling bird-of-prey alarm call as equivalent to their own eagle alarm call. Exposing vervets repeatedly to recordings of the starling bird-of-prey call so that they became habituated to it, reduced their response to the vervet eagle alarm call; and similarly vice versa. In contrast, the starling terrestrial predator alarm call was not treated in quite the same way as the vervet leopard alarm call, probably because the starlings make their call in the presence of a rather wide range of terrestrial animals, including snakes; the starling's call is therefore often produced in the presence of animals not threatening to vervets, and even if there is a threat the appropriate response is not obvious since what is appropriate is different for snakes and mammals.

Besides alarm calls, vervets make various kinds of grunt or chutter in different social circumstances. A vervet grunts when approaching a dominant individual, or approaching a subordinate, or watching (or initiating) a group movement into an open area, or spotting members of another group. These grunts sound much the same to us, and do not differ much in their frequency spectra, but recording the grunts made by the same monkey in different situations and then playing them back to other monkeys produced consistently different responses. For example playing a 'grunt to a dominant' caused the hearer to look towards the loud-speaker, whereas playing a 'grunt on seeing another group' caused the hearer to look in the direction towards which the loud-speaker was facing. After examining the consequences of naturally occurring grunts, Cheney and Seyfarth concluded that the 'grunts to a dominant' and the 'grunts to a subordinate' served a useful social purpose, making it less likely that the two would separate, and more likely that they would forage, feed or sit together. It seems clear then, that, like the alarm calls, the grunts refer to objects or events or situations* outside the originator, and serve to transfer information that is useful to the hearer.

In learning to cope with these vocal social signals, the young vervet has two problems. It has to acquire a good deal of knowledge of the social structure of those around it, and it has to learn to produce and understand a variety of grunts. It has to know, for example, who is dominant and who subordinate to itself, and the appropriate grunts for each; and if it meets an adult male it needs to know whether he is a long-term member of the group – and so deserving a 'grunt to a dominant' – or a recent immigrant – who should be given a 'grunt to another group'. According to Cheney and Seyfarth infants initially divide social situations into broad categories, and gradually learn to subdivide them. Infants begin grunting on the day they are born, and the grunts gradually become more adult-like – some taking only a few months, others up to two to three years.

The situation faced by the infant vervet is in some ways like the situation faced by the human baby who is proceeding from babbling to talking. Both have the

---

* More precisely, what the vervet perceives as objects or events or situations. Occasionally it will be mistaken.

double job of learning about the structure of their environment, including its social aspects, as well as learning how to impart and receive information by the vocal route. But how comparable is the use of alarm calls and grunts and chutters by an adult vervet with the use of language by humans? There are obvious differences. First, though vocalizations of the vervet can, like words, refer to specific objects and events, the vervet's vocabulary, despite being bigger than it was once believed to be, is minute compared even with that of an eighteen-month old child. Secondly, except perhaps in the use of alarm calls to deceive (if they really are used in this way), the references that are made are always to objects that are present or events that are current. Thirdly, there is no evidence of the use of syntax by the vervet – no evidence, that is, of an ability to use rules to combine a limited set of words to produce a much larger set of meaningful sentences. With these limitations, the level of vocal communication of the vervet falls well below what would conventionally be called a language. But vervets are old world monkeys whose ancestors separated from ours perhaps twenty million years ago. Can we find a closer approach to human language in our closer relations? What about the apes?

ACADEMIC APES

In her book, *The Chimpanzees of Gombe,* Jane Goodall lists about thirty calls – grunts, pants, barks, roars, cries, screams, squeaks, whimpers, laughs, lip smacks, tooth clacks – that are used by chimpanzees in the wild.[13] And a result of the many thousands of hours spent in watching wild chimpanzees is that the use of these calls, as well as of non-vocal signals, in different situations or different emotional states is at least partly understood. But the most critical evidence concerning the linguistic abilities of apes has come from experiments – there are now more than a dozen of them – in which attempts have been made to teach some kind of language to apes in captivity.[14]

The early experiments, usually with chimpanzees, all sought to teach the animals to converse in English; and they failed almost completely. After more than six years, Viki, a home-reared chimpanzee, could say only three or four words; her pronunciation was unclear, and it was not proven that she knew what the words referred to.[15] Comprehension was better, but it is well known that dogs can learn to recognize and understand a small number of words.[16] Because of the lack of success in these experiments, later experiments made no attempt to get the apes to talk. Instead, they were encouraged to communicate either by gesturing or by using mechanical devices – arranging plastic chips, or striking keys labelled with arbitrary geometrical patterns ('lexigrams') on a huge keyboard. In some experiments, the apes were also encouraged to listen to, and learn to understand, spoken words.

The first experiment to meet with some success was by Allen and Beatrice

Gardner, who used American Sign Language to teach a young wild-born female chimpanzee, 'Washoe'.[17] Washoe lived in a trailer at the Gardners' home (in Washoe County, Nevada), but she had extensive areas in which to play, she was well provided with games and toys, she was taken on trips, and she had at least one human companion all the time. The general aim was to give her an environment as stimulating as that of a young American child. All those taking part in the project learned American Sign Language, and this was the only form of verbal communication normally used in Washoe's presence.

To teach Washoe to interpret signs made to her, she was exposed to a wide variety of activities and objects, together with their appropriate signs; using sign language, her companions also chatted to her, making comments and asking questions in situations in which the meaning might be guessed. To teach her to make signs, the most effective method was for the teacher to hold Washoe's hands and guide them so that they adopted the right shape and position and made the appropriate movement. Relying on imitation, and rewarding increasingly close approaches to the desired pattern, proved much less effective. Washoe repaid all this attention by making her first sign at three months. By seven months she had learnt four signs; by twenty-two months, thirty; and by fifty-one months, when the projected ended, she was using well over a hundred; but these figures include many natural chimpanzee gestures. The earliest signs to be acquired were simple requests. Most of the later signs were names of objects, which Washoe used either as requests or as answers to questions – for example to name objects or pictures of objects. (For the sign for an object to count as part of her vocabulary, Washoe had to show that she could use it to refer to objects of the appropriate class, and not just one particular object.) Occasionally Washoe ran two, or occasionally more, signs together. Usually the combinations involved natural gestures ('come-gimme, hurry'), or had been taught, or could have been learned by imitation; but her use of 'water bird' for swan when asked the question 'what that?' in the presence of a swan seems to have been her own ingenious invention – though, as Herbert Terrace and his colleagues point out, she might simply have been identifying first the water and then the bird.[18]

Slightly later attempts to teach apes English using American Sign Language were made by Francine Patterson, using a one-year-old female gorilla, 'Koko', in San Francisco, and by Herbert Terrace using a two-year-old chimpanzee, 'Nim Chimpsky' – in New York.[19] Koko was taught by simultaneous exposure to American Sign Language and spoken English. After four years her sign vocabulary was about 250 words (though this included natural chimpanzee gestures), and her new coinages included 'white tiger', for a toy zebra, and 'bottle match' for a cigarette lighter. Spoken words were said to be understood about as well as signed words. Koko also became very familiar to the public, featuring in the *National Geographic* and other magazines, being the subject of children's books, and appearing on television. Nim was taught a simplified version of American

Sign Language. He learnt his first sign (drink) at four months, and by the end of the project (when he was three years and eight months old) he was using 125 signs (including many natural gestures). He also produced short multisign sequences, but if they contained more than two signs they were often repetitive. A detailed analysis of Nim's utterances, and particularly of three-and-a-half hours of video-tapes of nine sessions recorded over a period of sixteen months, led Terrace and his colleagues to conclude that Nim's multisign sequences lacked any grammatical structure, and that certain regularities that had been taken to suggest syntactical rules were in fact the result of his tending to copy what his teacher had just signed.

The alternative approach – to teach apes to communicate using plastic chips or keys labelled with lexigrams – at first appeared to be dramatically successful. It seemed to be possible to train chimpanzees to understand and create simple sentences – 'Mary give Sarah apple',[20] 'Please machine give piece of cabbage',[21] – but it was later realized that there was no proof that the individual chips or 'lexigrams' in such 'sentences' each conveyed a separate meaning. Pigeons can be trained to earn a reward by pecking in turn spots of four different colours,[22] but there is no suggestion that each spot means something to the pigeon. Many (though probably not all) of the strings of chips, or of lexigrams, that the experimenters translated into the kind of sentence just quoted could have been achieved by rote learning of a few stock phrases, with the substitution of particular words depending on clues such as the presence of a particular object or a particular person.[23]

Criticism of this kind does not apply, though, to the later work by Sue Savage-Rumbaugh and her colleagues at the Yerkes Regional Primate Research Centre in Atlanta; and their experiments on pygmy chimpanzees have provided some of the most convincing results of all the ape experiments.[24] The star pupil was a young male called Kanzi. The work started with unsuccessful attempts to teach Kanzi's mother to communicate using lexigrams, though her teachers also addressed her in English, and spoke English to one another in her presence. For two years the infant Kanzi was present during his mother's training sessions, and would occasionally press a random key and expect a reward from the machine. It was only at the end of this period, when his mother was temporarily away, that the two-and-a-half-year-old Kanzi revealed how much he had profited from the training directed at his mother. Now, when he went to the keyboard, he would spend time looking for a particular lexigram to press, and when, having pressed it, he was offered a choice of objects he would select the one designated by that lexigram. Conversely, if shown an object he could choose the corresponding lexigram. Later he began to put two lexigrams together, or to combine lexigrams with gestures – either natural chimpanzee gestures or, less commonly, the gestures of American Sign Language, which his teachers used to some extent though they themselves were not fluent.

In the course of these experiments, it was discovered that Kanzi appeared to understand some English words. This understanding was quite unexpected, but to encourage it the machine was modified so that pressing a key, not only made the lexigram light up, but also activated a speech synthesizer to say the appropriate English word. Kanzi's comprehension of English was tested either by requesting him to perform particular acts – fetching his ball, for example – that he would be unlikely to have performed just then had the request not been made, or by asking him to select from a number of photographs the one corresponding to a particular spoken word. In general, he tended to learn to recognize spoken words before he learnt to recognize the corresponding lexigrams. By the time he was five-and-a-half years old, he had a vocabulary of about 150 words, counting only those words that he had learnt both to understand and to use spontaneously.

It is impossible to read accounts of all these experiments without being impressed by the achievements of the apes and the devotion and efforts of their trainers. But what they tell us about the language potential of apes is both less clear and more controversial than the early reports suggested. The straightforward interpretation of the experiments on chimpanzees, pygmy chimpanzees and gorillas, is that members of all three species are capable of learning to use signs of some sort – sign-language signs, lexigrams, or heard spoken words – as references to objects, actions, attributes and people. Like children gradually generalizing the use of a new word, the apes seem to learn to extend the reference of a particular sign from a single object to a class of objects, and like children they sometimes over-extend the reference in a rational way that happens not to be correct. Unlike children, though, the apes tend to use 'words' very largely to express demands or desires rather than merely to comment, and they rarely refer to things that are not present or connected with current stimuli. These two features often make it difficult to be certain that a sign is being used to represent a particular object, rather than merely as part of a routine set of actions that the animal has learnt is effective in achieving a desire – like the pigeon seeking a reward by pecking in order at four colours, or the dog who brings his lead to his master when he wants to go for a walk.

The argument that the learnt signs or lexigrams really are being used *referentially* rather than *instrumentally* – to use the current jargon – is clearest in the behaviour of Kanzi, when he is presented with a photograph of an object and chooses the lexigram corresponding to that object. Unless he routinely conceives a desire for whatever is in the photograph, it is difficult to interpret his action as expressing a demand. On the other hand, a devil's advocate might point out that to learn that a symbol and an object are associated is not the same as learning that the symbol refers to the object; and he might then suggest that Kanzi had learnt to associate the object in the photograph with a lexigram, and had also learnt that choosing the associated lexigram when presented with a photograph of an object is apt to be rewarded.

But if there is a moderately strong case that apes, like young children, can use 'words' to refer to things, there is only a very weak case that they can produce anything equivalent to grammatical sentences. Despite early claims, it now looks very doubtful that the multisymbol utterances have any syntactical structure – that is to say, it is doubtful if the arrangement of the individual symbols contributes to the meaning. I have already mentioned the conclusions of Terrace and his colleagues that the detailed analysis of the utterances of Nim showed that his multisign sequences lacked any deliberate grammatical structure. Sue Savage-Rumbaugh has argued from Kanzi's ability to respond to sentences of spoken English (occasionally supplemented by the use of lexigrams), that he must be making use of syntactical rules because his response to a given word varies with its function in the sentence.[25] Her interpretation has, though, been questioned by Joel Wallman, who argues that the different responses can be explained by assuming that Kanzi simply chooses the most appropriate meaning, and that no syntactical analysis need be invoked.[26] The arguments on this question and, more widely, on the relations between the linguistic behaviour of apes and of children, are intricate and contentious, and further discussion would be out of place here. The state of play, as seen by someone at the sceptical end of the spectrum, has been nicely summarized by Wallman. He explains that ape researchers convinced of the linguistic abilities of apes:

> have protested that detractors employ rubber rulers and moving targets in comparing the apes and children ... Criteria are revised "on the run", new criteria are imposed, or different methods of imposing a given criterion are used for apes and for children ... The ape researchers, on the other hand, could be described as having fixated on certain of these criteria, generally the most mechanical and readily quantified, and then having either expressly trained them or searched them out in their data. This is, in itself, not an objectionable practice. One is put in mind, however, of Diogenes, the Athenian cynic who, in response to Plato's definition of man as "a featherless biped", produced a plucked chicken. In both cases, the claimed identity may be valid on the narrow grounds used, but there is still a profound difference between the alleged equivalent and the genuine article. The difference between the cases – aside from the fact that, as I have argued, the apes are *not* equivalent to children on the various indices used in the literature – is that Diogenes knew he was holding a chicken.

THE EVOLUTION OF THE 'LANGUAGE ORGAN'

> The language process ... confronts ... a truly boundless area, the scope of
> everything conceivable. It must therefore make infinite use of finite
> media...
> – Wilhelm von Humboldt (elder brother of Alexander) in 1836[27]

Even if Sue Savage-Rumbaugh is right in believing that Kanzi must have had
some primitive notion of syntactical rules to respond to English sentences in the
way he did, his use of language could hardly be described as 'making infinite use
of finite media'. To do that needs a flexible grammar, and one implication of the
work on apes that we have been discussing is that the evolution of such a
grammar is very unlikely to have occurred before our ancestral line diverged
from that of the apes, something like six million years ago. Given all the other
differences between us and apes that must have developed during that period, the
development of grammar in six million years does not seem too surprising, but
when and by what steps it occurred is far from clear.

Oddly, Chomsky, while recognizing the enormous selective advantages that
language must confer, sometimes writes as though he believes that the evolution
of human language cannot be explained by natural selection; he suggests that the
explanation is more likely to lie in molecular biology or in the unknown ways in
which 'physical laws apply when 10^{10} neurons are placed in an object the size of a
basketball, under the special conditions that arose during human evolution'.[28]
This is puzzling. As Steven Pinker and Paul Bloom point out, since language supplies information about the world (and particularly about other individuals with
whom we may need to interact) that could otherwise only be obtained first-hand,
it is obviously adaptive; and 'natural selection is the only scientific explanation of
adaptive complexity'.[29] If we adopt Chomsky's nomenclature and speak of the
machinery responsible for human language as the 'language organ', that organ is
a splendid example of Darwin's 'organs of extreme perfection and complication'.
For it to have resulted from natural selection, it must have evolved by a series of
steps 'each grade being useful to its possessor', but that requirement would seem
to present no more difficulty in the case of language than in the case of other
organs of extreme perfection and complication. Neither vocabulary nor
grammar are monolithic.

The real difficulty in elucidating the evolutionary history of 'the language
organ' is that we know very little about the organization and details of the relevant neural machinery as it now exists, and almost nothing about the stages by
which human language itself reached its present state. Accepting the view that
human language evolved by natural selection as a method for transferring particular kinds of information between individuals, does not, of course, imply that

all parts of the relevant machinery evolved because of their linguistic role, or even that the origin of every part was necessarily adaptive. Language involves complex handling of representations, but so do many other activities in the brain, and there has long been controversy about the extent to which the linguistic machinery is different and specialized. We know that, after brain damage in children not more than ten years old, other parts of the brain can sometimes take over the linguistic role of the damaged parts, but we know very little of the nature or the extent of the reorganization that makes this possible. The discovery by Myra Gopnik (see p. 291) of an inherited disorder caused by a single dominant gene, and resulting in the selective loss of particular grammatical abilities without affecting other aspects of speech or other mental abilities, implies that modern humans possess some machinery that is especially concerned with grammar.

Bickerton has speculated that the evolution of our linguistic ability occurred in two stages.[30] He draws attention to the paradox that only a slight improvement in tools occurred during the long period of hominid evolution in which the size of the brain was increasing from little more than that of an ape to the lower part of the range for modern humans, yet the rapid improvement in tools and other artefacts – cave paintings, decorated objects, personal ornaments – that started less than 50,000 years ago has been accompanied by no change in brain size. We therefore need an explanation for the slow but prolonged increase in brain size, and a different explanation for the rapid changes in artefacts beginning less than 50,000 years ago. He suggests that the increase in brain size was associated with the development of a protolanguage, consisting of a lexicon of symbols representing categories of objects, but totally devoid of any syntax, and that the rapid recent development was the result of the acquisition of syntactical ability. He supposes that the 'slow, clumsy, ad-hoc stringing together of symbols' that gradually became the hypothetical protolanguage would have provided sufficient advantage to account for the evolutionary increase in brain size, but would not have led to radical innovations in behaviour. Later, the addition of syntax, by making it possible to construct complex propositions – 'cause-and-effect, if ... then, not ... unless' – would have allowed our ancestors to escape from 'doing things the way we always did but a bit better' into 'doing things in ways no one thought of before'.

Bickerton also makes a more far-reaching point. It is that language has roles other than its role in communication. Because language is a representational system, it can store information and is involved in many thought processes, and these roles, too, are relevant in its evolution. He does not claim that thought is impossible without language – pointing out that that would lead to the absurd conclusion that 'no nonhuman (and for that matter, no infant human) could think' – but he does, rather startlingly assign the linguistic machinery of the brain an unexpected role.

This is best illustrated by the example that he himself gives. There are presumably, in the brain, several ensembles of neurons whose respective appropriate firing represents different aspects of, say, the concept 'cat'. There may, he says, be an 'auditory cat', a 'visual cat', and a 'linguistic cat'. The linguistic cat, he says, is a 'holistic cat', because if he (Bickerton) is asked about the word 'cat' he can tap his auditory knowledge, his visual knowledge, and so on. So far there is no problem. But then he goes on to suggest, albeit tentatively, that the linguistic cat is the *only* holistic cat because 'it takes some kind of arbitrary symbol to tie together all the representations of all the attributes that make up our idea of "cat"'. This is, of course, another aspect of the binding problem that we referred to in Chapter 14, but even if we accept that the 'linguistic cat' helps *us* to link together the 'auditory cat' the 'visual cat' and the 'olfactory cat', linguistic representations cannot be the only way of solving that particular binding problem, since it must be solved by animals without language, such as mice and other cats.

The case for an essential role of language in allowing us to make use of abstract concepts is perhaps stronger, but even here there are difficulties. As Steven Pinker has pointed out, we do not need to know the word *Schadenfreude* to understand the idea of pleasure in the misfortunes of others.[31] Animals without language obviously do not attach labels to abstract concepts and categories, but they use them. A dog will distinguish between people it trusts and people it does not trust, and to make such a distinction it must have appropriate representations in its brain. The point that, in animals with language, the representations used for linguistic communication can also be used in other ways, and that they play a large part in abstract thought, is both valid and important, but a monopolistic role for such representations in creating concepts is unlikely.

As well as accounting for the origin of language, we need to account for the origin of languages. Can we improve on the story of the Tower of Babel as an explanation of the multiplicity of languages in the world? The immediate explanation is that we can only inherit information stored in our parents' genes, and human genes store neither the information in the lexicons we use nor most of the information in the grammars we use. Instead, they store instructions for making a neural machine capable of two linguistic tasks. Exposed to speech at an early age and in an appropriate social setting, we are able to compile a lexicon; and, with the help of an inbuilt set of rules (Chomsky's 'universal grammar'), we are also able to deduce and apply the rules of the grammar employed in the speech we hear. This method of acquiring vocabulary and grammar works because what is crucial is that we should use roughly the same vocabulary and the same grammar as other members of our community, rather than a vocabulary or a grammar specified in some other way. (Indeed, as we progress from babbling to speaking, we actually lose the ability to make many sounds that do not occur in the language we are hearing.) But a consequence of the method is that languages gradually change – hence the multiplicity.

As to why the language machinery evolved in this fashion, the most plausible explanation, so far as the lexicon is concerned, is that it is both economical and flexible. It is economical because, by possessing a mechanism that builds up a lexicon from the speech that we hear, we avoid the need to store a great deal of information in our genes – the information is, as it were, stored in that part of the environment that we use as a source of inputs.[32] It is flexible because it allows the vocabulary to be expanded or modified as circumstances change. Unlike the elements of our vocabulary, the elements of our grammar are neither arbitrary nor independent, so some inherited rules are necessary; but because, with the help of those rules, we are able to deduce the rules of the grammar in the speech we hear, not everything has to be specified in the genes. Again, information is stored in the environment. How the inherited 'universal grammar' actually evolved, and how it or the lexicon are represented in the neural machinery, are questions to which we have no answers. Our understanding of the language areas of the brain lags far behind even our limited understanding of the visual areas.

V THINKING ABOUT THINKING

We can imagine what life would be like without vision; it is much more difficult to imagine what it would be like without language; it is almost impossible to imagine living without memory or emotions or the ability to switch attention between one thing and another or to plan. It is these features of the mind, and the machinery behind them, that I now want to turn to.

19 Memory

*forms, figures, shapes, objects, ideas, apprehensions, motions, revolutions: these
are begot in the ventricle of memory, nourished in the womb of pia mater, and
delivered upon the mellowing of occasion.*
—Shakespeare, *Love's Labour's Lost*

Since the pia mater is the fibrous sheath that encloses the whole brain,
Shakespeare was on safe ground in nourishing memories within it. 'Ventricle of
memory' is less happy; we now know that the ventricles of the brain – the cavities
inside it – contain nothing more exciting than a dilute solution of salts, sugar,
proteins, and a few other substances. So where are memories stored? And, more
interestingly, how are they stored?

In September 1953, William Scoville in Hartford, Connecticut, operated on a
twenty-seven-year-old mechanic in an attempt to relieve intractable epileptic
seizures.[1] The mechanic, HM, had been knocked unconscious by a bicycle at the
age of nine; he had had minor epileptic attacks up to ten times a day from the age
of ten, and major attacks with tongue biting, urinary incontinence, loss of consciousness and prolonged somnolence from the age of sixteen. Despite almost
toxic doses of anticonvulsants, the attacks had become more frequent and more
severe, so that by the time of his operation he was having about fifty major attacks
a year, forcing him to give up his job and making life intolerable. Surgery seemed
the only possible treatment, and because of the known tendency of epileptic
seizures to start deep in the temporal lobes, parts of both temporal lobes lying near
the midline were removed. The individual structures removed on each side
included the front two-thirds of the hippocampus, the amygdala, and some of the
adjacent areas of the cerebral cortex.[2] (Because the lateral parts of the temporal
lobe were not interfered with there was no risk of producing an aphasia.)

Following the operation, the seizures were less incapacitating (and eventually
very infrequent), and HM's intelligence seemed to be slightly improved – probably because he was less drowsy. But these improvements were bought at a
terrible price. The main effect of the operation was an unexpected and devastating loss of the ability to form new memories – what neurologists call an *anterograde amnesia*. There were other problems too. Apart from Dr Scoville, whom he
had known for many years, HM could no longer recognize any of the hospital

staff, and he could not find his way about. He also suffered a partial loss of memory of events before the operation: a *retrograde amnesia*. He did not recall anything of the pre-operative period in hospital, or the death of a favourite uncle who had died three years earlier; from time to time he would ask when his uncle was coming to visit, and each time he was told of his uncle's death he would be dismayed as if he were hearing the news for the first time. Yet his earlier memories seemed to be both vivid and intact, and he even remembered some trivial events just before he had come into the hospital.

In a detailed examination by Brenda Milner nineteen months after the operation, HM thought that the date was still 1953; he had no memory of having talked with a doctor just outside the consulting room; he constantly reverted to events in his boyhood, and he scarcely seemed to remember that he had had an operation.[3] Formal intelligence testing showed a slight improvement, particularly in arithmetic, compared with his performance before the operation, and a whole battery of tests failed to show any deficit in perception, abstract thinking or reasoning ability. His motivation was said to be excellent, and his personality unimpaired.

The very severe anterograde amnesia, and partial retrograde amnesia, persisted. A few years after the operation, when HM's family moved to a new address a few blocks away on the same street, he was unable to learn the new address – though he remembered the old one without difficulty – and he could not find his way home without help. He would do the same jigsaw puzzle, or read the same magazines, repeatedly without showing any familiarity with them. Half an hour after lunch he would not remember what he had had, or even that he had had lunch. An inkling of what such severe amnesia seems like to the sufferer is given by some remarks made by HM many years later.[4] In an interval between psychological tests, he looked up and said rather anxiously,

> Right now, I'm wondering. Have I done or said anything amiss? You see, at this moment everything looks clear to me, but what happened just before? That's what worries me. It's like waking from a dream; I just don't remember.

In the now classical paper in which Scoville and Milner first reported their findings – 'partly for their theoretical significance, and partly as a warning to others' – they also described the results of psychological tests on eight psychotic patients who had had similar, though generally less radical, surgical operations in attempts to improve the quality of life following the failure of more conservative methods of treatment.[5] They found that the degree of memory loss was related to the extent to which the hippocampus had been removed, and they concluded that 'the anterior hippocampus and [para]hippocampal gyrus, either separately or together, are critically concerned in the retention of current experience'. (On the other hand, the limited extent of HM's retrograde amnesia showed that the hippocampus – or at any rate the anterior two-thirds that had been removed on each side – was not essential for access to memories formed several years earlier.) Removal of a large part of the hippocampus and adjacent cortex on only one side did not affect memory. Scoville and Milner could not say whether interference with the amygdala contributed to the amnesia, since it had been removed on both sides in all the cases they studied, but nearly thirty years later an almost pure anterograde amnesia was described in a patient with lesions on both sides limited to the hippocampus.[6] This provides strong evidence that the hippocampus plays a crucial role in the formation of new memories; it does not, of course, imply that those memories are stored in the hippocampus, or that other parts of the brain are not involved.

Surgical interference and problems with the blood supply to the brain are not the only causes of the kind of *amnesic syndrome* experienced by HM. A century ago the Russian neuropsychiatrist Sergei Korsakoff described chronic alcholism as the commonest cause, though we now know that the amnesia is caused not by the alcohol but by the alcoholic's poor diet, and a consequent lack of thiamine, a vitamin essential for normal carbohydrate metabolism. Other causes of thiamine deficiency can also cause *Korsakoff's syndrome*. Examination of the brains of patients suffering from Korsakoff's syndrome shows no hippocampal damage, but instead damage to certain collections of cells in the between-brain that are closely connected with the hippocampus.[7] There is, too, often damage in the frontal lobes. The amnesic syndrome can also be caused by strokes, head injury, and the rare but frightening encephalitis that can follow herpes simplex infections if the virus travels from the nose up the olfactory nerves to reach the brain. Amnesia is, too, an early feature of Alzheimer's disease, generally preceding the dementia.

Some drugs tend to interfere with the long-term recall of events that occur while the drug is present.[8] Both hyoscine, a substance found in thorn apples, and the benzodiazepines, (substances such as Valium and Mogadon [nitrazepam]) act in this way. Hyoscine decreases the efficacy of the excitatory transmitter acetyl choline by competing for the receptor sites at which acetyl choline acts, and it may be significant that, in patients with Alzheimer's disease, there is a marked decrease in the number of nerve cells whose terminals release acetyl choline in the cerebral cortex. Benzodiazepines increase the efficacy of the inhibitory transmitter GABA by increasing the affinity of its receptors, and it may be significant that GABA receptors are found mainly in the cerebral cortex and limbic structures including the hippocampus. A footnote in my favourite textbook of pharmacology tells of the ninety-one-year-old grandmother of one of the authors, who 'was growing increasingly forgetful and mildly dotty having been taking nitrazepam for insomnia regularly for years'. When a 'canny practitioner' diagnosed the problem and stopped the drug there was a dramatic improvement.[9] On the other hand the amnesic effect of benzodiazepines is sometimes exploited to protect patients from potentially unpleasant memories. I remember, after undergoing an endoscopy for which I had been given a substantial dose of Valium, being told by the physician that he had found nothing wrong and that he was sorry I had had a 'rough ride'. Not only did I have no memory of any discomfort, but I felt on top of the world.

HOW MANY KINDS OF MEMORY?

At the time that HM was operated on, and throughout the nineteen fifties, memory tended to be thought of as just one kind of process, though several years earlier Donald Hebb had suggested that there might be different mechanisms for short-term memory and long-term memory; and back in 1911 the Swiss psychologist Edouard Claparède had described what was to become known as implicit memory – a sort of covert memory which affects the behaviour of a subject without the subject being aware that the memory exists. By the late 1960s it was clear both from studies of normal behaviour and from studies of patients suffering from neurological diseases that memory could be of several different kinds.

Short-term memory (also called working memory), lasts a matter of seconds or at most a few minutes; long-term memory can last anything up to a lifetime. (Although people with senile dementia who forget what they have recently said or recently heard are often described as 'having trouble with their short-term memory', it is in fact the loss of the ability to form new long-term memories that accounts for their distressing behaviour.) We looked at the role of verbal short-term memory in Chapter 16, and at the use of PET scanning to locate the relevant machinery. Despite HM's complete loss of long-term memory for events subsequent to his operation, his short-term memory – as judged by the number of digits he could repeat forwards or backwards immediately after hearing them spoken, or by his ability to report the sequence of coloured lights he had just seen displayed – remained within normal limits.[10] And like normal subjects, in the absence of any distraction he could prolong the memory by continuous rehearsal. Although he could not remember a conversation afterwards, while it was going on he could remember the words he heard long enough to converse sensibly; indeed he could repeat and transform sentences with complex syntax, and get the point of jokes even when these depended on double meanings.[11]

The division of long-term memory into explicit and implicit categories really started with what Claparède himself describes as a curious experiment on one of his patients.[12] The patient was a forty-seven-year-old woman who had already been in an asylum for five years suffering from Korsakoff's syndrome. Claparède concealed a pin in his hand when he shook hands with her. She soon forgot the prick, as she forgot all recent events, but whenever he subsequently attempted to shake hands with her she withdraw her hand.

This experiment of Claparède's can be regarded simply as an example of classical 'conditioning', but more interesting examples of implicit memory are provided by a number of memory-dependent tasks that patients with the amnesic syndrome are still able to do. HM, for example, was able to learn several motor skills including tracking a moving spot with a stylus, and drawing round the outline of a star-shaped figure when all he could see was the reflection of the figure in a mirror. With these quite tricky tasks, his ability improved as he went through many trials spread over several days, yet each day he was unaware that he had attempted the task before.[13]

These examples of the ability of some patients to use information they do not know they possess is reminiscent of several of the kinds of behaviour that we met in the chapter on *Disordered Seeing with Normal Eyes* – the behaviour of the patient DF, who could post a card through a slot without difficulty though she could not indicate the orientation of the slot; the covert recognition of faces by de Haan, Young and Newcombe's patient, PH, and Sergent and Poncet's patient, Mme V.; and, most striking of all, the uncanny performance of patients with blindsight. What all these kinds of behaviour have in common is that the patients are able to make use of information that they are not conscious of possessing, *but which normal subjects would be conscious of possessing.*

Neither short-term memory nor implicit memory is, of course, what most people have in mind when they say 'I used to have a good memory', or 'I seem to be losing my memory'. What they are referring to (and what is so conspicuously lost in patients like HM or patients with Korsakoff's syndrome) is *explicit* memory – the sort of memory that allows you to say 'I remember this' or 'I don't remember that'.

Explicit memory itself can be of two kinds;[14] *episodic memory* – memory of individual events personally experienced; my seeing the Erecthion for the first time, in the spring of 1997 – and the oddly named *semantic memory* – oddly named because the term is used to refer not only to memories of the meanings of words but also to memories of facts, of concepts and of general knowledge of the world; my knowing, for example, that each caryatid on the Erecthion supports two and a half tons of marble. The elements of semantic memory must each have been acquired, directly or indirectly, from some past personal experience, but that experience is generally forgotten. Where it is remembered, a memory may contain both episodic and semantic elements. My curious and useless knowledge about the loads borne by caryatids is the result of such a memory: I happen to remember reading a sentence by Osbert Lancaster (who thought caryatids were a lapse from the usual classical Greek good taste) describing them as 'elegant flower maidens [who] simper as unconcernedly as if they had never been called upon to balance two and a half tons of Pentelic marble on their pretty little heads'.[15]

In patients such as HM, it is the loss of episodic memory that is most conspicuous, since most of his semantic memories would have been established before the period covered by the retrograde amnesia. As we have seen, language and the rules of arithmetic caused him no difficulty. He did, though, learn the meanings of new words less easily than normal people.

Losses of episodic and of semantic memory do not always occur together. We have already met many examples of loss of semantic memory, though not under that name. The patients with visual agnosias for objects or faces (Chapter 13) and the patients who could not remember particular words or their meanings (Chapter 16) were all suffering from the loss of different kinds of semantic memory, but their episodic memory – their memory for past events in which they had been personally involved – was not usually affected. The converse pattern – loss of episodic memory without loss of semantic memory – is rare, but examples have recently been reported by F. Vargha-Khadem, and her colleagues in London and Bethesda.

They described three patients, in their teens or early twenties, with severe anterograde amnesia resulting from brain injuries that occurred at birth or in early childhood.[16] All were unable to learn their way about, to say how they had spent the day, or to remember telephone conversations, television programmes, visitors or holidays; and forty minutes after spending several minutes copying a complex drawing they could reproduce almost none of it (Figure 19.1). The remarkable thing about these patients is that all three went to mainstream schools and learnt to speak, to read, to write, and to learn about the world, at levels described as 'within the low to average range'. Because the brain injuries occurred so early, most of this semantic knowledge must have been acquired after the brain injury, and after loss of episodic memory was first noticed.

Using magnetic resonance imaging, Vargha-Khadem and her colleagues found that both hippocampuses in all three patients were only about half the normal size, whereas the temporal lobe cortical areas were normal or almost normal. They therefore suggest that the hippocampus is crucial only for the acquisition of episodic memories, and that it is cortical areas adjacent to the hippocampus that largely support the acquisition of context-free semantic memories. Where both hippocampus and adjacent cortical areas are destroyed – as in HM – the patient can acquire neither episodic nor semantic memories.

TOPOGRAPHICAL MEMORY

Learning our way about a house or a town or a tract of countryside is something we are rather good at – perhaps not surprisingly given the survival value of not getting lost. Anyway, it is something we manage without any formal instruction and at quite an early age. Initially, learning our way must depend on episodic memory, but as the area becomes familiar we do not remember the individual journeys that we have made, and our knowledge of the layout would seem to be just part of our knowledge of the world – in other words, part of our semantic memory.

copy   delayed recall

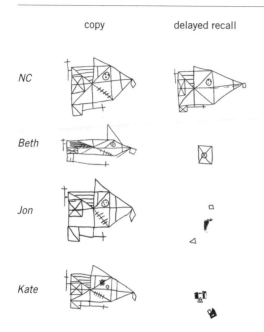

NC

Beth

Jon

Kate

FIGURE 19.1 Three patients in their teens or early twenties, who had hippocampuses about half the normal size, spent several minutes copying a complex drawing. Their copies were fairly accurate, but forty minutes later, when they were asked to reproduce the drawing from memory, they could produce almost nothing. The performance of a normal person in this task is shown at the top of the figure. (From F. Vargha-Khadem and her colleagues.)

In the 1970s John O'Keefe and his colleagues recorded the activity of nerve cells in the hippocampus of a rat that could move freely around in an enclosure, and they made the startling discovery that, while the firing of some cells depended on what the rat was doing, the firing of others depended only on where in the enclosure it was.[17] These *place cells* therefore seemed to represent a cognitive map of environmental space. Cells with similar properties were later found in the rat's posterior parietal lobes. Experiments like those by O'Keefe and his colleagues cannot, of course, be done on humans, but it is well known that lesions in the posterior parietal cortex cause profound disturbances in the appreciation of space in both humans and monkeys; and there are recent imaging studies that show a dramatic increase in blood flow in certain areas, and particularly in the right hippocampus, of humans who are recalling knowledge of the topography of the local environment.[18]

In the first of these studies Eleanor Maguire and her colleagues at the Institute of Neurology, in London, used PET scanning to see which areas of brain became active when experienced London taxi drivers described in detail how they would get from – to give just one example – Grosvenor Square, in Mayfair, to 'The Elephant and Castle', in Southwark, dropping a passenger at the Bank of England on the way.[19] London taxi drivers were chosen because, before they are licensed, they study for three years and must pass stringent tests of their knowledge of London topography. It turned out that the right hippocampus along with several other areas (including areas in the parietal lobes) lit up as each driver described his route. A similar increase in the blood flow through the right hippocampus was not produced simply by recalling and describing a sequence of events, or unconnected landmarks or scenes: it seemed to occur only when the task required that a fairly complex series of landmarks be remembered in sequence in their spatial context.

In a second series of studies, Maguire and her colleagues made use of a commercially available computer game that presented the viewer with a coloured three-dimensional 'fully textured' first-person view of a virtual-reality town.[20] The town had streets, shops, bars, a cinema, church, bank, train station and video-games arcade, and subjects could move through the streets and into and through buildings using a keypad with forward, backward, left-turn and right-turn buttons. An editing program allowed streets to be blocked or doors into buildings to be opened or closed. Subjects were allowed up to an hour to explore the town and learn their way about, and they were then PET-scanned as they performed four different tasks. The first was to use their knowledge of the town to move from a starting point to a destination that was identified by being shown on a screen. The second was similar except that, between the learning period and the test period, a road was blocked and doors into certain buildings were locked. The third task was to move through the town simply by following a trail of arrows. And the fourth was just to look at a series of static pictures of the town. Only the first two tasks required navigation using topographical knowledge. Comparison of the regional blood flows during the different tasks, and (for the navigational tasks) comparison between successful and unsuccessful attempts, showed that activity in the right hippocampus was strongly correlated with successful navigation using topographical knowledge. Where the topography had been changed by blocking streets or locking doors, the subject had to switch strategy to work out detours, and in this situation there was also increased blood flow in the left frontal region.

THE RELATION BETWEEN SHORT-TERM AND LONG-TERM MEMORY

Because patients with the amnesic syndrome usually have normal short-term memories, it has long seemed likely that what is amiss in such patients is a failure in the process that is thought to convert short-term memories into long-term memories. That hypothesis assumes, of course, that entrance to the long-term memory store is necessarily via a short-term store – an assumption that seemed shaky when patients were discovered whose short-term memory was limited to one or two digits but who, in spite of this, seemed to have normal long-term memory.[21] It was later realized, though, that impairment of short-term memory could be selective, and restricted to memories of certain kinds, suggesting that the different kinds are stored separately.[22] A patient with impaired short-term memory of one kind might then succeed in remembering facts or events long-term by starting with short-term memories of other kinds. For example, a patient whose short-term store for *phonological* information was damaged would be unable to remember spoken words by their sounds but might be able to remember them by their meanings.[23]

Providing a route to long-term memory stores is not, of course, the only role of the short-term stores. They also hold information for immediate use, whether or not it is destined to enter a long-term store. The short-term store for phonological verbal information, for example, almost certainly serves as an aid to speech comprehension by providing a back-up while syntactic and semantic analysis proceeds. Patients with selective impairment of verbal short-term memory find it difficult to understand sentences with convoluted syntax, or instructions packed with detail – 'Put the red circle between the yellow square and the green square'.[24] More generally, the existence of short-term stores allows the brain to compare the visual or auditory information, or any other kind of information, that it is receiving at any moment with the information (not necessarily of the same kind) that it received a few moments ago, a crucial ability in a changing world.

There is evidence from observations on monkeys, made by C. F. Jacobsen at Yale over sixty years ago, that the ability to hold information of some kinds for immediate use involves the so-called prefrontal cortex,* probably in association with other areas. He discovered that normal monkeys who see food being placed under one of two identical cups can find it if they are given immediate access but their performance gets steadily worse with delay, becoming no better than chance after two minutes. Monkeys lacking prefrontal cortex perform no better than chance even if the delay is as little as two seconds.[25]

More recently, work by Joaquín Fuster and his colleagues, at the University of California at Los Angeles, showed that during the period in which a monkey is remembering the position of a hidden object there are cells in the prefrontal cortex (and also in parts of the parietal and temporal cortex) that fire continuously.[26] And Patricia Goldman-Rakic and her colleagues at Yale have shown that, for different positions of the hidden object, it is different cells within the prefrontal cortex that fire, so there is a kind of map.[27] Similar delayed-response tasks can be used to test a monkey's short-term memory for the shape or colour of objects, and it turns out that such memories are also accompanied by the firing of nerve cells in the prefrontal cortex. Because these cells are segregated from those that fire while the monkey remembers position, it looks as though the

---

*This is the part of the cortex of the frontal lobes that lies in front of the areas concerned with movement and with speech. The name is illogical because the prefrontal cortex is part of – not in front of – the frontal lobes, but it was introduced by Ferrier more than a century ago and it is now thoroughly entrenched.

prefrontal cortex contains separate processing mechanisms for remembering 'what' and 'where' an object is.[28]

SO HOW DOES MEMORY WORK?

Like almost every other activity of the nervous system, remembering depends on the subtleties of the interactions between nerve cells. In Chapter 9 we saw how the peculiar properties of certain glutamate receptors (the so-called NMDA receptors), enabled them to produce long-lasting changes in synaptic responsiveness. In Chapter 15 we saw how modification of synaptic strengths in an artificial neural net enabled the net to recognize photographs of faces. We also saw how modification of synaptic strengths in line with Hebb's principle – that a synapse becomes more effective if the presynaptic and postsynaptic cells fire simultaneously or almost simultaneously – allowed a neural net to learn to associate two patterns that were repeatedly presented together. Work with neural nets can, of course, only be suggestive, but there is persuasive evidence that the suggestion is along the right lines.

Strangely enough, the first clear-cut direct evidence that learning depends on specific changes in synaptic effectiveness came not from mammals or even other vertebrates but from experiments in the 1970s on the marine snail *Aplysia californica*. This little animal has a variety of defensive reflexes for withdrawing its tail, its respiratory gill, and the spout that it uses for discharging waste and sea water. These responses can be strengthened or weakened under different circumstances, and in a long series of elegant experiments Eric Kandel and his colleagues have elucidated the mechanisms involved in great detail, with the deliberate aim of analysing the machinery of learning in an animal with a central nervous system small enough (and individual nerve cells large enough) to make investigation of the contribution of individual cells less daunting.[29] Initially, touching the siphon causes withdrawal of the gill, but with repeated touches the response declines, and the animal is said to be *habituated* to the stimulus. This habituation is accompanied by a diminution in the amount of transmitter (glutamate) released by the terminals of sensory cells at their synapses with motor cells (or with intermediate nerve cells that themselves form synapses with motor cells). Noxious stimuli applied to the head or tail of the snail increase the withdrawal responses to harmless stimuli subsequently applied to the siphon. This *sensitization* is accompanied by modification of the properties of a number of synapses in the neural circuit concerned with the gill-withdrawal reflex, including those that are involved in habituation. Both habituation and sensitization can last anything from a few minutes to several weeks depending on the amount of training, but the same set of synapses seem to be involved irrespective of the duration. Another important feature of the picture that emerges from the work on *Aplysia* is that the synapses whose modification is responsible for learning are the synapses that are involved in the behaviour that the learning will change. In other words, learning is not done by separate 'memory cells' but by the cells engaged in the task with which the learning is concerned.

In the early 1970s – shortly after the first papers by Kandel and his colleagues on memory in marine snails, and twenty years after HM had had his unfortunate but informative operation – Tim Bliss, from the National Institute for Medical Research in London, and Terje Lømo, from the University of Oslo, provided evidence for an increase in synaptic efficacy following intense synaptic activity in the hippocampuses of anaesthetized rabbits.[30] They showed that, in line with Hebb's prediction (p. 253), a short burst of rapidly repeated electrical stimulation of nerve fibres that formed synapses with the dendrites of particular cells in the hippocampus increased the efficacy of those synapses – as judged by the effects of subsequent single stimuli – and that the 'potentiation' sometimes lasted as long as ten hours, which was as long as the preparation could be maintained. In later experiments, 'long-term potentiation', as they called it, sometimes lasted for days or weeks. Since the nerve fibres that Bliss and Lømo were stimulating formed one of the main inputs from the cerebral cortex to the hippocampus, the possible relevance of the long-term potentiation to memory was obvious.

In the last quarter-century, a good deal has been learnt about the mechanism of long-term potentiation, and its likely role in learning and memory. The 'Hebbian' behaviour is nicely explained by the peculiar properties of the glutamate receptors of the NMDA type described in Chapter 9. If the postsynaptic cell has just fired when the presynaptic cell releases glutamate, the potential difference across the postsynaptic membrane will be reduced, and the binding of glutamate to the receptor will open the channel that allows calcium ions to enter and initiate the various changes that lead to long-term potentiation.

A number of subtleties about long-term potentiation and the events that produce it have been revealed by experiments on slices of hippocampus, which are easier to work with than a whole brain. *Single* brief bursts of repetitive stimulation (say 100 shocks per second for one second) increase the synaptic strength, but the effect declines spontaneously over the next couple of hours. In contrast, several bursts in rapid succession cause a similar increase in synaptic strength which is maintained over hours or days. Significantly, the difference in behaviour does not occur if the slice of hippocampus is pre-treated with drugs that inhibit protein synthesis: the long-term potentiation occurs but it then spontaneously declines. The implication is that two processes are involved: one that follows either single or repeated bursts of stimulation, that does not involve the synthesis of new protein, and that presumably results from a change in the receptor that is slowly but spontaneously reversed; and another that is initiated only by repeated bursts, that *does* require the synthesis of new protein, and that causes a more or less permanent change in the receptor, so consolidating the altered behaviour of the synapse.

Since the protein molecules that consolidate the change in the synapse are synthesized in the nucleus of the nerve cell, a long way from the synapses on the dendrites, there is a problem: how does the new protein get to the right synapses? The answer seems to be that it binds to any synapse in the post-synaptic cell that has been active fairly recently, not just the synapses whose activity triggered the protein synthesis. The evidence comes from some elegant experiments by Uwe Frey in Magdeburg and Richard Morris in Edinburgh.[31] They showed that a *single* brief burst of repetitive stimulation to a nerve forming one synapse on a cell could cause the *maintained* type of long term potentiation at that synapse *provided that within the previous ninety minutes or so there had been repeated bursts of stimulation to another nerve that formed a different synapse on the same cell.* It looks, then, as if the reversible change at a recently active synapse not only strengthens the synapse for an hour or two, but also enables it to pick up the new protein that makes the increase in strength more permanent. Because their results show that proteins whose release is triggered by vigorous activity at one synapse can consolidate synaptic changes at other synapses, Frey and Morris have wondered whether our tendency to remember the incidental details accompanying momentous events – Where were you when you heard that Kennedy had been shot? – might be explained by the consolidating effect of a burst of protein synthesis initiated by synaptic events caused by the news.

This pattern of interaction between different synapses on the same cell obviously represents a departure from the original Hebbian assumption that modification of a synapse depended only on events at that synapse, but there is, anyway, independent evidence that the assumption is wrong. It turns out that interactions between synapses, leading to their modification, can also occur between synapses on neighbouring cells,[32] probably as the result of the release of nitric oxide (NO) which readily passes through cell membranes and acts as a diffusible messenger.[33]

The recent experimental work on synapses is impressive, and the notion that consolidated long-term potentiation plays a key role in memory is very attractive, but the evidence I have discussed so far is all rather indirect. If the notion is right, it ought to be possible to interfere with learning by interfering with the hypothetical mechanisms; and it is.

Evidence for a role for NMDA receptors (that is, glutamate receptors of the NMDA type) comes from experiments by Richard Morris and his colleagues in Edinburgh.[34] They showed that a selective and reversible inhibitor of NMDA receptors, which (as expected) prevented long-term potentiation in hippo-campal slices, also affected the ability of rats to find a submerged platform in a water maze. Rats given the inhibitor took longer to find the platform; and a comparison of the effects of different doses of inhibitor showed that the concen-trations of inhibitor that were needed, in the rats' tissues, to affect performance in the maze were similar to the concentrations needed, in the hippocampal slices, to inhibit long term potentiation.

Although the obvious interpretation of this result is that the inhibitor interfered with 'spatial learning' by acting on NMDA receptors on cells in the hippocampus, it could have been inhibition of NMDA receptors elsewhere that was responsible. The link between *hippocampal* NMDA receptors and topographical memory has, though, been dramatically confirmed by more recent experiments by Susuma Tonegawa, Eric Kandel, and their colleagues at Columbia University. Using sophisticated molecular biological techniques, they developed a method which makes it possible, at a chosen time in the development of an embryo, to knock out a selected gene in a selected region of the brain or in a specific cell type.[35] With this method, they were able to knock out the gene for a subunit of the NMDA receptor, in the small region of the mouse hippocampus that is known to contain the nerve cells whose activity is related to the mouse's position in space (see p. 321). The mice grew up without obvious abnormalities, but when the adult mice were tested in a water maze similar to that used by the Edinburgh workers they were found to have great difficulty in learning to find the submerged platform. Other kinds of learning were not affected.

We have seen that *lasting* changes in the behaviour of synapses in the hippocampus depend on the synthesis of new protein, so if learning depends on such changes we should expect it to be affected by substances that inhibit protein synthesis. And it is. When mice, for example, were injected with a suitable inhibitor of protein synthesis at least five minutes before they were trained on a very simple maze, they remembered the correct path for about three hours but not longer. Normal mice remembered fairly well even after a week.[36]

A role for nitric oxide in memory is suggested by experiments showing that rats that had been injected with a drug that prevents nitric oxide synthesis were unable to find their way in a water maze that normal rats mastered without difficulty.[37] And recently, inhibition of the synthesis of nitric oxide was shown to reduce glutamate release in the olfactory bulbs of a sheep, and also to prevent the sheep from learning the smell of its new-born lamb, so that it accepted its own lamb or other lambs indiscriminately.[38] Once the smell had been learnt, inhibiting the synthesis of nitric oxide did not interfere with recognition or with the sheep's normal response.

Not all changes in synaptic strength are brought about by long-term potentiation acting through NMDA receptors; and learning can involve changes in the number of synapses, and even of dendrites, as well as changes in synaptic strengths. But what is clear is that at the cellular and sub-cellular level machinery exists that is capable not only of simple logical operations but also of being modified by previous experience so that its behaviour changes. It is this machinery that forms the basis of the ability of networks of nerve cells to learn and to remember.

Whizzing them over the net
—John Betjeman, *Pot Pourri from a Surrey Garden*

If the last paragraph sounds upbeat, let me say at once that the way in which the extensive networks of nerve cells that are involved in memory actually work is much less well understood than are the events at the synapses or within the cells. We have seen that, starting with the unfortunate case of HM, there is a great deal of evidence that structures in the hippocampal region of the temporal lobe (and related structures in the between-brain) are necessary for the creation of new memories about events or facts. We have seen, too, that they are also necessary for the retention of past memories, if those memories are of events that occurred only a short time before the damage, but not if they occurred in the remote past. But how short is short and how remote is remote?

HM could not remember the death of his uncle three years before his operation, though he had a normal memory of his early life. There are other cases of retrograde amnesia resulting from damage in the hippocampal region, or in the related structures in the between-brain, in which memory is impaired for events that occurred twenty or more years before the brain damage, usually with the greatest impairment in the most recent years.

Graded impairment of memory, with older memories being better retained, has often been described but it is difficult to be certain about. To know what a patient with retrograde amnesia has forgotten, you need to know what a normal person with similar experiences would remember, and that is not easy. In one case, though, this problem was solved in a rather dramatic fashion. Nelson Butters and Laird Cermak in Boston studied an eminent sixty-five-year-old alcoholic scientist who was unfortunate enough to get amnesia associated with Korsakoff's syndrome but serendipitous enough to have completed his autobiography just two years earlier.[39] By comparing what he remembered with what he had previously written, Butters and Cermak were able to show that his memory of personal events got steadily better as the events chosen became more distant, and that this relation held all the way back to childhood. Here, then, the gradient of impairment existed. But as Butters and Cermak point out there is still a problem: were the older memories remembered better because they were older or because they had become memories of a different kind? Even in people who don't write autobiographies, there is a tendency for old memories to become favourite tales and to be produced over and over again. With continual rehearsal the memories become 'independent of specific temporal and spatial contexts', in other words they become semantic memories. And Cermak and Butters suggest that the gradient they observe may simply be the result of the relatively greater vulnerability of episodic memories.

Whether or not this suggestion is right, the selective preservation of older memories in patients with retrograde amnesia needs an explanation in neural terms. The increasing stability of memories with age is sometimes referred to as the result of consolidation, but the timescale is so much longer than the timescale of the consolidation associated with long-term potentiation (and the conversion of short-term to long-term memories) that it is extremely unlikely that similar mechanisms are involved. The explanation usually given is that, as time passes, there is a gradual reorganization of the storage of long-term memories so that the hippocampal region is no longer involved.

Since the episodes that give rise to memories generally involve a variety of perceptions, it seems likely that the laying down of such memories involves nerve cells in the association areas and in secondary or higher order cortical areas concerned with the different senses. (I shall refer to all these cortical areas collectively as 'the neocortical zone'.)[40] It is also likely that recalling memories involves recreating something like the original pattern of activity in those same sets of cells, or at least some of them. But we know that laying down new episodic memories also depends on a contribution from the hippocampuses and adjacent regions – 'the hippocampal zone'. Initially then, both the hippocampal zone and the neocortical zone must act together. Eventually, when consolidation is complete, the memories are stored in such a fashion that they are available without the involvement of the hippocampal zone, implying that storage is then wholly in the neocortical zone. We therefore want to know the answers to two questions: why is the hippocampal zone necessary for laying down new memories? And what goes on during consolidation that eventually makes the hippocampal zone unnecessary for retaining old memories?

There are no sure answers to these questions, but a popular hypothesis is this.[41] Memory storage occurs as patterns of synaptic strengths (and perhaps also of the firing thresholds of nerve cells) distributed across nerve networks. The neocortical zone and the hippocampal zone together form a single network. Although the storage capacity of the hippocampal zone is much less than that of the neocortical zone, it is supposed that changes in synaptic strengths can occur much more readily in the hippocampal zone than they can within the neocortical zone. So, for recent experiences, the synaptic modifications in the neocortex are not adequate to ensure that the activation of neocortical neurons forming *part* of the pattern of activity representing a memory is able to recreate the *whole* pattern. At this stage, the recreation of the whole pattern depends on a contribution from the hippocampal zone, which, through its widespread connections is able to link together the separate areas of neocortex involved in a particular memory. Each time the memory is recalled, the pattern in the neocortical zone becomes better established and, with time, the contribution from the hippocampal zone becomes superfluous; consolidation is complete.

Whether the hypothesis is right or not, it is of course true that recalling memories repeatedly helps us not to forget them. It would, though, be unsatisfactory if the only way to consolidate memories were to recall them repeatedly, and one possibility, first suggested by David Marr in 1971, is that consolidation involves the replaying of memories stored in the hippocampus back to the neocortex during sleep.[42] Indeed, this might be an important function – perhaps even *the* important function – of sleep. So it is interesting that during phases of 'slow-wave sleep' – so-called because during these phases the electro-encephalographic records show slow changes in voltage – groups of cells in the rat hippocampus discharge in synchronous bursts that lead to increased activity in adjacent areas of cortex that connect with many areas in the neocortical zone.[43] There is also evidence, in rats, that hippocampal cells that tend to fire together while the rat is exploring also tend to fire together during slow-wave sleep that follows the exploration.[44]

An obvious question about the proposed roles of the hippocampal zone and the neocortical zone in memory is why this curious arrangement should have evolved. A possible answer has been suggested by James McClelland, Bruce McNaughton and Randall O'Reilly, in Pittsburgh and Arizona.[45] Using the hippocampal zone as a buffer for rapid storage of new information allows that information to be fed slowly into the neocortical zone, and they argue from computer simulations that this slowness makes it possible for new patterns to be incorporated with less disruption of existing patterns.[46] That this is likely to be a real problem in the brain is suggested by the ease with which our memories can become confused by later events, particularly if those events resemble, but are not identical with, those we are trying to remember.

The idea of shifting representations of memories from one part of the brain to another has received some direct support from imaging studies, though the relevant memories are not any kind of explicit memory but are memories of motor skills. There are a number of reasons for believing that the acquisition of motor skills starts in the prefrontal cortex but that, as the task becomes more automatic, the cerebellum takes over and probably becomes the site of the motor memory. Recently, Reza Shadmehr and Henry Holcomb, at Johns Hopkins, used PET scanning to measure local changes in blood flow in subjects learning a skilled motor task.[47] The results were complicated, but the significant feature in the present context is that when they compared the patterns of blood flow during two performances of the task, one immediately after acquisition of the skill and the other five and half hours later, they found that though the performances of the task were indistinguishable the circulatory pattern had changed. And the change indicated a shift from prefrontal regions of the cortex to a number of regions including the cerebellum. As this change in circulatory pattern occurred after the skill had been established, it suggests that the site in which the memory is represented must also have changed.

FORGETTING

one condition of remembering is that we should forget.
– Théodule Ribot, *Les Maladies de la Mémoire*, 1884

We tend to regard forgetting as a failure of memory, the result of an inadequacy in the machinery for remembering, but a moment's reflection shows that that cannot always be right. If we were to remember every telephone number we have ever used, every scene we have ever looked at, every conversation we have ever heard, every sentence we have ever read, we should be overwhelmed with memories. In fact we transfer to our long-term memory stores only a tiny fraction of the contents of our short-term stores. We retain only what is likely to be useful.

In rare individuals, the mechanism that so effortlessly prevents us from being swamped appears to be inadequate. In Moscow in the mid-1920s the neuropsychologist Alexander Luria was consulted by a young man, S., who had come to see him at the suggestion of the editor of the paper on which S. worked as a reporter – a job he had drifted into without any particular plan of what he wanted to do in life.[48] The Editor had noticed that when he assigned tasks for the day, including lists of addresses and detailed instructions, S. never took any notes, yet when challenged he could repeat the entire assignment word for word. Luria's first impression was of a 'rather ponderous and at times timid person ... who couldn't conceive of the idea that his memory differed in some way from other people's'. Other impressions accumulated over nearly thirty years.

Luria began by giving S. a series of words, then letters, then numbers, reading to him slowly or presenting them written on a blackboard. Provided there was a gap of three-to-four seconds between items, he could remember seventy or more words or numbers, and it did not matter whether the words were real words or nonsense syllables. He could reproduce the series in reverse order, or state what preceded or followed any given word or number. If the numbers were presented as a table, he could, as it were, 'read off' diagonal rows. As if this were not remarkable enough, Luria later found that S., though he needed a little more time in which to do it, could recall these series fifteen years later, together with the context in which they had been given:

> "Yes, yes... This was a series you gave me once when we were in your apartment... You were sitting at the table and I in the rocking chair... You were wearing a grey suit and you looked at me like this... " And with that he would reel off the series precisely as I had given it to him at the earlier session.

S. told Luria that there were two things that helped him remember. The first was an ability to continue to 'see' the numbers he had imprinted in his memory, just as they had appeared on the blackboard or the piece of paper. Seeing the pattern of numbers it was just as easy for him to read off diagonals as to read off horizontally or vertically, though a number carelessly written might later be 'misread'. The second thing was an involuntary habit of associating visual images with words that he heard. And not just visual images. S. turned out to be one of those rare people who experience synaesthesia; that is to say, a stimulus that normally causes a sensation of one kind simultaneously (and consistently) causes a sensation of another kind.[49] (This seemingly improbable phenomenon can be mimicked in normal people with the aid of drugs. When, in an experiment, the psychologist Richard Gregory was given ketamine – an NMDA channel blocker related to phencyclidine – he found that, with his eyes shut, stroking the bristles of a brush or being stroked on the hand with a comb gave him a sensation of brilliant red and green images.[50] Perceiving sounds as visual sensations has also been reported by people taking LSD, mescalin or psilocybin. And in *Captain Corelli's Mandolin* the hero makes the heroine wonder whether 'the wires in his brain were connected amiss' by disliking a certain tune 'because it was a particularly vile shade of puce'.)*

S.'s synaesthesia took the form of seeing particular visual images, and sometimes also experiencing particular sensations of taste or touch, when he heard particular words or voices or noises. These additional sensations meant that S. could more easily translate a series of words into a series of visual (and other) images. To remember the order of the images, he would imagine some well-remembered street – usually the street in which he had lived as a child, or Gorky Street in Moscow – and starting at one end he would imagine walking down the street placing the visual images corresponding to the successive words against the imaginary backdrop.

Luria reports a conversation he had with S. as they were walking in Moscow:

'You won't forget the way back to the Institute?' [Luria] asked, forgetting whom [he] was dealing with. 'Come now,' S. said. 'How could I possibly forget? After all, here's this fence. It has such a salty taste and feels so rough; furthermore, it has such a sharp, piercing sound…'

There was, though, a tiresome side to the intrusion of synaesthetic sensations and the incessant and spontaneous recall of images that contributed to S.'s

---

*The composers Alexander Scriabin and Olivier Messiaen are sometimes said to have experienced synaesthesia but this seems doubtful. They certainly used visual metaphors involving colour to describe music, and Scriabin, in his *Prometheus (The Poem of Fire)*, included a silent keyboard which controlled the play of beams of coloured light; but it is not clear that they themselves had consistent and involuntary visual experiences when hearing certain sounds. For a discussion of the claims of various artists, writers and musicians to synaesthesia, and of the difficulty in distinguishing true synaesthesia from metaphor and analogy, see Harrison, J. E. & Baron-Cohen, S. (1997) in *Synaesthesia* (eds S. Baron-Cohen & J. E. Harrison), pp. 3–16, Blackwell, Oxford.

tenacious memory. He describes how the sound of the z in the Russian word for mayonnaise ruined the taste; how, if he read while he ate 'the taste of the food drowns out the sense ...' And there were more serious problems. Although his ability to think in visual images sometimes enabled him to solve problems in his head that others could solve only by algebraic arguments or by manipulating objects, the incessant involuntary surfacing of visual images made it difficult for him to extract the essence of a sentence. Halfway through hearing or reading a description he would already be imagining something that might well be incompatible with what followed. You might think that this difficulty would be avoided when he was reading about abstract concepts that did not obviously lend themselves to visualization, but S. repeatedly told Luria: 'I can only understand what I can visualize'. In conversation, the irrepressible images would make him digress, his speech would become cluttered with irrelevant details, and he would find it difficult to retain the thread of the argument.

After being unsuccessful in dozens of jobs, S. finally became a professional mnemonist, giving stage performances – the solution to the problem of making a living found by many of those, some *idiots savants*, some of normal intelligence, gifted with prodigious factual, visual or musical memories.[51] Even here, though, S.'s inability to forget caused problems. Giving successive performances in the same hall on the same evening, he would sometimes begin to 'see' the chart of random numbers from a previous performance. Eventually, he learned how to prevent the appearance of these unwanted images, but how he did it is a mystery. All Luria has to say is: 'It seems pointless to conjecture about a phenomenon that has remained inexplicable.' Thirty years later there is nothing to add.

20 The Emotions

And Joseph made haste; for his bowels did yearn upon his brother: and he sought where to weep...
— Genesis xliv. 30

Oh Caesonia, I knew that men felt anguish, but I didn't know what that word, anguish, meant. Like everyone else I fancied it was a sickness of the mind – no more. But no, it's my body that's in pain. Pain everywhere, in my chest, in my legs and arms. Even my skin is raw, my head is buzzing, I feel like vomiting. But worst of all is this queer taste in my mouth.
— Caligula, in Camus's play *Caligula*

In 1884 William James published a provocative article, which is still quoted today.[1] It was entitled *What Is an Emotion?*, and in it he argued that the common-sense view of the sequence of events that leads to the bodily expression of emotions is wrong. The common-sense view holds that a perception of some object or event gives rise to an emotion, which then gives rise to the bodily expression of that emotion. 'We lose our fortune, are sorry and weep; we meet a bear, are frightened and run; we are insulted by a rival, are angry and strike.' What really happens, James says, 'is that we feel sorry because we cry, angry because we strike, afraid because we tremble'. The 'bodily changes follow directly the PERCEPTION of the exciting fact', and 'our feeling of the same changes as they occur IS the emotion'. He accepts that, crudely stated, the hypothesis will meet with immediate disbelief, but in twenty pages he makes out a plausible case.

The vital point of his theory, he says, is this: 'If we fancy some strong emotion, and then try to abstract from our consciousness of it all the feelings of its characteristic bodily symptoms, we find we have nothing left behind ... ' And later: 'What kind of emotion of fear would be left if the feeling neither of quickened heart beats nor of shallow breathing, neither of trembling lips nor of weakened limbs, neither of goose-flesh nor of visceral stirrings, were present, it is quite impossible to think.' And he is equally eloquent about rage and grief. It is crucial for his hypothesis that the bodily effects occur quickly enough to precede the arousal of an emotion, and he reminds us of the 'cutaneous shiver which like a sudden wave flows over us' when we listen to 'poetry, drama or heroic narrative', and of how, when we suddenly see a dark form moving in the woods, 'our hearts stop beating, and we catch our breath instantly, before any articulate idea of danger can arise'. The bodily effects associated with an emotion are, of course, involuntary, but if his hypothesis is right, the voluntary production of these bodily effects 'ought', he says, 'to give us the emotion itself', and he reminds us 'how the giving way to the symptoms of grief or anger increases those passions'. Conversely, 'whistling to keep up courage is no mere figure of speech'. Finally he describes patients in whom loss of sensibility over the surface of the body is accompanied by emotional insensibility. He accepts that his article does not prove his case, but concludes: 'The best thing I can say for it is, that in writing it, I have almost persuaded *myself* it may be true.'

The obvious weakness of James's view is that emotions, as distinct from moods, are generally *about* something – a loss, a success, a danger, an obstacle, a person – and to say that 'the feeling of the [bodily] changes as they occur IS the emotion' is to claim that an emotion excludes any representation of what it itself is about. This doesn't square with our experience of emotions and the way we refer to them. We are familiar enough with the bodily changes and the feelings associated with them – the flushed face, dilated nostrils and uncontrollable tremor of rage; the cold sweat, dry mouth, loose bowels, gooseflesh and clammy palms of fear; the sinking feeling and nausea of sudden anxiety – and we accept that they form a striking component of the relevant emotion, but only a component. We also recognize what you could call a cognitive component; our anger includes thoughts about the cause of the anger, our fear about the cause of the fear, and so on.

Another difficulty with the James theory is that the same bodily effects may occur in very different emotional states, and even in non-emotional states, so that something more is needed to account for particular emotions. In the early 1960s, Stanley Schachter and Jerome Singer at Columbia University suggested that it was the cognitive component that gave direction to the emotion. To test this hypothesis they injected normal volunteers with what was said to be a vitamin supplement but was actually either adrenaline (epinephrine) or an inert substance, and then placed them in situations that were either emotionally neutral or were slightly cheerful or slightly irritating. They found that all those given the inert substance and all those placed in the emotionally neutral situation had little change in mood. A marked change in mood required *both* the bodily changes caused by the adrenaline and the appropriate social situation.[2]

THE SEAT OF THE EMOTIONS

Which parts of the brain are involved in emotions? Before the days of PET scanning, the only way to answer that question (like so many other questions about what goes on where) was to observe the behaviour of animals or people after selective destruction of different parts of the brain by surgery, trauma or disease.

In the last decade of the nineteenth century, and in the decade following the First World War, experiments in Germany, the Netherlands and the USA showed, surprisingly, that animals in whom the whole of the cerebral cortex had been removed could still display the outward signs of anger and fear; indeed they did so in response to slight stimuli such as would be ignored by a normal animal.[3] Thinking it unlikely that an animal lacking the whole of its cerebral cortex could be conscious, the investigators referred to the response as 'sham rage' and they interpreted the abnormal ease with which the response was elicited as a 'release phenomenon'. Normally, they supposed, the 'centre' in the brain initiating or coordinating the response was under the control of the cortex, which allowed the response to occur only when it was appropriate. Released from that control, the same 'centre' fired off at the least provocation. (This kind of approach had been rather successful in explaining many observations about movement and posture, so it was natural to try to extend it to the study of emotional responses.)

Further experiments pointed to the hypothalamus – the collection of nerve cells in the floor of the between-brain – as the relevant 'centre'. In the absence of the hypothalamus, 'sham rage' did not occur; conversely, weak electrical stimulation of a part of the hypothalamus caused signs of rage.[4] And this made sense. It was known that the hypothalamus controls the sympathetic nervous system – the part of the nervous system particularly concerned with preparing the body for fighting or fleeing – and many of the changes in sham rage can be produced by stimulation of the sympathetic nervous system.

Switching off fear – the Klüver–Bucy syndrome

> and fears shall be in the way,
> and the almond tree shall flourish...
> — Ecclesiastes, xii, 1

There is, though, another collection of nerve cells, in the forebrain but outside the cerebral cortex, that is important in the generation of emotional responses. In the years just before the Second World War, Heinrich Klüver and Paul Bucy, in Chicago, were investigating regions of the brain that were thought to be involved in drug-induced hallucinations. In the course of their experiments they discovered that the removal of both temporal lobes in monkeys produced unexpected and bizarre changes in behaviour.[5] Monkeys previously wild and aggressive became tame and easy to handle, showing neither fear nor anger, and approaching people or other animals without hesitation. They explored and touched objects restlessly, but although their visual acuity was normal and they could locate objects accurately, they seemed unable to recognize familiar objects by sight and they tended to put everything – a comb, a sunflower seed, a screw, a live snake, a piece of banana, a live rat – into their mouths, discarding objects found to be inedible. The monkeys also became sexually overactive, attempting to copulate with other monkeys of either sex, and even with animals of other species or inanimate objects.

The visual agnosia component of the Klüver–Bucy syndrome is the result of the loss of those secondary visual areas that lie in the temporal lobes, but the tameness, the restlessness, the tendency to put things in the mouth, and probably also the abnormal sexual activity, turned out to be associated with the loss of the amygdala – the almond shaped collection of nerve cells buried in the white matter of the temporal lobe in each hemisphere.[6]

A remarkable experiment by J. L. deC. Downer at University College, London, suggested that sensory information had to reach the amygdala if its emotional significance was to be registered.[7] In a 'split-brain' monkey (with the optic nerve fibres that cross from one side to the other – Fig. 11.2 – also divided), each amygdala receives visual information only from the eye on the same side. In such a monkey, who also had only one functioning amygdala, Downer found that if he covered the eye connected to the functioning amygdala the normally wild and aggressive monkey appeared very tame to all visual stimuli and would accept raisins from the experimenter's hand. If he covered the other eye, so that the monkey saw the world only through the eye with access to the functioning amygdala, it behaved in its normal wild fashion.

But the amygdala is not only involved in handling *current* stimuli with emotional content; it seems to play a central role in emotional memory. The evidence for this comes mainly from a series of studies of one of the ways in which animals learn to recognize clues to impending danger. Just as Pavlov's dog learnt to associate the sound of a bell with food, and would salivate on hearing the bell even in the absence of food, so a rat can learn to associate auditory or visual stimuli of no intrinsic significance with, say, a mild electric shock or the appearance of a cat, and will then react appropriately in response to the auditory or visual stimulus alone. And this *defensive conditioning*, as the psychologists call it, is not peculiar to animals. We have already met a human example in Claparède's patient, who would not shake hands with him after the incident with the pin. In fact, Claparède's patient illustrates nicely two features characteristic of defensive conditioning: the ease with which the association is formed, and the persistence of the effect. This makes evolutionary sense. You don't want to have to experience a threatening situation repeatedly to learn to avoid it, and you don't want to forget what you have learnt.

What happens in conditioning is that the animal learns to produce a response (the 'conditioned response') to a stimulus (the 'conditioned stimulus') that would not previously have caused that response. The rat or rabbit or baboon who has learnt to associate a sound with a mild electric shock will react to the sound alone by 'freezing', by altering its heart rate and by increasing its blood pressure. There may, too, be reduced sensitivity to pain, and increased secretion of hormones that prepared the body for activity. (It is this reduced sensitivity to pain – the result of the release of natural morphine-like substances at nerve endings in the pain pathways – that is thought to account for the frequent reports of severely wounded soldiers who did not recognize the severity of their wounds until much later.) All these responses are innate (and useful) responses in states of fear, and what is new is the trigger that elicits them. In the 1960s and 1970s, experiments on defensive conditioning in a variety of mammals showed that destruction of the amygdala on both sides interfered with the acquisition of new conditioned responses, and led to the loss of conditioned responses that had already been acquired.[8] This was true not only where the conditioned stimulus was a sound, but also where it was a flashing light.

How does the conditioned stimulus influence the amygdala? We know that auditory information is routed via the thalamus to the auditory region of the cerebral cortex, so it seemed natural to suppose that if the conditioned stimulus were a sound the amygdala would receive information about it from the cortex. But, in experiments on rats conditioned to respond to a sound, Joseph LeDoux and his colleagues, at Cornell and at New York University, found that the freezing and the rise in blood pressure that followed the sound were quite unaffected by removal of the auditory cortex.[9] On the other hand, both responses were abolished by destroying the thalamus or the auditory pathways leading to it. Clearly, then, the information about the sound had to get as far as the thalamus, but it did not have to reach the cortex. Using modern techniques for tracing nerve pathways, LeDoux and his colleagues found that nerve fibres from the auditory part of the thalamus went to several subcortical regions including the amygdala; and by selectively damaging the different routes they showed that it was the fibres from the thalamus to the amygdala that were essential – see Figure 20.1.

This explained how the sound managed to influence the amygdala, and it also seemed to imply that the emotional learning did not involve the cerebral cortex. But here the story has an interesting twist. In real life, most of the stimuli involved in defensive conditioning are more complex than a simple sound, and it seemed unlikely that such stimuli could be recognized without the involvement of the cerebral cortex. Theodore Jarrell and his colleagues at the University of Miami therefore trained rabbits by exposing them repeatedly to two tones differing widely in pitch, one always followed by a shock and the other not. After this training, the tone associated with the shock, given alone, caused a change in heart rate, whereas the other tone did not. The question was, how would such rabbits behave without their auditory cortex. It turned out that, after removal of the auditory cortex, the rabbits no longer responded differently to the two tones, *and both tones caused the change in heart rate*.[10] This implied two

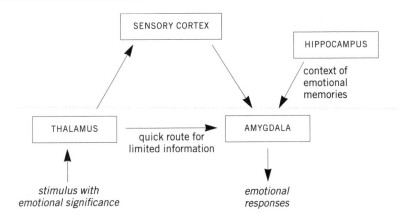

FIGURE 20.1 Information leading to emotional responses reaches the amygdala by several routes.

things. In the first place, there must be two routes carrying information about sound from the thalamus to the amygdala: a direct route, and a route via the auditory cortex (Figure 20.1); as LeDoux puts it, there must be a 'low road' and a 'high road'. Secondly, it is only the cortex that supplies information enabling the amygdala to respond differently to notes of different pitches. (At first sight this second conclusion is surprising, since all the auditory information reaching the cortex comes from the thalamus, but recordings from individual nerve cells in the thalamus show that those cells that send information directly to the amygdala are very poor at discriminating between notes of different pitch.)[11]

Why should such an arrangement have evolved? The answer is, almost certainly, that there is an advantage in having a 'low road' that carries limited information but gets it to the amygdala as quickly as possible, as well as a high road that carries more detailed and 'processed' information that inevitably arrives later. As LeDoux points out, for the hiker who has been walking through the woods 'it is better to have treated a stick as a snake than not to have responded to a possible snake'. In rats, it is reckoned that the cortical route takes nearly twice as long as the twelve milliseconds taken by the direct route.[12] As every Scot knows: '... ye'll take the high road, and I'll take the low road, And I'll be in Scotland afore ye, ... '

A road from the hippocampus

There is also a third route by which sensory information reaches the amygdala (Figure 20.1). Threatening events in real life come with a context, and it is often worth remembering that context. It pays to avoid the place where you nearly trod on a viper or were nearly run down by a car coming fast round a bend. So you might expect that in experiments on defensive-conditioning, the animals might remember not only the significance of the conditioned stimulus but also the significance of the surroundings in which the training took place. And they do. Sometimes the response occurs as soon as the rat is put in the training box, without any specific stimulation. But we saw in the last chapter that episodic memories – memories of events in their contexts – depend on the integrity of the hippocampus; so is the hippocampus involved in remembering the threatening nature of particular places? This question was answered by examining the effects of hippocampal damage, first, on conditioning to a tone coupled with a shock, and then on conditioning to the chamber that was used for the experiments.[13] The upshot was that the conditioning to the tone was unaffected, but the conditioning to the chamber – the place where the tone was heard – was prevented. It seems that information from the hippocampus is not required for an animal to learn to associate a salient stimulus with danger, but it is required for the animal to remember that a particular place is dangerous.

Coping with change

We saw earlier that it makes evolutionary sense that defensive conditioning should not be easily reversed. But situations change and it is desirable to be able to reassess risks; the path on which we nearly trod on a viper may have compensating advantages, or the viper may have departed. Repeated exposure to the conditioned stimulus *not* followed by the noxious unconditioned stimulus does, in fact, eventually lead to loss of the conditioned response. This loss seems to be less the result of passive forgetting than of further learning. In the first place, even when they have been extinguished, conditioned defensive responses both in rats and humans can recur spontaneously or as the result of stress, suggesting that though the response is suppressed the record of the events that gave rise to it has not been totally erased.[14] Secondly, there is evidence in rats that the suppression of a conditioned response can be prevented by blocking the NMDA receptors in the amygdala – suggesting that what is going on during suppression is new learning.[15]

Defensive conditioning has been studied in great detail, but the world is full of carrots as well as sticks, and animals can also be trained to associate a neutral stimulus with a reward. This association too seems to involve the amygdala. A normal monkey can easily learn that food is under one of two objects, and quickly relearn the association if from time to time the food is transferred between the two objects; a monkey without amygdalas cannot do this.[16] Edmund Rolls and his colleagues showed that the amygdalas of monkeys who had been conditioned to respond to a neutral visual stimulus associated with a reward, such as food, contained nerve cells that responded to the visual stimulus alone.[17] When the conditions of the experiment were then changed so that the neutral visual stimulus became associated with a stimulus the monkeys dislike (saline) instead of a rewarding stimulus (food), the responses of the cells in the amygdala to the visual stimulus alone did *not* change; but the overall response of the monkey changed, and so did the responses of certain nerve cells in the inferior part of the prefrontal cortex. It looks, then, as if the known ability of monkeys to readjust their behavioural responses when the significance of a neutral visual stimulus is repeatedly changed depends on nerve cells in the prefrontal cortex as well as on cells in the amygdala.

THE HUMAN AMYGDALA

The general picture that emerges from experiments on animals is that the amygdala receives information of different kinds from the thalamus, from many parts of the cortex and from the hippocampus, and that it produces the bodily responses associated with emotion partly through the hypothalamus, and partly through its projections to other parts of the brain, including back-projections to the cortex.[18] In humans, these back projections to the cortex, particularly to parts of the prefrontal cortex, are thought to influence current perceptions of the environment, and to play an important part in the conscious aspects of emotion.

As a guide to what goes on in our own brains, animal experiments have two obvious limitations: extrapolating from animals to humans is uncertain; and animals cannot tell us what they are feeling. We need to complement the animal work by considering the behaviour of patients in whom, as a result of disease or of surgery to treat disease, the amygdala has been selectively damaged, or prevented from feeding information to the cortex in the normal way. We can also look at the effects of electrical stimulation of the amygdala reported by patients during surgery.

Unlike monkeys, humans with bilateral lesions limited to the amygdala do not suffer from the Klüver-Bucy syndrome – though something like that syndrome, together with severe memory loss, was produced by the removal of both temporal lobes in a desperate attempt to treat a young epileptic who had made several attempts to strangle his mother and to crush his younger brother.[19] But selective removal of the amygdala on both sides does often have a calming effect – corresponding to the taming effect in monkeys – and in the 1960s and 70s, surgical operations to remove part of one or both amygdalas were performed on a large number of aggressive or hyperactive or psychotic patients.[20] A substantial fraction of these showed improvements that seemed to justify such 'psychosurgery', but because of the risks of the operation, unfortunate side effects, and a decrease in improvement rates when patients were followed up for longer periods, such operations are now virtually never done.[21]

It was not these operations, though, that introduced the term psychosurgery. In 1935, the Portuguese neurologist Egas Moniz was in London for the Second International Congress of Neurology.[22] Here he heard an account, by Carlyle Jacobsen and John Fulton from Yale, of experiments to investigate the effects of bilateral removal of the frontal lobes on the ability of chimpanzees to solve problems. An incidental finding was that one particularly emotional chimpanzee, who became upset during the preliminary testing, was cured of her anxieties by the operation, and subsequently approached the tests without distress. According to Fulton, after the talk 'Moniz arose and asked, if frontal lobe removal prevents the development of experimental neuroses in animals and eliminates frustrational behaviour, why should it not be feasible to relieve anxiety states in man by surgical means.' On his return to Lisbon, and with the collaboration of a surgical colleague, Moniz followed up his own suggestion. Initially they tried injecting alcohol into the frontal lobes, but soon they switched to cutting some of the nerve tracts connecting the prefrontal cortex with the rest of the brain. Because the tracts form part of the white matter in the brain, the operation was known as prefrontal leucotomy (from the Greek *leucos* = white, and *tomos* = a cutting).

The operation initially met with considerable success, probably because the desirable effect (the calming thought to be caused by interruption of the flow of information between the amygdalas and the prefrontal cortex) is more immediately apparent than the undesirable changes in personality that often also occur (see Chapter 21). In any event, it was taken up with some enthusiasm in the 1940s in the USA and the UK, particularly for treating patients with severe chronic anxiety, depression threatening to lead to suicide, and severe obsessive-compulsive disorders. In the UK alone over 10,000 seriously ill patients were treated in this way between 1942 and 1954.[23] Later the combination of unfortunate side effects and the availability of more effective and safer psychoactive drugs – chlorpromazine ('Largactil'), was licensed by the US Food and Drug Administration in 1954 – led neurosurgeons to abandon the operation almost completely. In the rare cases in which it is still done, a more precise technique is used which ensures that only selected parts of the prefrontal cortex on each side are affected.

A striking resemblance between the picture seen in patients and the picture derived from work with animals is shown in a study by Antoine Bechara and his colleagues in Antonio Damasio's laboratory in Iowa.[24] They compared the results of attempting *defensive conditioning* in three patients. The first, SM, was a young woman in whom the amygdala on both sides had been severely damaged but in whom the rest of the brain was undamaged.[25] The second patient had damage to the hippocampus on both sides but both amygdalas were intact. The third patient had severe damage to both the amygdala and the hippocampus on both sides.

For the conditioning, the unconditioned stimulus was a startlingly loud sound – a one-second, 100 decibel, blast on a boat horn. The conditioned stimulus was a square of monochromatic light projected onto a screen; the light could be red, yellow, green or blue, but only the blue light was ever followed by a blast on the horn. The response was a fall in the electrical resistance between two points on the skin of the palm – a measure of the nervous stimulation of the sweat glands.

The results were very clear. SM, the patient with selective damage to the amygdala on both sides, failed to acquire a conditioned response to the blue light, but she remembered without difficulty that it was the blue light and only the blue light that was likely to be followed by the blast of the boat horn. The patient with the selective damage to the hippocampus on both sides readily acquired the conditioned response to the blue light, but despite this he could not remember the association between the blue light and the boat horn. The patient with damage to both amygdala and hippocampus on both sides failed to acquire the conditioned response and also failed to remember the association between the blue light and the horn. Clearly, the amygdala was needed for an emotionally linked conditioned response, and the hippocampus was needed for the acquisition of an emotionally neutral explicit memory.

Patients with damage to the amygdala on both sides have further problems. The Iowa group found that SM had great difficulty in recognizing emotions – and particularly the emotion of fear – in other people's faces[26]. Yet her eyesight was good and she had no difficulty in recognizing identity from faces. Andrew Young and his colleagues, in the UK, described a patient with damage to the amygdala on both sides who not only found it difficult to recognize facial expressions (especially the expression of fear) but who also had difficulty in recognizing the intonation patterns, in voices, that convey emotion, or even those patterns that indicate whether a sentence is a statement, a question, or an exclamation.[27]

Evidence that the amygdala is important in the formation of emotional *memories* comes from a patient with amygdala damage on both sides studied by Larry Cahill and his colleagues in California and Germany.[28] The patient and suitably matched control subjects were told a short story, with accompanying slide-show, about a young boy walking with his mother to visit his father at work. At one stage in the story the boy is severely injured in a traffic accident and this was illustrated with graphic pictures. After the story, the patient was asked to rate his emotional reaction to the story on an arbitrary 1–10 scale, and his rating was close to the average of the controls – which does not, of course, imply that the intensity of his emotional reaction was normal. A week later, the patient and the controls were asked to recall events from the different stages of the story, and, as might be expected, the controls remembered the details of the emotional part of the story much better than the other parts. The patient with damaged amygdalas did not.

The effects of electrical stimulation of the amygdala have been observed by surgeons, particularly during investigations of intractable epilepsy. Fear was the most common response; anxiety, guilt, and anger were much less frequent; weak stimulation sometimes caused feelings that were described as 'good' or 'pleasurable'.[29] Other sensations that have been reported include thirst, smells, tastes, sexual feelings and a feeling described by the patient as 'being about to belch'. Stimulation may also cause both auditory and visual illusions, déjà vu experiences, and hallucinations.[30] The curious auras and dramatic evocations of old memories described by Penfield and his colleagues (see pp. 178–80) were elicited by stimulation of the temporal lobe cortex, but later investigations have shown that hallucinations in which the patient relives past experiences may be elicited more readily by stimulation of the amygdala. Surprisingly perhaps, this was true even when the stimulation did not cause an after-discharge spreading to neighbouring structures. It seems unlikely that hallucinations of this kind can happen without involving both visual and auditory areas of cortex, and the suggested explanation is that the various cortical association areas, the amygdala and the hippocampus, though differing in function, all form part of a single associative network, and that memories of past experience, encoded as patterns of synaptic strengths distributed widely across the network, can be recreated by activating a part of the pattern in one area.

SELF-STIMULATION

In 1953 James Olds made an odd observation. He and Peter Milner, working in Hebb's laboratory at McGill, were using a technique then new to psychologists. An electrode previously implanted (under anaesthesia) in a particular region of the brain of a rat was used to stimulate the animal with brief bursts of very weak electric shocks while it was conscious and unrestrained. What Olds noticed was that as the rat wandered about a table-top enclosure it tended to return to the point at which it had last been stimulated, as though the stimulation was in some way rewarding. To test this notion, he arranged that the rat could stimulate itself through the implanted electrode by pressing on a lever at the side of its enclosure, and he found that the rat pressed the lever persistently.[31]

Systematic investigation of this phenomenon by Olds and others produced results that were surprising in two ways. The first surprise was the number and variety of sites in the brain at which stimulation seemed to be rewarding – sites in the midbrain and hindbrain as well as in various areas of the forebrain, including the hypothalamus, the amygdala, and the prefrontal cortex. There were also other sites at which the effect of stimulation seemed to be aversive, since the rat, having pressed the lever once, would avoid pressing it again. The second surprise was the intensity of the apparent desire of the rat to stimulate itself when the implanted electrode was in certain areas. In one experiment of Olds's, with the electrode implanted in the front part of the hypothalamus the rat pressed the lever about 2000 times an hour, with only two brief breaks, for almost twenty-four hours. Even higher rates were sometimes seen for short periods. In other experiments, Olds found that, to get to the lever and start stimulation, the rat would run through a maze or even cross an electrified grid that gave shocks to its feet. In an experiment by Aryeh Routtenberg and Janet Lindy in which both feeding and the opportunity to self-stimulate were restricted to a single hour each day – a period long enough for normal rats to maintain their body weight – some rats with electrodes in the hypothalamus preferred to spend so much of that hour on self-stimulation that they lost weight and would have died had the experiment not been discontinued.[32]

It seemed likely that the reward to the animal from self-stimulation was related to rewards such as food or sexual contact associated with the normal function of the area being stimulated. In an experiment in which the implanted electrode was in the lateral hypothalamus – an area known to be involved in the control of feeding – the rate of lever pressing was greater when the animal was hungry than when it was not.[33] Lowering the animal's blood sugar level by an injection of insulin increased the rate of lever pressing; increasing the blood sugar level by injecting the hormone glucagon decreased the rate. Even more impressive was an experiment in which two electrodes, controlled by two different levers, were implanted in different sites in the hypothalamus. A rat that had had plenty of food and water showed no strong preference between the two levers. After being deprived of food for twenty-four hours it consistently chose one lever; after being deprived of water, it consistently chose the other.[34] In male rats that had had the stimulating electrode implanted in the back part of the hypothalamus, the rate of lever pressing was unaffected by depriving the animal of food, but it was greatly reduced by castration and it was then restored by the injection of androgens.[35]

Although it is natural to assume that persistent self-stimulation is associated with pleasure to the animal, proof of that association is not possible except in humans. The effects of stimulation and self-stimulation in humans have, of course, only been observed in the course of investigating or treating desperately ill patients, but a large number of such observations on patients with intractable epilepsy, psychoses, cerebral tumours or Parkinson's disease have been reported.[36] It is clear that stimulation at a variety of forebrain sites, as well as sites in the mid- and hindbrain, can cause pleasurable feelings; these may be of a rather vague kind – a feeling of well-being – or may be specifically sexual or take the form of pleasant tastes or smells.

A CLUE TO ADDICTION

There is a striking resemblance between the obsessive self-stimulation by rats with electrodes implanted in certain areas of their brains, and the self-administration of addictive drugs by people. This resemblance is not coincidental; there is evidence that habit-forming brain stimulation and several, though not all, habit-forming drugs act through the same neural machinery.[37]

The neural machinery most immediately affected by brain stimulation obviously depends on the site of the electrode, but the effects of stimulation at one point in a network can be extensive, and there is evidence for an underlying pattern. The sites that lead to the most consistent and strongest self-stimulation lie along a bundle of nerve fibres that runs from the forebrain downwards through the hypothalamus and the midbrain to the upper part of the hind brain. Electrical stimulation along the line of the bundle is thought to produce effects on the forebrain by a curiously indirect route. The nerve fibres most easily stimulated are large myelinated fibres which conduct impulses downwards to the mid brain. Here they excite nerve cells whose slowly conducting unmyelinated axons pass upwards, also in the bundle, to form synapses at various sites in the forebrain, including the amygdala, the nucleus accumbens (a small collection of nerve cells close to the basal ganglia), and the prefrontal cortex. These ascending fibres produce their effects by releasing the neurotransmitter dopamine, a substance closely related to adrenaline (epinephrine).*

There is, though, another way of increasing the concentration of dopamine at these forebrain sites, and that is by the use of drugs that *either* increase the release of dopamine from the nerve terminals *or* block the re-uptake mechanism that removes it from the synaptic clefts. Amphetamine, cocaine, some opiates, nicotine, phencyclidine and cannabis are all thought to increase dopamine concentrations at forebrain sites in one or other of these ways.[38]

If an increase in dopamine concentration at some forebrain sites is itself rewarding, that could account both for the addictive effects of self-stimulation and the addictive effects of these psychoactive drugs. The hypothesis could also account for several other observations.[39] It could explain why, in rats with electrodes implanted in areas in which stimulation is rewarding, all these drugs increase the frequency with which the rats press the lever that causes stimulation, whereas dopamine antagonists – drugs that compete with dopamine for dopamine receptor sites, but which do not act like dopamine when occupying the sites – reduce this frequency. And it could explain why injections of dopamine antagonists block the rewarding effect of intravenous amphetamine – an effect which can be measured by seeing how frequently rats press a lever activating a device that injects a shot of amphetamine into the rat's bloodstream. The effects of dopamine in the nucleus accumbens seem to be particularly important, since selective bilateral destruction of the dopamine-releasing nerve terminals at this site alone** blocks the rewarding effect of intravenous amphetamine.[40]

---

*The ascending fibres discussed here are different from the fibres ascending from the midbrain to the basal ganglia, whose failure to release sufficient dopamine is the cause of Parkinson's disease.

**The selective destruction is achieved by injecting 6-hydroxydopamine locally. It is taken up by the nerve terminals and there oxidized to a substance which destroys them.

THE DOPAMINE HYPOTHESIS FOR SCHIZOPHRENIA

Schizophrenia, of course, involves much more than the emotions, but this is a convenient place to mention one of the most exciting and frustrating developments in psychiatry in the last four decades – exciting because it has increased both our understanding of the disease and our ability to treat it, and frustrating because, in both respects, the increase is so limited.

Amphetamine, which as we have just seen releases dopamine from nerve terminals at sites in the forebrain, can produce in addicts who inject large doses into themselves a condition very similar to acute paranoid schizophrenia, with dreamlike states, delusions and hallucinations.[41] So, could excessive dopamine release at forebrain sites be responsible for schizophrenia?

In favour of this hypothesis, substances such as apomorphine, which are known to mimic the effects of dopamine, make the symptoms of schizophrenic patients worse, and drugs that are dopamine antagonists or that deplete the brain of dopamine tend to suppress the so-called 'positive' symptoms of schizophrenia – the delusions, hallucinations, and disordered thoughts, and the feeling that thoughts are being controlled by an outside agency. The *prima-facie* case provided by observations of this kind led the pharmaceutical industry to produce and investigate a large number of dopamine antagonists. It turns out that there are two main classes of dopamine receptor, and the striking finding is that there is a close correlation between the affinity of drugs for the so-called D_2-receptors and the efficacy of the same drugs in treating schizophrenia. This holds true for nearly a score of anti-psychotic drugs with affinities varying more than a thousand-fold.[42]

There is, though, an awkward exception. Clozapine is currently the most effective (as well as the most expensive) drug for treating the positive symptoms of schizophrenia, helping some of the 30 per cent or so of patients little affected by the more typical antipsychotic drugs,[43] yet its affinity for D_2-receptor sites is substantially less than its clinical efficacy would suggest. For a time, it looked as though this anomalous behaviour could be explained by two discoveries: first, that there is a subclass of D_2-receptor sites, known as D_4-receptor sites, with slightly different properties; and secondly that clozapine has a particularly high affinity for these sites.[44] If it is excessive dopamine activity at D_4-receptor sites that is responsible for schizophrenia, the efficacy of clozapine is to be expected. And this hypothesis looked even more attractive when Philip Seeman and his colleagues in Toronto demonstrated a sixfold increase in the number of D_4-receptor sites in the basal ganglia of patients with schizophrenia – though whether this increase was related to the disease or was caused by the drugs used to treat it was not clear.[45]

In any event the hypothesis collapsed.[46] There were drugs with similarly high affinities for the D_4-receptor sites that were not especially effective or were even ineffective, and other drugs that had particularly low affinities for these sites but which were just as effective as the typical antipsychotic drugs. There were even individuals who, as a result of gene mutations, lacked D_4-receptor sites yet appeared to be psychologically normal.[47]

A more likely explanation (at least in part) of the anomalous efficacy of clozapine, is that it reflects differences in behaviour of dopamine receptor sites in the basal ganglia and in the cerebral cortex. The classical studies of the affinities of antipsychotic drugs for dopamine receptor sites used tissue from the basal ganglia (of rat brains), because the concentration of dopamine receptor sites is much higher there than in other parts of the brain.[48] But if abnormal activity at dopamine receptor sites is important in causing the positive symptoms of schizophrenia, you might expect sites in the cerebral cortex to be more relevant than sites in the basal ganglia. And there is some evidence that sites in the cortex do behave differently. For example, Michael Lidow and Patricia Goldman-Rakic, at Yale, showed that in rhesus monkeys both typical antipsychotic drugs and clozapine cause an increase in the number of D_2- and D_4-receptor sites in the prefrontal and temporal cortex, whereas only the typical antipsychotic drugs had a similar effect in the basal ganglia.[49] (The increase in number is thought to be a compensatory response to the blocking of the receptor sites.) And Lyn Pilowsky and her colleagues at the Institute of Psychiatry and University College, London, have indirect evidence suggesting that, unlike the D_2-receptor sites in basal ganglia, D_2-receptor sites in the cortex have similar binding affinities for typical antipsychotic drugs and for clozapine.[50] If it is the sites in the cortex that matter for treating schizophrenia, there is no anomaly. And the low affinity of clozapine for sites in the basal ganglia explains nicely why the Parkinsonian-like side effects that are often so troublesome to patients treated with typical antipsychotic drugs are absent in patients treated with clozapine – see first footnote on page 346.

Encouraging though these findings are, it is still not clear why clozapine should be effective in some of those cases where other antipsychotic drugs fail, or why all dopamine-receptor blockers are ineffective for some patients with schizophrenia, and only partly effective for most. Even where they are effective at treating the delusions, hallucinations and disordered thoughts of a schizophrenic patient, they generally do little to improve the so-called negative symptoms – the withdrawal from social life and the emotional flatness. Indeed, when given to normal people, these drugs tend to cause apathy and reduced initiative. And, of course, even if the dopamine hypothesis is right, that still leaves the underlying cause to be discovered.

DEPRESSION AND ANTIDEPRESSANTS

Important though the dopamine pathway is thought to be, it is only one of several pathways ascending to the forebrain – pathways which are characterised by the different neurotransmitters that are released at the axon terminals. At least two of these pathways – the nor-adrenaline pathway and the serotonin (5-hydroxy-tryptamine) pathway – are thought to affect mood, though both also have other functions. Like a modest increase in the concentration of dopamine, an increase in the concentration of nor-adrenaline or of serotonin is associated with an improved mood.

At first sight the properties of the dopamine, nor-adrenaline and serotonin pathways explain rather satisfactorily the effectiveness of the three classical classes of antidepressant drugs: the *monoamine oxidase inhibitors,* the *tricyclic antidepressants* (so-called because of their molecular structure), and the *selective serotonin reuptake inhibitors.* The concentration of a transmitter in the synaptic cleft depends not only on how much is released from the presynaptic cell but also on how much is taken back into the presynaptic cell by reuptake mechanisms in the cell membrane. The monoamine oxidase inhibitors increase the release of dopamine, nor-adrenaline and serotonin by inhibiting an enzyme within the presynaptic nerve terminals that normally destroys these transmitters; more is therefore available for release. The tricyclic antidepressants inhibit the membrane transport systems responsible for the reuptake of nor-adrenaline and of serotonin. And, as their name suggests, the selective serotonin uptake inhibitors preferentially inhibit the reuptake of serotonin. The best known member of this class, sold under the name of Prozac, is said to be the most widely prescribed of all antidepressants.[51]

Unfortunately, there are problems with this tidy story.[52] Some drugs that increase transmitter release are not antidepressants, and some antidepressants seem neither to increase the release of transmitter nor to delay its re-uptake. There may be explanations for such exceptions, such as the existence of many different kinds of receptor for each transmitter. A more fundamental difficulty with the whole story is that the effects at the nerve terminals occur very quickly whereas the effects on behaviour take two to four weeks to develop. By the time the patient is improving, transmission at the relevant synapses may even be inhibited rather than facilitated. There is no doubt that the antidepressants work (though not in every case) and that they have the cellular and biochemical effects described; but, as Humphrey Rang and his colleagues have remarked:

> the gulf between the description of drug action [at the cellular and biochemical level] and the description ... at the functional and behavioural level remains, for the most part, very wide. Attempts to bridge it seem, at times, like throwing candy floss into the Grand Canyon.[53]

*   *   *

We seem to have come a long way since William James put forward his notion that, when we have some exciting perception, it is the feeling of the bodily changes as they occur that *is* the emotion, and Bard and others showed that stimulation of the hypothalamus caused 'sham rage'. There is, though, a faint and distorted echo of those days in a recent finding that one of the hormones produced by the hypothalamus has an effect on mood as well as effects on the body. In times of stress the adrenal cortex secretes a steroid, cortisol, which has a variety of appropriate metabolic, vascular and other effects. What makes the adrenal cortex secrete cortisol is a hormone secreted by the pituitary, and what makes the pituitary secrete this hormone is a hormone secreted by the hypothalamus. And it turns out that if this hypothalamic hormone – *corticotrophin releasing factor* or *CRF* – is injected into the brains of laboratory animals it causes insomnia, decreased appetite, decreased libido and anxiety – all features of depression in humans.[54] How important this effect is, how soon the pharmaceutical industry will produce blockers of the CRF-receptors, and whether such blockers will be useful, are all questions waiting for answers.

21 Planning and Attention

Doctor, here is business enough for you.

At half past four in the afternoon of a hot September day in 1848, Phineas Gage, the young and capable foreman of a gang working for the Rutland and Burlington Railroad in Vermont, was blasting away rock in the line of a proposed new track.[1] A hole had been drilled, gunpowder and a fuse had been inserted, and it was his assistant's job to cover the powder with sand which Gage would then tamp down with an iron rod. The procedure was utterly routine but something made Gage glance away. He then turned back to the rock and, without waiting for his assistant to pour in sand, began tamping. The explosion turned the tamping rod into a missile. Over three feet long, tapered at the upper end, and an inch and a quarter thick at the base, it entered Gage's left cheek, pierced the base of the skull, emerged from the top of his head and landed thirty metres away. Astonishingly, Gage was not killed, or even made unconscious for more than a few minutes. He was taken by ox cart to a hotel in the nearby town of Duttonsville (now Cavendish), sitting upright, talking, or writing in his notebook. At the hotel he sat in the verandah until the arrival of a local doctor whom he greeted with the words heading this paragraph. That same evening he was clear-headed enough and optimistic enough to say that he didn't want to see his friends as he would be back at work in a day or two.

Bleeding, profuse at first, was controlled within thirty-six hours. Infection, before the days of antibiotics, was more awkward to manage, but after twelve difficult weeks Gage was ready to go home. Six months after the accident he returned to Duttonsville, where Dr Harlow, the doctor who had been largely responsible for his care, said, 'His physical health is good, and I am inclined to say that he has fully recovered.'

This, as Harlow soon realized, was an overstatement. It was true that in spite of thirteen pounds of iron having been driven through the front part of his brain, Gage had neither paralysis nor anaesthesia; he could still see, hear and smell, and his memory, speech and comprehension all seemed to be unaffected; but he had been blinded in his left eye, and his personality had been radically transformed. In an account that was not published until twenty years after the accident, Harlow explains:

The equilibrium or balance, so to speak, between his intellectual faculties and his animal propensities, seems to have been destroyed. He is fitful, irreverent, indulging at times in the grossest profanity (which was not previously his custom) manifesting but little deference for his fellows, impatient of restraint or advice when it conflicts with his desires, at times pertinaciously obstinate, yet capricious and vacillating, devising many plans of future operation, which are no sooner arranged than they are abandoned in turn for others appearing more feasible... Previous to his injury, although untrained in the schools, he possessed a well-balanced mind, and was looked upon by those who knew him as a shrewd, smart business man, very energetic and persistent in executing all his plans of operation. In this regard his mind was radically changed, so decidedly that his friends and acquaintances said he was 'no longer Gage'.[2]

And that verdict fits with the way he spent the rest of his life. For a couple of years he worked with farm animals or in a livery stable, or he appeared with his tamping iron as an exhibit in T P Barnum's Circus. Later he went to Chile for seven years where he cared for horses or drove a six-horse coach. Finally he returned to the USA to live with his mother and sister, until he died from an epileptic seizure in 1860.

What sort of brain damage caused such a dramatic change in personality? There is no doubt that the front part of the left frontal lobe was severely damaged, but there was considerable disagreement among those who examined Gage during the year following the accident about the extent of damage to the right frontal lobe.[3] There was no autopsy when Gage died, but some years later Harlow persuaded the authorities to exhume the body, and both Gage's skull and his tamping iron are now in a museum at Harvard. A number of attempts have been made to deduce the injuries to the brain from the injuries to the skull and the dimensions of the tamping iron, the most recent and the most sophisticated by Hanna Damasio and her colleagues in Iowa.[4] What is clear from these studies is that severe damage to the front part of the left frontal lobe (including much of the left prefrontal cortex) was almost certainly accompanied by less severe damage to the right prefrontal cortex.

In the century and a half that has passed since Phineas Gage's accident there have been no cases quite like his, but many in which a similar combination of psychological changes has resulted from bilateral lesions in the prefrontal area. In 1932 a New York stockbroker developed a life-threatening tumour which pressed downwards on his frontal lobes destroying all but the areas involved in language and movement.[5] Surgery saved his life, but though his senses, movements, speech and memory were all unaffected, and he could still play checkers, calculate and do intelligence tests, he never worked again. Instead, he stayed at home making plans that he never put into effect, and upsetting his wife by his embarrassing or facetious remarks, his boasting, his lack of care for her, and occasionally his abuse – all features quite foreign to his normal personality. Similar results have been described following damage to the prefrontal cortex by trauma, by surgical removal of epileptic foci, by haemorrhage or thrombosis, by infections, or deliberately by psychosurgery.[6] A striking feature of all these cases is that the characteristic clinical picture occurs only when there is damage to the prefrontal cortex on *both* sides.

Although loss of the ability to organize daily life satisfactorily, or to make and execute effective plans for the future, can occur in patients who are able to perform at a high level in intelligence tests of various kinds, usually the impairment of planning is accompanied by some loss of intellectual ability; and almost always there is an emotional flatness – what psychologists like to call 'a lack of affect'.* This emotional flatness contributes both to the lack of drive characteristic of such patients and to the deterioration in personal relationships that may lead to the neglect of family responsibilities and the break-up of families. (Emotional flatness, though, is not incompatible with occasional brief flare-ups. A medical colleague tells me of an unfortunate patient who had had a prefrontal leucotomy and who used to like sitting quietly on a particular seat in the crypt of a London church. One day he found someone else on the seat and politely asked him to move. Receiving a refusal, he promptly strangled the usurper.)

Although it is clear that damage to the prefrontal cortex leads to faults in the formation and execution of plans and to unsatisfactory social behaviour, both planning and social behaviour are complex activities involving many different processes, and it is far from clear what these processes are and which of them are disrupted. The sort of approach that proved so useful in investigating, say, the visual cortex, is not readily applicable because many of the processes are likely to be found somewhere in a causal chain far removed from both the stimulus input and the response output.

In this situation, one way ahead is to try to break down the complex behaviour into simpler stages and to devise tests for looking at these stages. For example, one of the requirements for the effective execution of plans is the ability to change behaviour when the predicted response is not obtained. We saw in the last chapter (p. 341), how some nerve cells in the prefrontal cortex of monkeys changed their responses to a neutral stimulus when the associations (pleasant or unpleasant) of that stimulus were changed. This kind of experiment cannot be done in humans, but it is worth asking whether damage to the human prefrontal cortex is associated with a similar loss of the ability to change responses when conditions change; and the answer is that it is – as shown by a very simple test, the Wisconsin card-sorting test. In this test, the criterion for choosing cards is changed by the experimenter from time to time and the subject has to notice that the previous criterion is no longer working and to decide what the new criterion

*From the Latin *affectus*, meaning disposition.

is. The test is designed so that most normal subjects find the task reasonably easy, but Brenda Milner found that patients with frontal lobe damage had great difficulty, tending to stick to one choice – to 'perseverate', as the neuropsychologists say – instead of making use of the new information.[7] (In striking contrast, Scoville and Milner's unfortunate patient HM (pp. 315–7), who had lost hippocampus and amygdala on both sides and was almost unable to form new memories, performed well.) Other tests in which the situation faced by the patient is changed, and the patient's ability to change response appropriately is assessed, show a similar loss of flexibility in patients with frontal lobe lesions, particularly when the test requires the patient to take several factors into account.[8] Sometimes the patient is even aware of the correct response but just doesn't make it.

Comparisons of the effects of damage to the prefrontal cortex in different cases have suggested that there is some correlation between the precise area damaged and the effect on behaviour,[9] though full-scale double-dissociation studies have not been made, [10] and it seems that the characteristic loss of the ability to make and execute effective plans can be associated with damage in any part of the prefrontal cortex. Very roughly, the consensus seems to be that patients with damage to the region on the underside of the brain lying above the eyes tend to be impulsive and emotionally labile, to have poor social judgement and to be easily distracted.* Sometimes they are euphoric, though the euphoria is not incompatible with irritability. In contrast, patients with damage to the upper and lateral surface of the prefrontal cortex tend to be apathetic unless driven by external stimuli, and to have difficulty in suppressing wrong responses. These patients often lack insight into or concern about their own disabilities; they have difficulty in imagining alternative scenarios, and they may have problems with tasks that depend on the ability to recall the order of recent or remote events. All of these problems contribute to the difficulty in planning.

ATTENTION

'Everyone knows what attention is', says William James; yet he goes on to tell us: 'It is the taking possession by the mind, in clear and vivid form, of one out of what seem several simultaneously possible objects or trains of thought... It implies withdrawal from some things in order to deal effectively with others...'[11] A century later, James's definition wears well, and though we tend to use novel phrases – parallel processing, winner-takes-all – the crucial questions remain the same. Given the abundance of stimuli affecting our sense organs at any moment,

---

*Loss of social skills and social status following damage to the parts of the frontal lobes overlying the orbits has also been reported in rhesus monkeys – see Butter, C.M. & Snyder, D.R. (1972) *Acta Neurobiol. Exp.*, 32, 525–65.

and the richness of the supply of memories, images and ideas available to us, it is clear that the production of a single stream of thought must involve selection; so what are the principles on which selection is based, and what is the mechanism for selecting?

The answer sometimes given to the first question is that we pay attention to what is important for determining current behaviour; but since our behaviour at any moment is more likely to be influenced by what we are attending to than by what we are ignoring there is an element of circularity in that answer. Given our origin as products of natural selection, we might expect to have evolved in such a way that we most readily give our attention to matters that promote our survival or reproduction; and in a general way we do. But many matters that are vital for our survival – the level of acidity in our blood, for example – are controlled by the brain without any conscious awareness; and the close attention of the orchid enthusiast as he pollinates an orchid has only the most indirect connection with the survival and reproduction of the enthusiast whatever it may do for the orchid. This should not be surprising. Natural selection has given us brains and minds that promote our survival and reproduction, but the curiosity, understanding and ingenuity that serve those ends are not monopolized by them. As Steven Pinker has warned us, we must not confuse our goals with our genes' goals.[12] And as T. H. Huxley said a century earlier, 'What is often called the struggle for existence in society... is a contest, not for the means of existence, but for the means of enjoyment.'[13]

The mechanisms involved in selecting objects of attention are ill understood, though interesting information has come from observations on normal people, on patients with disorders of attention caused by brain damage, and on animals.

Switching of attention can obviously be 'stimulus driven' ('bottom up' in current jargon), as when we are startled by the cry of a baby, or 'centrally driven' ('top down'), as when we decide to see if the baby has a fever.

Top down control of attention can also be involuntary. Each time we make a sudden movement of the eyes – and we are forever making them as we glance round the room or read or walk or talk or eat – we *involuntarily* pay no attention to the image on the retina during the time the eye is moving. If you find this difficult to believe, ask a friend to stand about two feet in front of you and to look alternately at your left and your right eye, switching about every second or two. You will have no difficulty in detecting the rapid flick of your friend's eyes at each switch. Now stand about a foot in front of a mirror and look alternately at the left and right eye of your image in the mirror. You will see no movement, because, while your eyes are making sudden movements, you are not paying attention to the images on your retinas. The advantage of ignoring information from the retinas during sudden movements of the eyes is that, because the movement of the images on the retinas is caused by the eye movement, it gives no extra information about the objects being looked at, and merely complicates analysis. How the suppression of the information occurs is still not understood, though one suggestion is that it involves some sort of 'gating' in the visual thalamus that prevents incoming visual information from proceeding to the cortex.

Neglect

Ignoring visual information during sudden eye movements is normal and useful. But ignoring information involuntarily can also be disadvantageous, and then neurologists tend to talk about neglect. Patients with brain damage in any of a number of areas – but particularly in the parietal lobes, and usually the right parietal lobe – sometimes show a bizarre neglect of stimuli on the opposite side of the body. This one-sided neglect may affect vision, hearing or touch and may even be accompanied by failure to wash or dress the affected side of the body or even to acknowledge that it is part of the body. If the neglected half of the body is paralysed, the patient may deny that there is anything wrong. The visual neglect is sometimes accompanied by blindness in half of the visual field, but even where such blindness exists it is unlikely to explain the neglect, since loss of half of the visual field does not usually lead to neglect so long as the patient is free to move his eyes. When patients with left-sided neglect are asked to copy a drawing they tend to copy only the right-hand half, though if the drawing contains several separate elements side by side they may copy the right-hand half of each element – see Figure 21.1.

One-sided neglect does not only apply to 'on-line' information, but also to information stored in the memory. Edoardo Bisiach and Claudio Luzzatti, in Milan, showed this dramatically in an elegantly simple experiment on two patients who were suffering from left-sided neglect following strokes. The patients were asked to imagine themselves in the Piazza del Duomo, and to describe what they would see, first, if they were looking at the front of the cathedral from the opposite side of the square, and then if they were looking in the opposite direction from the front doors of the cathedral.[14] For each view, both patients gave a good description of what they would see in front and on their right, and ignored completely or almost completely what they would see on their left. And the patient who did not completely ignore objects on his left mentioned central and right-side items 'in a lively manner', sometimes commenting on them, whereas the few objects on the left were mentioned 'in a kind of absent-minded, almost annoyed tone'.

Patients with one-sided visual neglect often have difficulty in reading because they tend to ignore the beginning or the end of the word or line, often substituting wrong letters to make a wrong word. A patient with a right parietal lesion and left-sided neglect, for example, might read *train* as *rain* or *willow* as *pillow*. If the word is presented in mirror-reversed form, it is still the left part of the word – now the *end* of the word – that usually causes difficulty. This is what you would

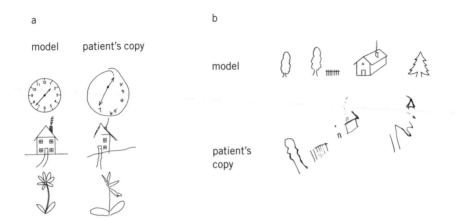

FIGURE 21.1 Examples of left-sided spatial neglect. In (a) the patient was asked to copy drawings of single objects. The left-hand part of each drawing was largely omitted. In (b) a different patient was asked to copy a single drawing of a scene containing several separate objects arranged horizontally. The left-hand part of *each* object (or, in one case, pair of objects) was omitted. ((a) is from F.E. Bloom and A. Lazerson; (b) from R.A. McCarthy and E.K. Warrington.)

expect. Similarly, Alfonso Caramazza and Argye Hillis, at Johns Hopkins University, described a seventy-seven-year-old woman, NG, with a *left*-sided brain lesion and *right*-sided neglect, who, as expected, had difficulties with the ends of words – reading *hound* as *house, stripe* as *strip*. Unexpectedly, though, when words were presented in mirror-reversed form, NG continued to have difficulty with ends of words though these were now on the left side of the word.[15] She also had difficulty with the ends of words when the letters were arranged vertically. The implication of this finding – and it is a startling one – is that however the printed words are presented (normal horizontal; mirror-reversed horizontal; vertical) they must be represented – re-presented makes the meaning clearer – somewhere in her brain as if they were written from left to right. And because most patients with unilateral neglect do not behave like NG, it also follows that the fault responsible for the difficulty in reading in NG must occur at a different level of processing from the fault in the majority of these patients.

   As if the behaviour described in the last couple of pages were not bizarre enough, there is an extra twist to the story. Patients with left-sided visual neglect improve dramatically while ice-cold water is trickled slowly through the left ear, or warm water is trickled slowly through the right ear.[16] This sounds like witchcraft, but the explanation (so far as the effect can be explained) is more mundane. Trickling water colder or warmer than blood temperature through the ears causes convection currents in the horizontal semicircular canals, which reflexly affect the direction of gaze. In the acute stage of visual neglect, patients tend to look away from the neglected part of the field, and even to turn their heads away from it. Appropriate stimulation of the horizontal semicircular canals not only counteracts these tendencies but also transiently relieves the neglect. Altering the direction of gaze cannot be the whole story, though, since the remedial effect of trickling cold water through the outer ear is not restricted to visual neglect; it has also been described in patients with right-sided lesions who neglect sensations from the skin on the left side of their bodies.[17] The suggested explanation is that the oddly asymmetric pattern of flow in the horizontal semicircular canals on the two sides alters the orientation of a spatial frame of reference in the right parietal lobe.[18]

Attention at the cellular level

Because paying attention to a selected small area of the visual field makes us more likely to notice what is going on there, you might wonder whether nerve cells in the cerebral cortex whose receptive fields include both the area under attention and other areas are particularly responsive to stimuli occurring in the area under attention. It turns out that they are. The effect was first looked for in the parietal lobe of a monkey,[19] because of the tendency for neglect in humans to be associated with lesions in that lobe, but similar effects were later also seen in the areas of the monkey cortex concerned with seeing colour and form, with recognizing faces and hands, and with detecting movements.[20] In some experiments, the difference between the responsiveness of the area under attention and other areas within the cell's receptive field was so marked that it was almost as if the receptive field had shrunk right down to the area under attention. Indeed, because the receptive fields of nerve cells in the secondary visual areas tend to be large, it is possible that this effect of attention helps to localize stimuli. In the monkey's primary visual cortex, where receptive fields are small, no effect of attention could be demonstrated.

Recording from individual nerve cells is not normally possible in humans, but recently techniques have been developed that use electrical or magnetic changes detected at the surface of the head to calculate the location of large populations of active nerve cells. Because the time resolution of these techniques is very good, they can be used to get some idea of the changes of activity in different regions following auditory or visual stimuli, and they can therefore be used to see how attention affects the responses. For example, flashes of light in an attended area of the peripheral visual field cause greater potential changes than similar flashes in an unattended area.[21] Analysis of these changes suggests that the increased activity is always outside the primary visual cortex, in agreement with the results of direct recordings from monkeys.

What these experiments on monkeys and people show is that paying attention to a small part of the visual field is accompanied by significant changes in the behaviour of nerve cells in the parietal lobes and in the secondary visual areas, and that these changes are of a kind that might be useful both in noticing events and in localizing them within the visual field. But we have no idea how attention produces these effects or what its neural nature is. The effects of attention are not, by the way, directly connected with eye movements. Both we and monkeys can fix our gaze on one point and our attention on a point a little way off – we do it whenever we are interested in some event but don't wish to stare – and the effect of attention on the responsiveness of cortical nerve cells is still present.

Attention and the binding problem

We know that the later stages of analysis of the shape, colour and motion of the objects we are looking at are handled by separate neural machinery in different areas of the brain. The question therefore arises: how can we tell when several edges, or features of other kinds (textures, colours, patterns of movement) correspond to the same object or part of an object? The obvious answer – that if the different features all relate to the same place in the visual field they must refer to the same object – is true but of limited help, because the size of the receptive fields of nerve cells tends to become large in the visual areas responsible for the later stages of processing. Each individual cell in these areas will therefore give only a rough indication of position, and it is difficult to know how accurate an indication can be derived from the population of cells that are active at any one time.

In spite of this, we can normally solve such binding problems effortlessly and promptly. How do we do it?

Since, as we have just seen, attention can dramatically reduce responses in the *unattended* part of a nerve cell's receptive field, one suggestion is that binding is achieved by scanning the scene with a 'spotlight' of attention, and assuming that features that are simultaneously in the small area 'under the spotlight' are features of the same object.[22] As Anne Treisman and Garry Gelade put it, 'focal attention provides the "glue" which integrates the initially separable features into unitary objects'. This hypothesis does not fit very well with our subjective impression that we are immediately aware of the conjunction of features that correspond to an object, but that might be because we only become aware of the final outcome of what could be quite complicated information processing. A more serious weakness is that, even if the hypothesis is true, it still leaves the mechanism of attention and the control of its direction unexplained.

A strong suggestion that attention *can* play a crucial role in binding is provided by the dramatic failure in the binding of different visual features that was observed by Anne Treisman and her colleagues in a 58-year-old patient, RM.[23] A few years earlier RM had had two strokes, which left him with damage to both parietal lobes. Selective damage to both parietal lobes is rare and gives rise to a disorder of visual attention that leads to a strange collection of symptoms. Of these, the two most striking are a loss of the ability to localize objects in space (either by pointing at them or describing their positions), and a difficulty in seeing more than one object at a time. RM suffered from both of these difficulties, and he also had a problem in linking the colours and shapes of objects. If you are shown two letters printed in different colours and are asked to describe what you see, you will have no difficulty in doing this unless *either* you are distracted while the letters are being displayed *or* the display is extremely brief (less than 200 milliseconds). When RM was shown, say, a red X and a blue O, he would often attribute the wrong colours to the two letters even if he looked at the display for ten seconds in the absence of any distractions. And this inability to bind shape and colour seemed to be the result of a difficulty in localization. Asked to say whether the X was on the left or the right of the O, he performed no better than chance. Asked to compare two shapes and say which was the taller, he often gave the wrong answer; and, significantly, he found the comparison easier if he were shown the shapes one after the other rather than simultaneously. These observations show that a very severe impairment of visual attention can interfere with spatial localization sufficiently to prevent binding; but though this is what would be predicted by Treisman and Gelade's hypothesis, it does not, of course, prove it.

A quite different hypothesis to account for binding is that nerve cells responding to the different features of the same object fire synchronously, and that it is this synchrony that indicates that the features are those of the same object.[24] Synchronous firing of nerve cells situated in the different parts of the visual cortex of cats or monkeys certainly happens, over short periods of time,[25] and the assemblies of synchronously firing cells sometimes include cells in cortical areas dealing with different visual features, or even in cortical areas dealing with different sensory modalities. The hypothesis is therefore plausible; but is it true?

Charles Gray, at the University of California at Davis, and his colleagues recorded from nerve cells in two places in the cat's primary visual cortex about 7 mm apart.[26] The two nerve cells were chosen so that they had similar orientation preferences, and their receptive fields did not overlap and were spatially displaced along the line of preferred orientation. This arrangement made it possible to stimulate the cells with retinal images of bars (aligned in the preferred orientation, and moving perpendicular to that line) in three ways: two bars moving in *opposite* directions; two bars moving in *the same* direction; and *one long bar* moving in that direction – see Figure 21.2. What they found was that with the two bars moving in opposite directions there was no synchronization; with two bars moving in the same direction there was slight synchronization; with one long moving bar there was strong synchronization. This is just what you would expect if the synchronization reflects binding. This experiment involved only cells in the primary visual cortex, but a similar result was obtained by A. K. Engel and his col-

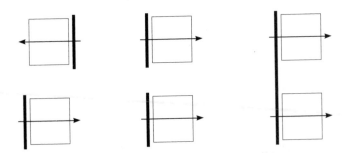

FIGURE 21.2 The three stimuli used in the experiment by Charles Gray and his colleagues. The rectangles represent the receptive fields of the nerve cells whose activity was being recorded. (From C. M. Gray, P. König, A. K. Engel and W. Singer.)

leagues in experiments in which they looked for synchronization between cells in the primary visual cortex and in the part of the visual cortex concerned with movement.[27] It is crucial to the hypothesis that individual nerve cells are able to change the partners with which they become synchronized when the visual image changes, and further experiments along the lines of those just described have shown that they are.[28]

A weakness of the hypothesis is that, even if it is true that synchronous firing of cells indicates that their activity is the result of stimulation by the same object or part of an object, it is not at all clear how the synchronization is established in the first place. The achievement of synchronization would, though, provide a plausible role for the reciprocal connections between nerve cells in different areas that is such a marked feature of the cerebral cortex.

The neural correlates of consciousness

Of the many neural processes going on at any moment while we are awake, some are presumably essential if we are to have the thoughts or feelings that we have at that moment, while others are not. To know which processes are essential would seem to be a necessary first step – though only a first step – towards knowing why we are conscious; yet it is a first step that is surprisingly difficult to take. Stimulating various sense organs gives us various sensations, and we know a good deal about many of the neural events that are the links in the causative chains involved; but our knowledge of these neural events tends to become increasingly vague as we progress along the chains, and always peters out before we reach our destination. Presumably, somewhere along each chain there must be some neural activity which is necessary, and perhaps sufficient, for us to have the feelings or thoughts we are having, but though neuroscientists are happy to talk about the *neural correlates of consciousness*, these correlates seem as elusive as the crock of gold at the end of the rainbow.

There are two quite different reasons for this. The first is the practical difficulty presented by the overwhelming complexity and inaccessibility of the neural events in any animal that we can be confident has sensations and thoughts. The second is a strong gut feeling that even if we knew that the occurrence of particular neural events was a necessary and sufficient condition for having a certain sensation or a certain thought, we still wouldn't understand why we had the sensation or thought. The crock would be empty. I shall come back to this gut feeling in the last section of the book. Here I want to look at approaches to finding the neural correlates of consciousness.

The direct approach would be to follow along the chain of events initiated by some sensation-causing stimulus, in the hope that at some stage it would lead to a pattern of neural events that not only correlated well with the current 'contents of consciousness' but which could be shown to be both necessary and sufficient for causing those particular thoughts or feelings. That is a very tall order indeed, not just because of the difficulty of elucidating the neural machinery likely to be involved, but also because of the need to monitor the contents of consciousness during the investigation. If there were a separate 'consciousness module',* a clue to its whereabouts might be provided by those strange cases in which brain damage seems to dissociate conscious awareness from otherwise normal or partly normal responses to stimuli – patients with blindsight; the patient, DF, who could post cards through slots and pick up plaques normally but had no idea of the orientation of the slot or the width of the plaque; patients with covert recognition of faces, amnesic patients with intact implicit memory. In fact, there is no evidence for a separate consciousness module. Even for the limited part of consciousness that is connected with vision, the different deficits seem to depend on damage in different regions; and the selectivity of the deficits is itself awkward to explain in terms of damage to a single consciousness module. DF, for example, was at a loss when she was asked about orientation or width, but seemed to have a normal awareness of colour and surface texture.

Because there is a close connection between attention, short-term memory and consciousness, you might expect the neural events responsible to be closely connected, and they probably are. But with attention causing changes in nerve cells in the parietal lobes and in various visual areas in the temporal lobes, and short-term memory causing changes in nerve cells in the prefrontal cortex, the connection is hardly a guide to where to look for neural correlates of consciousness.

A more subtle approach to this search is, first, to expose a monkey to a stimulus that is ambiguous and can give rise to two different conscious perceptions, then to train the monkey to indicate which perception it currently has and when it

---

*There are, of course, regions of the brain whose activity determines whether we are awake or asleep, but that is a different problem.

changes, and, finally, to look for cells in the monkey's brain whose behaviour changes when the perception changes. The point of this procedure is that, because the stimulus remains constant, all the early processing should remain constant, and cells whose behaviour changes when the perception changes are likely to be part of the machinery that makes conscious perception possible.

An ingenious experiment along these lines was done by Nikos Logothetis and Jeffrey Schall, at the Massachusetts Institute of Technology.[29] They took advantage of the phenomenon known as binocular rivalry. When, with the aid of a suitable optical device, two dissimilar patterns are presented to your two eyes, as if from the same place, you might expect to see the two superimposed; and for a moment you do. But, very soon, what you see alternates between the two patterns, changing every few seconds. For example, if your left eye sees a horizontal grating drifting upwards, and your right eye sees a similar grating drifting downwards, what you see is a grating whose direction of drift alternates. Similar binocular rivalry occurs in rhesus monkeys, as can be shown by training them to report the direction of movement when both gratings drift in the same direction (so that there is no rivalry), and then exposing them either to gratings drifting in the same direction or to gratings drifting in opposite directions. When the gratings drift in opposite directions, trained monkeys indicate frequent changes in what they are perceiving, just as normal people do. The interesting question is: in this situation, how do the monkey's nerve cells behave in the part of the visual cortex concerned with movement (Zeki's area V5)? What Logothetis and Schall found was, first, that in all situations the responses of many cells changed only when the real direction of drift changed, but, secondly and more interestingly, *in the rivalrous situation the responses of some cells alternated in phase with the (indicated) perception although the direction of drift seen by each eye remained constant.* In other words, the behaviour of these cells was in line with what the monkey reported it was seeing, rather than with what it was looking at.

In experiments in which a *stationary* pattern consisting of parallel horizontal bars is presented to one eye and a *stationary* pattern consisting of parallel vertical bars is presented to the other eye, people and monkeys see only one pattern at a time, changing between horizontal and vertical every few seconds. In experiments of this kind with monkeys trained to indicate the orientation they are perceiving at any moment, David Leopold and Nikos Logothetis have shown that cells which modulate their activity in line with the changing perception, despite the constant stimulus, can be found both in the primary visual cortex and in the part of the visual cortex particularly concerned with colour and orientation (Zeki's area V4).[30] It had always been thought that the alternation of perception between the two orientations in this kind of experiment was the result of binocular rivalry, which was itself the result of reciprocal inhibition between nerve cells in the primary visual cortex that received information from the right or the left eye. But Leopold and Logothetis noticed that the cells in area V4 whose firing correlated with the perceived orientation could be activated by either eye; and this made them wonder whether the rivalry was not the result of competition between the two eyes but of competition between alternative perceptual interpretations at a higher level of analysis. To test this possibility, they and their colleagues at Baylor College of Medicine, using themselves as guinea-pigs, tried the effect of rapidly alternating the rival stimuli (i.e. the stationary patterns oriented in the two directions) between the two eyes.[31] What they found was that the *perceived* direction of the pattern continued to alternate in the normal way (changing on average every 2.3 seconds) although the switching of the stimuli between the two eyes was much more frequent (every third of a second). Clearly, the rivalry was not between the two eyes but between two alternative perceptions.

Another ingenious approach to finding neural correlates of consciousness is to look for neural activity associated with illusory perceptions. We have seen that PET scanning experiments of Semir Zeki and his colleagues showed that normal people perceiving the continual illusory rotation in Isia Leviant's pattern of concentric circular blue bands superimposed on alternate black and white spokes (p. 235), showed increased blood flow in areas surrounding and overlapping with the V5 area of the visual cortex – the area rich in cells that are excited by retinal images that are really moving. Similarly, Roger Tootell and his colleagues at the Massachusetts General Hospital, using MRI scanning, found evidence of increased activity in the V5 area in people experiencing the waterfall illusion – the sense of seeing something moving when looking at a stationary scene after prolonged exposure to movement in one direction (see p.235).[32] What is more, the time course of the increased activity fitted well with the time course of the illusory sensation.

Perhaps the most intriguing suggestion about neural correlates of consciousness came from Francis Crick and Christof Koch in 1990.[33] They argued that being conscious of objects and events is obviously connected with attention, and attention is connected with the binding problem; and they accepted the suggestion of Christof von der Malsberg that it is the synchronous firing of nerve cells concerned with the different features of an object that achieves binding.[34] But they then went on to suggest a startling further role for this firing – *that the synchronous firing of nerve cells concerned with the different features of an object also makes us conscious of that object.* A feature of synchronous firing is that it is usually accompanied by oscillations in voltage with a frequency of between thirty and eighty cycles per second,[35] and Crick and Koch pointed out that 'the likelihood that only a few simultaneous distinct oscillations can exist happily together might explain, in a very natural way, the well known limited capacity of the attentional system'. Their hypothesis does not explain why we are conscious; but, given that we are, it offers a plausible reason for the narrowness of the stream.

VI THE PHILOSOPHY OF MIND
– OR MINDING THE PHILOSOPHERS

Philosophy is a battle against the bewitchment
of our intelligence by means of language
— Wittgenstein, *Philosophical Investigations*

Most of my scientific and medical colleagues tend to be dismissive of philosophy. Their attitude, of course, reflects a realistic assessment of the likely relevance of philosophy to their immediate professional preoccupations, but there is more to it than that. It may partly reflect the dichotomy of British education and the persistence of C. P. Snow's 'two cultures' – philosophy goes with 'arts' and 'humanities' on the other side of the great divide. I suspect, too, that it is encouraged by an exaggerated view of the lack of progress in philosophy, compared with medicine or any of the sciences. And, paradoxically, it may also reflect the success of Wittgenstein and the Vienna Circle in persuading us that the main role of philosophers is to release us from the linguistic tangles in which earlier philosophers have trapped us. If (changing the metaphor) the 'aim of philosophy is', as Wittgenstein says, 'to show the fly the way out of the fly-bottle',[1] it would be better not to get into the bottle in the first place.

Whatever their reasons, I feel that the attitude of my colleagues is misguided, and that, in the field that is the subject of this book, the philosophical problems – or, at any rate, some of them – are both fundamental and interesting. I therefore want to devote this last section to considering a few of them. Of the three chapters that follow, two are concerned with aspects of the 'mind–body' problem; the third discusses the problem of free will.

I am acutely aware of Hilary Putnam's remark, 'Any philosophy that can be put in a nutshell belongs in one', but that is a risk I have to take.

22 The 'Mind–Body Problem' – a Variety of Approaches[1]

In the second chapter of this book we saw that a combination of common sense and a proper respect for physics got us into a quandary. Common sense suggested that there were obvious and frequent mutual interactions between the mainly public world of physical states and events and the private world of (conscious) mental states and events. On the other hand, a proper respect for physics suggested that the explanation of all physical events lay in antecedent physical events. At first sight, epiphenomenalism – the notion that mental events are caused by physical events but cannot themselves cause either physical or further mental events – seemed an ingenious way out of the quandary; but it still left the mental events (the epiphenomena) dangling unexplained from the physical events that caused them; and, as we saw, epiphenomenalism made it difficult to answer the two evolutionary questions asked by William James in 1879.[2] If mental events are inefficacious, why have we evolved brains that make them possible? And, assuming that they are inefficacious, how can we account for the striking correlation between the nature of our conscious sensations and the survival value of the activities associated with them?

Epiphenomenalism is, anyway, no longer fashionable,[3] but I am going to argue that the notions that have replaced it are not much better at coping with the philosophical and scientific questions that any satisfactory theory of the mind needs to answer. Let us look at some of these notions.

BEHAVIOURISM

One way of dealing with an awkward problem is to ignore it and concentrate on easier problems. At the beginning of the 20th century, psychology as an academic discipline had got itself into a sterile phase. It had started as an offshoot of philosophy, much of it was rather abstract, and great emphasis was placed on the results of introspection – results which often led to disagreements which there was no objective way of settling. Then, just before the First World War, John B. Watson, the thirty-five-year-old Professor of Psychology at Johns Hopkins University, published a series of lectures that he had given at Columbia University the previous summer.[4] He argued that the proper subject of psychology was not consciousness or mental states but behaviour, and that the scientific way to study psychology was to give up introspection and concentrate on the way people or animals behaved in different situations. A few years later, in the situation of a male professor with a female research student, Watson's own behaviour was such that he had to retire from academic life.[5] Despite his departure, from the early 1920s until well after the Second World War psychology in America and England was dominated by what became known as *behaviourism*.[6] The behaviourists did not, on the whole, deny the existence of consciousness but, reacting to the years of sterile introspection, they considered that only overt behaviour was worth studying. The switch of attention to psychological phenomena that could be studied objectively proved to be highly fruitful, and the contribution of the so-called 'methodological' behaviourists to the development of the subject was considerable; and of course their methods are still used. But the behaviourists could not be expected to illuminate what they chose to ignore, and they ignored consciousness.

Philosophers sympathetic to behaviourism argued that beliefs, desires, emotions, being in pain, and so on, are best regarded as dispositions to behave in particular ways, a view that was expounded with extraordinary elegance and power by the Oxford philosopher Gilbert Ryle, in his *Concept of Mind*.[7] Even the most elegant advocacy, though, cannot persuade us for very long that the significant thing about toothache is our being disposed to say that we have toothache, to snap at our colleagues and use clove oil and go to the dentist. And if beliefs, desires, fears, and so on, are to be diagnosed solely on the basis of dispositions to display the appropriate behaviour, it is difficult to deny the possession of these mental states to robots designed to display such behaviour. Partly for reasons of this kind, and partly because behaviourists wilfully ignored the need to explore or even consider the mechanisms linking events-that-induce-behaviour to the behaviour induced, behaviourism as a philosophy has been out of favour for the last forty years.

FUNCTIONALISM

By deliberately ignoring linking mechanisms, the behaviourists prevented themselves from regarding the dispositions that they equated with mental states as causes of behaviour, or the possession of a disposition to behave in a particular way as evidence that the possessor was in a particular physical state. This seems odd, but Ryle is quite explicit: 'To possess a dispositional property is not to be in a particular state, or to undergo a particular change; it is to be bound or liable to be in a particular state, or to undergo a particular change, when a particular condition is realized.'[8]

The so-called *functionalists* recognized that failure to consider the linking mechanisms was a fatal weakness in behaviourism, and they proposed instead that mental states were causative – that they were, in the words of the Australian David Armstrong, 'functional states of the organism apt for bringing about certain sorts of behaviour'. [9] Each mental state was also supposed to be apt to be caused by particular sorts of events, and apt to have particular effects on other mental states. This triple characterization of a mental state – What is its effect on behaviour? What causes it? And what are its effects on other mental states? – is attractive because we seem to use criteria of just these kinds to recognize particular mental states in ourselves or other people. I recognize that you believe it is going to rain because I see you look often and apprehensively at the lowering sky, because lowering skies are a recognized cause of a belief that it is going to rain, and because your shutting of the novel in which you seemed to be totally engrossed a moment ago suggests that a dominant desire to know who dunnit has been replaced by a dominant desire to keep dry. But defining mental states in this way says nothing about what they are states of. They might be physical states of the brain or they might be states of some other kind – even, conceivably, states of some mysterious 'thought substance'. In fact, functionalists almost invariably regard mental states as physical states of the brain, but we shall come back to this later.

IDENTITY THEORIES

Ryle's attitude to the common-sense view of the mind – which he called with 'deliberate abusiveness' (his phrase not mine) 'the dogma of the ghost in the machine' – was that it is a category mistake: it represents the facts of mental life as if they belonged to one logical category, when they actually belong to another. (It is the sort of mistake, he explains, that the foreigner visiting Oxford or Cambridge makes when, having been shown a number of colleges, libraries, playing fields, museums, scientific departments and administrative offices, he then asks. 'But where is the University?')

The group of theories that I now want to discuss also regard the common-sense view as being a category mistake, but they regard mental states not just as dispositional states or functional states, but as physical states of the brain associated with behavioural propensities – just as brittleness in glass reflects both the physical state of the glass and its propensity to shatter. According to these theories, mental processes are not something *caused* by physical processes in the brain but are actually *identical* with these processes. For this reason such theories are called *identity theories*.[10]

One of the ways in which philosophers make life difficult for normal people is by using ordinary words to convey curiously restricted meanings. To understand identity theories you need to understand three words in philosopherspeak: *identical, type* and *token*. 'Identical' is used in the sense in which Persia and Iran are identical (there is only a single object) and *not* the sense in which identical twins or a pair of candlesticks are identical (there are two objects, each being, notionally at least, an exact copy of the other). 'Type' is used to denote a category or set of instances, as opposed to the individual instances – the 'tokens' – which belong to the category. Take the word 'pomegranate'. It contains eleven letter *tokens* but only nine letter *types*, because the letter types 'a' and 'e' are each represented by two tokens. The phrase 'instances of a mental state of a given type' therefore refers to instances of a mental state which are indistinguishable in character, though occurring at different times or to different individuals. You are now equipped to understand identity theories.

These theories come in several varieties. The most straightforward assumption is that all the different instances of a mental state of a particular type – a feeling that there's too much garlic in the gazpacho, say – are individually identical with instances of a physical brain state of a particular type – hence the label *type/type* identity theory. That assumption, though, is more demanding than is necessary to avoid the philosophical impasse that results from trying to believe both common-sense dualism and orthodox physics. And many people find it implausible that each instance of a given sensation, say, in different individuals, or in the same individual at different times, is identical with an instance of precisely the same physical state of the relevant part of the brain. To avoid the philosophical impasse, an adequate and less demanding assumption is that each instance of a mental state of a given type is identical with an instance of *some* physical brain state, but the different instances of the mental state of the given type need not all be identical with instances of the same physical brain state (*token/token* identity theory). This then raises the question: if you look at a series of instances of a mental state of a given type, can you say anything about the nature of the brain states with which, in each instance, the mental state is identical?

One very strange answer to this question, suggested by the American philosopher Donald Davidson, is that though each instance of a mental state is identical with an instance of some physical brain state, there are in general no laws describing the connections between the mental and physical characteristics.[11] This extraordinary theory has the attraction that it avoids the philosophical impasse of dualism without 'reducing' mental states (including moral beliefs) to physical brain states, a reduction which leaves many people feeling uncomfortable. But its basic premise is difficult to reconcile with the experimental findings of psychophysicists, which point to the existence of laws of precisely the kind the theory excludes. And, as Tim Crane and Hugh Mellor have pointed out, whole industries making anaesthetics, scents, narcotics, sweeteners, coloured paints and lights, loudspeakers and soft cushions could not exist in the absence of such laws.[12]

A more plausible, and a much more fashionable, answer to the question is that the different instances of a given mental state are individually identical not with different instances of a unique physical state of part of the brain, but with different instances of a unique *functional* state – that is, the one-to-one correspondence is not between particular types of mental process and the activity of particular sets of neurons, but between particular types of mental process and particular patterns of neuronal behaviour responsible for particular computational processes.[13] In computer terminology, what matters is the program and not the involvement of any particular piece of hardware. Those who hold this view are sometimes known as *functional-state identity theorists.*

You may be thinking that this is all nonsense. Wouldn't claiming an identity between two things as apparently different as, say, a feeling of remorse and neuronal events of a particular pattern be making a category mistake of the grossest kind? After all, information about feelings, sensations and mental images comes from questioning or introspection; information about neuronal events or brain states comes from experiments of the kind done by physiologists or biochemists. But, as the Oxford psychologist U.T. Place pointed out in 1956, though such differences prevent the identity from being obvious, they do not make it necessarily untrue.[14]

Statements of identity which are true by definition, such as 'a square is an equilateral rectangle', are obvious provided that we are aware of the meaning of the words. But statements of identity that are empirical (that are known as the result of experience) – a cloud is a mass of water droplets in suspension in air; lightning is a movement of electric charge through the atmosphere – may or may not be obvious. And they are likely to be far from obvious where the methods for observing the characteristics of each of the two members of the pair are very different. A cloud is not obviously a mass of suspended water droplets because to recognize a cloud we have to be a good way off, whereas to see the water droplets we have to be very close; indeed, as Place points out, if we are close enough to see the water droplets we tend to talk of fog or mist rather than cloud. It needed Benjamin Franklin and his kite to prove the identity of lightning with the movement of electric charge through the atmosphere. It was not obvious that the temperature of a gas is identical with the mean kinetic energy of the gas molecules, though it is. And a century elapsed between the discovery of genetic packaging by Mendel and the identification of genes with stretches of DNA with a particular function.

Such arguments, though, only reassure us part of the way. Note that, with each of these pairs, recognition of the identity explains why the first member has the properties it has. The properties of clouds are those to be expected of large collections of water droplets suspended in air; in experiencing a thunderstorm we are simply experiencing on a very large scale the sparks and crackling associated with charge movements through the atmosphere on a small scale; heat flows from a gas at high temperature to a gas at a lower temperature because, on average, the faster moving molecules of the hotter gas will accelerate the more slowly moving molecules of the cooler gas; and the abilities of genes to determine inherited characteristics follow elegantly from the replicating and information-storing properties of DNA. It is the explanatory nature of these identities that makes us happy to regard them as identities and not merely examples of invariant correlation. We say that genes are identical with stretches of DNA, rather than that the existence of genes is invariably correlated with the existence of stretches of DNA. As Hilary Putnam has pointed out (using a different example), the difference between the two kinds of statement is important because, though both may lead to the same predictions, they open and exclude different questions.[15] If we say that the existence of a gene is invariably correlated with the existence of a stretch of DNA, we leave open the questions: 'What is a gene then if it is not a stretch of DNA?', and 'Why is the existence of a gene invariably correlated with the existence of a stretch of DNA?', questions which are excluded by the statement of identity.

Unfortunately, when we move from considering the examples we have just been discussing to considering the hypothetical identity of mental states and brain states (or mental processes and brain processes), the satisfying explanatory character of the identity seems to be missing. Nothing that we know about brain states seems able to account for the existence of conscious awareness, of feelings of pain or pleasure or fear or anger, or even of the qualities associated with simple sensations. This lack of any obvious explanatory character makes us reluctant to assume we are dealing with identities rather than mere correlations. The excluded question 'why are conscious mental states invariably correlated with brain states?' is just the question we want to ask. You might say that the supposed identity of the conscious mental states with physical states of the brain provides an answer to that question, but it is an answer that fails to satisfy us. We don't feel we understand conscious mental states better in the light of it, in the way we did feel we understood clouds, lightning, temperature and genes better in the light of the identities we discussed earlier.

It may be, of course, that the failure of hypothetical identities between conscious mental states and brain states to have any satisfying explanatory value is simply a consequence of our ignorance of the brain states, but in clinging to that view most of us find it difficult to avoid an uncomfortable feeling of whistling in the dark.

The currently recommended therapy for that uncomfortable feeling is to consider the analogy of computers. If, following Daniel Dennett we ask of a chess-playing computer, 'Why did it move its queen?' we can be answered in two ways.[16] Following an exposition of the rules of chess if we don't know them already, it can be pointed out to us that the queen was being threatened by the white bishop, and that this particular move by the queen not only removed that threat but also put the white king in check. Or we can be given a complete account of the computer's circuitry and program and initial state, which will enable us – if we can understand such matters, and can bear the tedium – to understand why the queen was moved in the way it was. It is a far cry from talk of black queens being threatened by white bishops to talk of flip-flop circuits and electrical pulses, but both kinds of answer provide an accurate explanation of the same event. 'Don't you see', we are urged, 'that the difference between talk of bishops and talk of flip-flop circuits and electrical impulses, is like the difference between talk of mental events and talk of neurons and nervous impulses?' The relation between brain and mind, we are asked to believe, is like the relation between the computer and the program that is running through it.[17] Those who argue in this way admit that the brain differs from most computers in doing a great deal of processing in parallel, and in continually making small changes in its hardware, but these seem to be unimportant differences.

Up to a point, the computer analogy works remarkably well. The chess-playing program has goals and sub-goals and strategies, just as we do. And though the goals of the computer program originated with the programmer, whereas ours are the result of natural selection and our own past experience, they are equally goals and are equally effective in explaining behaviour. There is, of course, no mystery in chess-playing computers. They are immensely clever, and few of us can understand them in detail, but we can understand them in principle – or at least we can understand in principle how a machine might play noughts and crosses, or how a 'smart' missile could fly along the streets of a mapped town, and recognize and destroy its target. Similarly, we can see in principle how the working of neural machinery – the computations performed by the brain – could give rise to overt behaviour; could, say, enable us to react differently, though unconsciously, to different visual scenes. Neuroscientists can already tell us something about the computational strategies used for limited purposes by parts of the brain. What we cannot even begin to see is how the working of any machinery – artificial or neural – can give rise to consciousness.

Where we seem to have got to, then, is that functional-state identity theory provides us with a plausible way of avoiding the impasse that results from trying to reconcile common sense dualism with orthodox physics, though, at least in the present state of our knowledge of brain states, the supposed identity of mental states and brain states fails to account for the characteristics of conscious mental states in the way in which, say, the supposed identity of genes and stretches of DNA accounts for the characteristics of genes, or the supposed identity of the temperature of a gas and the kinetic energy of its molecules accounts for the effect of temperature differences on heat flow.

There are two other problems with identity theories of mind that I want to talk about. The first is an argument by the American logician and philosopher Saul Kripke. The second is concerned with William James's evolutionary questions.

KRIPKE'S ARGUMENTS

In two fascinating essays published in the early 1970s, Saul Kripke put forward elegant and ingenious arguments which, he claimed, made identity theories of mind untenable.[18] I believe that that claim is wrong and I want to explain why. But irrespective of that claim, Kripke's arguments are well worth looking at because of the intense light they throw on the subtle differences of meaning that can be conveyed by sentences proclaiming identities between things. Following long philosophical arguments, though, is not to everyone's taste, and if you are sure that you lack that taste you will not lose the thread of this chapter if you skip the next few pages and go straight to the William James heading on page 382. If you don't know whether you have the taste, try it. You may get hooked.

Kripke's argument depends crucially on the distinction between necessary and contingent truths. A necessary truth is a statement which, first, is true, and secondly, could not have been false. The statement '2 + 2 = 4' is an example. A contingent truth is a statement which happens to be true but which might not have been; the world could have been such as to make it false. The statement 'Bifocal glasses were invented by Benjamin Franklin' is an example. It has been widely believed that necessary truths are always truths which are known to be true *a priori*, whereas contingent truths are always empirical truths (i.e. statements known to be true only as the result of experience), but Kripke points out that this belief is incorrect. Among several counter-examples that he gives, perhaps the most straightforward is the identity of a 'star' that the Greeks pointed to in the evening and called *Hesperus*, and a 'star' that they pointed to in the morning and called *Phosphorus*. Since both 'stars' were in fact the planet Venus, *Hesperus* and *Phosphorus* are necessarily identical, but that was not knowable *a priori*. It follows that there exists a class of statements which if true are necessarily true but whose truth is not knowable *a priori*; and Kripke argues that theoretical identity statements such as [non-radiant] heat is molecular motion', or 'my being in pain at such and such a time is my being in such and such a brain state at such and such a time', fall into this class. We will consider the way he reaches this conclusion in a moment, but if it is correct it is significant because identity theorists have always assumed that the identity of mental states and brain states (not being knowable *a priori*) is a contingent identity – indeed, the theory is often referred to as the 'contingent identity theory'. And a contingent identity theory is clearly incompatible with Kripke's conclusion that identities of this kind, if they exist, are necessary.

Let us look at Kripke's argument that identity statements such as 'heat is molecular motion' or 'a particular mental state is identical with a particular brain state' are, if true, necessarily true.

There is a well known paradox about identity statements which arises in this way:

If x and y are identical* then every property of x is a property of y.

Everything is necessarily identical with itself, so x has the property of being necessarily identical with x.

But y shares all the properties of x, so y has the property of being necessarily identical with x.

It follows that if two things are identical they are necessarily identical.

This conclusion is paradoxical, since it seems to exclude the possibility of contingent identity statements, yet there is no doubt that such statements exist. 'The first Postmaster General of the United States is identical with the inventor of bifocals' is such a statement. (The first Postmaster General, who was in fact Benjamin Franklin, might have been someone other than Benjamin Franklin; and someone other than Benjamin Franklin might have invented bifocals.) Kripke deals with this paradox by introducing a distinction between what he calls *rigid designators* and *non-rigid designators*. By rigid designator, he means a term that (in *our* language as *we* use it) designates the same object in all possible worlds (all possible states of this world; all counterfactual situations) in which the object exists. 'The inventor of bifocals' is a non-rigid designator because one can readily imagine states of this world in which bifocals were invented by someone other than the man who actually invented them – Benjamin Franklin. 'Benjamin Franklin', however, is a rigid designator, since although there have doubtless been other men called Benjamin Franklin, when we talk about Benjamin Franklin in this sort of context we are designating a particular man in all possible situations – the man who, among other things, happened to be the inventor of bifocals and the first Postmaster General of the United States. Kripke points out that *contingent identity statements are possible provided that at least one of the designators is non-rigid*. (This is because by using a non-rigid designator – 'the inventor of bifocals' – instead of a rigid designator – 'Benjamin Franklin' – the speaker is identifying the object only via one of its contingent properties – that the man we know as 'Benjamin Franklin' invented bifocals.) It follows that the identity of things does not entail the *necessary* truth of statements of that identity which use non-rigid designators.

Kripke goes on to claim that theoretical identity statements such as 'Heat is the motion of molecules' are, if true, necessarily true, since only rigid designators are used. He acknowledges that theoretical identity statements of this kind have always been thought of as contingent, but that, he says, is largely because their origin in scientific discoveries was thought (wrongly) to imply that they were contingent. And Kripke draws attention to a fallacy in another piece of reasoning which can give the illusion that the statement 'heat is the motion of molecules' is contingent. We can readily imagine a group of people with peculiar

---

*The word *identical* is here being used in the sense in which Persia and Iran are identical; there is only a single object.

physiology who do not get the sensation that we call the sensation of heat from increased molecular motion but who get it from some other stimulus, or perhaps do not get it at all. In such people the normal association between molecular motion and the sensation of heat would certainly be lacking. But, he points out, this merely proves the contingency of the identity statement that molecular motion is the thing that produces in us the sensation that we now refer to as the sensation of heat. And it is the rigid designator 'heat', not the non-rigid designator 'the thing that causes in us the particular sensation that we now refer to as the sensation of heat',* that we are concerned with in the statement 'Heat is the motion of molecules'. The possible existence of the group of people with peculiar physiology is therefore not an argument that 'Heat is the motion of molecules' is a contingent identity statement.

Kripke pays so much attention to the nature of statements of the identity between heat and molecular motion because of the alleged parallelism between statements of that kind and statements such as 'My being in pain at such and such a time is my being in such and such a brain state at such and such a time'. Identity theorists have generally regarded both as examples of contingent identity statements. Since Kripke has shown that the statement 'Heat is molecular motion' is not contingent but necessary, the identity theorist is forced to choose between two alternatives. He can maintain the parallelism by changing tack, and claiming that statements of the identity of a mental state and a brain state are not, as had previously been thought, contingent, but are in fact necessary. Or he can discard the parallelism and continue to assert that such identity statements are contingent; but to be in a position to do that he must show that the term used to designate the mental state *or* the term used to designate the brain state, *or* both, are non-rigid designators. Kripke finds neither alternative viable, and concludes that identity theories of mind are untenable. It is this part of his argument that seems to me open to question, and that I wish to examine.

Kripke rejects the change in tack because he feels that the truth of a *necessary* identity statement linking a particular mental state – say, a pain – and a particular brain state is not compatible with our being able to imagine the brain state existing without the pain, and the pain existing without the brain state. And we cannot, he says, dismiss this ability as an illusion – 'something we can apparently imagine, but in fact cannot, in the way we thought erroneously that we could imagine a situation in which heat was not the motion of molecules'. That illusion depended on our picking out heat by a certain contingent property – its giving us the sensation of heat – but we cannot pick out pain in this way. It is not a contingent property of pain that it gives us the sensation of pain. For a sensation to be felt as pain is for it to be pain.

*The phrase 'the thing that produces in us the sensation that we now refer to as the sensation of heat' is a non-rigid designator because the world might have been such that something other than heat was the cause of that sensation.

The alternative of discarding the parallelism, and maintaining the long-held view that statements of the kind 'My being in pain at such and such a time is my being in such and such a brain state at such and such a time' are contingent, is possible only if the designation of either the mental state or the brain state is non-rigid. And Kripke argues that the designation of neither is non-rigid, because the way we pick out a mental state – say, a pain – or a brain state, is by referring to essential properties not contingent properties. 'Whenever the molecules *are* in this configuration, we *do* have such and such a brain state. Whenever you feel *this*, you do have a pain.'

I want to examine, first, Kripke's argument that the truth of a *necessary* identity statement linking a particular mental state, such as a pain, and a particular brain state is not compatible with our being able to imagine the brain state existing without the pain, and the pain existing without the brain state. Kripke argues that the seemingly analogous situation, in which the necessary identity statement 'Heat is molecular motion' *is* compatible with our apparent ability to imagine heat without molecular motion, is not relevant, because what we really imagine is not heat without molecular motion but the *sensation of heat* being caused by something other than molecular motion. And the confusion between heat and the sensation of heat has no counterpart in the case of pain because to have a pain and to have a sensation of pain are not distinguishable. This, of course, is true, but what the argument proves is merely that the (mistaken) feeling that statements of identity between mental states and brain states are contingent cannot be accounted for in the same way as the mistaken feeling that the identity statement 'Heat is molecular motion' is contingent. There could, however, be other reasons for having the feeling. I suggest five, of which one is applicable only to statements of the kind made by token/token identity theorists. The four more generally applicable are, in decreasing order of importance:

(i) With most theoretical identities, e.g. heat and molecular motion, or light and electromagnetic waves, or genes and DNA, the identity can be seen to explain important properties of one of the pair. In contrast, the explanatory power of a supposed identity between mental states and brain states is very limited. Even if the identity theory is true, (conscious) mental states appear to *emerge* inexplicably from brain states. And not knowing how this comes about it is perhaps natural to say that they *happen to emerge*, and from there it is only a small step to say that the supposed identity between mental states and brain states is contingent.

(ii) The naïve model that Kripke and others use – the notion that pain is identical with stimulation of particular nerve fibres called C fibres – is apt to mislead. We feel that it ought to be possible to block the C fibres beyond the point at which they are stimulated and so prevent the pain. Or we could interfere beyond the terminations of the C fibres to produce the same effect as C-fibre activity and hence have pain without the stimulation of C fibres. As soon as these suggestions are made explicitly they are obviously beside the point, because they are merely saying that the state of having C fibres stimulated is not the brain state that is identical with pain; but if they are not made explicit this is not obvious and one is left with the feeling that the mental state and the brain state are separable.

(iii) Although, when it comes to discussing philosophy, few of us are Cartesian dualists, in everyday life nearly all of us are *de facto* Cartesians. So we start with the notion that mental states and brain states are separable.

(iv) For people brought up to believe that necessary truths must be knowable *a priori*, and unaware of Kripke's argument for the existence of empirical but necessary truths, the observational/experimental origin of the hypothesis that there is an identity between a pain and the corresponding brain state suggests that a statement of this identity must be contingent. And a tendency to assume that a statement of an empirically established identity is contingent is not instantly abolished by the acceptance of Kripke's argument for the existence of empirical but necessary truths.

These four reasons seem to be adequate to account for our (mistaken) intuition that mental states and brain states are separable, and there seems no good reason why the *type/type* identity theorist should not simply reconcile his theory with Kripke's proof that statements of identity involving rigidly designated objects are necessary rather than contingent by changing tack and saying that, contrary to earlier views, statements that mental states are identical with brain states are, if true, necessarily true.

For the *token/token* identity theorist there is a further, and cogent, reason why identities between mental states and brain states, though necessary, seem to be contingent. Such theories hold that each instance of a mental state is identical with an instance of some physical brain state, but the *different* instances of a mental state of a given type need not be identical with instances of the *same* physical brain state. This implies that, though the twinge I felt in my second upper left premolar at 4.15 pm last Thursday must have been identical with a particular physical brain state (a state that existed in my brain at 4.15 pm last Thursday), twinges indistinguishable from that twinge – as judged by introspection – could be identical with different physical brain states. It is true that the world might have been such that, at 4.15 last Thursday, instead of the twinge I actually had, I might have had an indistinguishable twinge identical with a different brain state, but (if we assume that token/token identity theory is correct) the statement: 'The twinge I had at 4.15 pm last Thursday is identical with a brain state I had at 4.15 last Thursday' would still have been true. That statement (still on the assumption that token/token identity theory is correct) is therefore necessarily true. However, the fact that I might have had a twinge not discernibly different – as judged by introspection – and a different brain state gives a strong illusion of contingency.* To see why it is an illusion consider the genuinely contingent identity of 'the first Postmaster General of the USA' and 'the man who invented bifocals'. That identity is contingent because we can envisage counterfactual situations (other states of this world; other worlds) in which we suppose someone other than Benjamin Franklin to have been 'the first Postmaster General of the USA' without having to assume that that person was 'the inventor of bifocals', or in which we suppose that someone other than 'Benjamin Franklin' invented bifocals without having to assume that that person was the first Postmaster General – in more general terms, the objects designated by the two phrases can vary independently. In considering the identity between mental states and brain states, we are prevented from envisaging independent variation of that kind by the assumption that token/token identity theory is true.**

So again, Kripke is right in stating that such identities, if true, are necessary; but again there are satisfactory explanations of our intuitive feeling that they are contingent.

Finally, let us look at the fashionable functionalist form of *token/token* identity theory. This holds that the different physical brain states that are individually identical with different instances of a given mental state all have the same functional role – i.e. they are functionally indistinguishable. (It is the common functional characteristics of the different physical states that is supposed to account for the indistinguishability of the different instances of the given mental state.) But introducing the concept of functional brain states does not affect the identity relations between the individual instances of the mental states of the given type and the corresponding physical brain states; these will be as described in the paragraph before last, and do not involve a breach of Kripke's rule.

The relations between mental states, functional brain states and physical brain states believed to exist by those who hold this kind of identity theory, and the failure of Kripke's argument to make such theories untenable, may be made clearer by an analogy. Consider a fictitious disease – let's call it the 'S' syndrome – which is caused by any member of a family of rare dominant genes (so that the presence of any one of these genes causes the disease). These genes are identical with pieces of DNA with slightly different base sequences, but let us suppose that,

* This feature of token/token identity theories, and its relevance to Kripke's argument, was pointed out by Colin McGinn in an article primarily concerned with Davidson's *anomalous monism* – McGinn, C. (1977) *Analysis*, 37(2) 78–88.

**More generally, a true identity statement using non-rigid designators will be necessarily true if the references of the non-rigid designators change in a parallel fashion in counterfactual situations. Nevertheless, the use of non-rigid designators may give the illusion that the truth of the identity is contingent.

because of redundancy in the genetic code, they all code for precisely the same abnormal protein,[19] and that it is the presence of this protein (in place of a normal protein) that causes the disease. The pieces of DNA with which the different members of the family of genes that can cause the disease are individually identical are therefore *physically* distinct but *functionally* indistinguishable. In possessing these features, these pieces of DNA are analogous to the physically distinct but functionally indistinguishable brain states that are supposed to be individually identical with the different instances of a mental state of a given type. The different members of the family of genes that cause the disease are individually identical with these pieces of DNA, and in possessing this feature they are analogous to the different instances of the mental state of the given type, which are supposed to be individually identical with physically different but functionally indistinguishable brain states.

Consider a particular patient suffering from the 'S' syndrome. The cause of the disease in this patient is the presence (in each cell nucleus) of a particular member of the family of 'S' genes – a member identical with a piece of DNA with a particular structure. The world might have been such that the cause of the disease in this patient was another member of the family, identical with a piece of DNA with a different structure, so the designator: 'the gene that is the cause of the 'S' syndrome in this patient' is not a rigid designator. For that reason, we are tempted to believe (wrongly) that any true identity statement involving that designator will be only contingently true. Yet, unless we reject the standard gene/DNA identity theory, there is no doubt that the statement: 'The gene that is the cause of the 'S' syndrome in this patient is identical with the piece of DNA that is the cause of the 'S' syndrome in this patient' is necessarily true.

Of course, the failure of Kripke's argument that identity theories of mind are untenable does not prove that some kind of identity theory of mind is correct, but it does remove a serious objection to such theories.

WILLIAM JAMES'S QUESTIONS

Let us now turn to William James's awkward evolutionary questions. If mental events cannot cause physical events, why did consciousness evolve? And how can we account for the striking correlation between the nature of our conscious sensations and the survival value of the activities associated with them?

If some kind of identity theory is correct, the mental aspects of the brain states get an evolutionary ride on the coat-tails of the physical aspects, and the question 'Why did consciousness evolve?' reduces to the apparently simpler question 'Why did certain physical features evolve?' But the gain in simplicity is illusory. Instead of the question faced by the believer in epiphenomenalism 'Why did those physical brain states or processes that give rise to conscious mental phenomena get selected?', the believer in identity theory has to face the question: 'Given that not all physical brain states or processes are identical with conscious mental states or processes, why did those brain states or processes that are identical with conscious mental states or processes get selected?'

Our difficulty in answering this question is that, not only are we unable to say why (assuming some form of identity theory to be true) some brain processes are also conscious mental processes, but we do not even know what features of a brain process are necessary for it also to be a conscious mental process; so it is difficult to decide what it is about brain processes possessing those features that is likely to have conferred a selective advantage. As we saw in the last chapter, neuroscientists are happy to talk about the neural correlates of consciousness, but they have made rather little progress in identifying them. It is not even known whether different kinds of consciousness – say, consciousness of a visual scene, consciousness of pain, and consciousness of our own thoughts – involve a common mechanism or type of mechanism.

The enormous amount of work that has been done on the visual system makes it possible for us to understand in principle how we might recognize objects, and respond to them, as a robot might; but the question: 'What extra neural processes must occur for us to be conscious of objects?' remains unanswered.

There are, of course, people who believe – perhaps correctly – that talk of extra processes is mistaken, and that any machine with powers of analysis and recognition comparable with our own would necessarily be conscious. For such people, James's main question – why did consciousness evolve? – presents no difficulty: it evolved because it is a necessary feature of processes naturally selected because of the powers of analysis and discrimination that they confer. The notion that consciousness appears automatically in this way is not too easy to accept, however, partly because it is not clear why it should be a feature of such processes, and partly because, as we have seen in various neurological conditions, there is evidence that analysis and discrimination by the brain of a kind that is normally associated with consciousness can sometimes occur without consciousness – though, admittedly, the analysis and discrimination are at an extremely low level. And, of course, some kinds of consciousness – a violent twinge of lumbago, for example – are not obviously connected with any complicated analysis or discrimination. In fact, feelings such as these show in the starkest way the dilemma we face. Unless we revert to interactionist dualism, we must either believe that the pain of lumbago is a necessary feature of neural processes that have some other and beneficial effect that could account for their evolution through natural selection – interrupting activity to prevent further injury, for example – or we have to admit that we are unable to answer James's main question: we do not know why pain evolved.

At first sight the existence of pain, and the apparent lack of connection between pain and complicated analysis and discrimination, might suggest that consciousness does not necessarily involve neural processes concerned with analysis and discrimination. And that might suggest that to determine the neural correlates of consciousness it would be better to look at pain than at vision, since the problem of distinguishing between nervous activity needed for consciousness, and nervous activity needed for the analysis and discrimination involved in the unconscious recognition of the subject of attention, would be largely avoided. But to argue in this way could be wrong. It is characteristic of pain that it leads, or at least tends to lead, to the interruption, and perhaps cessation, of whatever activity the subject experiencing the pain is engaged in. In humans and the higher animals most activities require the central nervous system to engage in a substantial amount of analysis and discrimination, the results of which continuously modify the activity. A mechanism for interrupting or stopping activity is therefore bound to interact with mechanisms for analysing and discriminating. So if it is true that consciousness is a necessary feature of machinery capable of analysis and discrimination at the level at which our nervous systems perform these functions, it would not be altogether surprising that changes in consciousness should accompany the operation of devices that interrupt activity and that have been selected, for example, because such interruption prevents further injury.

Even if, without understanding how it happens, we accept that particular neural events, selected by natural selection for their physical effects, can somehow give rise to consciousness, that still leaves us with James's subsidiary question. In the language of identity theorists: Why should events promoting survival – eating, drinking, mating, moderate exercise – tend to cause brain states identical with *pleasant* mental states, and events threatening survival – lack of food or drink, injuries – tend to cause brain states identical with *unpleasant* mental states? You might suggest that all brain states associated with events promoting survival have in common sub-states associated with actions aimed at prolonging those events, and that, similarly, all brain states associated with events threatening survival have in common sub-states associated with actions aimed at terminating those events. You could then suppose that the common sub-states lead to resemblances in the mental aspects of the respective brain states. But while this could explain why there tend to be similarities between the sensations accompanying different survival-promoting events, and similarities between sensations accompanying different survival-threatening events, it would not explain why the former sensations tend to be pleasant and the latter unpleasant – though it is, of course, extremely fortunate for the quality of our lives that they are.

A CHOICE OF QUANDARIES

It seems that all the theories that allow us to escape from the quandary that a combination of common sense and a proper respect for physics got us into land us in other quandaries. In this awkward situation different philosophers react in different ways.

At the pessimistic extreme, some philosophers who, as Herbert Feigl puts it, 'like to wallow' in the perplexities of the mind–body problem, finally declare it insoluble.[20] 'Perhaps', he says, 'this is an expression of intellectual masochism, or a rationalization of intellectual impotence. It may of course also be an expression of genuine humility.' Genuine humility is hardly an occupational hazard of philosophers, but rationalization of intellectual impotence is an accurate description of the recent work of Colin McGinn, who considers why we might be incapable of forming the concepts necessary to explain how consciousness arises from the activities of the brain.[21] It is not, he says, that there is anything super-natural about the process, but simply that our cognitive competence is inade-quate for the task of understanding it; just as armadillos cannot solve problems of elementary arithmetic and a five-year-old child cannot understand relativity theory. Chomsky had earlier pointed out that 'while man's mind is no doubt adapted to his requirements, there is no reason to suppose that discovery of scientific theories in particular domains is among the requirements met through natural selection'; successful theories could exist, inaccessible to us but possibly accessible to a differently structured mind.[22] There is then, McGinn argues, no reason for expecting us to have evolved the intellectual equipment needed to grasp how our own minds ultimately operate. He therefore suspects that, for us, the nature of the dependence of consciousness on the brain is an insoluble problem. The inability of philosophers over two millennia to solve this problem encourages his suspicion.

It is both a strength and a weakness of McGinn's hypothesis that it cannot be disproved – except by the solving of the problem. Clearly we are not cognitively omnipotent, and the solution of the mind–body problem could conceivably be beyond our reach, even when that reach is extended by the information-storing and information-processing devices that we have invented or may yet invent. But the argument that we can be expected to lack competence for such problems because they were irrelevant to our hunter-gatherer ancestors (or their ancestors) is not convincing, since so many problems that we have been able to solve – for example, those concerning the laws of motion, the nature of electromagnetic radiations, the origin of species, the basis of inheritance – must have been equally irrelevant. Equally unconvincing is the argument from the premise that so many acute minds have thought for so long and have made so little progress. Until rel-atively recently the same could have been said about the problem of the origin of life.

An odd feature of the line McGinn takes is the different way it affects different individuals. In the preface to his book *The Problem of Consciousness*, he tells us: 'The new perspective on the problem of consciousness that I advocate... had the welcome effect of dissolving the nagging intellectual discomfort that I had long associated with the problem. For the first time I felt able to *relax* in the face of the mind–body problem – as if an incessant torment had been finally laid to rest.'[23] This is strikingly reminiscent of Wittgenstein's remark: 'The real discovery is the one that makes me capable of stopping doing philosophy when I want to. – The one that gives philosophy peace, so that it is no longer tormented by questions which bring itself in question.'[24] (Whether Wittgenstein would have regarded McGinn's 'new perspective' as a real discovery is, of course, another matter.) Steven Pinker, in the last few pages of *How the Mind Works*, finds McGinn's hypothesis 'exhilarating, a sign of great progress in our understanding of the mind'.[25] Sadly, for me, and I suspect for most biologists, the suggestion that there are grounds for thinking that we may be constitutionally unable ever to solve what is now the most challenging problem in biology does not exhilarate, though it may chasten. And since there is a good chance that the hypothesis is wrong, the invitation it offers to set the problem aside is an invitation that it would be unwise to accept.*

An extreme view of a very different kind is that of the *eliminative materialists* – Paul Feyerabend, Richard Rorty, Stephen Stich, and Paul and Patricia Churchland.[26] What the supporters of this view want to eliminate is the common-sense view of psychological phenomena, which they like to call *folk psychology*, and which Paul Churchland describes as 'a radically false theory, a theory so fundamentally defective that both the principles and the ontology of that theory [i.e. the entities and properties postulated by the theory] will eventually be displaced, rather than smoothly reduced, by completed neuroscience'. And the theory, he says, is not only false; it gives us 'a radically misleading conception of the causes of human behavior and the nature of cognitive activity'.

This is strong language, and Churchland attempts to justify the strength by pointing out various ways in which, he claims, folk psychology has failed. *First*, it has not provided any insights into such mental phenomena as 'the nature and dynamics of mental illness, the faculty of creative imagination... the ground of intelligence differences between individuals... the nature and psychological functions of sleep...the miracle of memory...the nature of the learning process'. *Secondly*, it has been 'stagnant for at least twenty-five centuries'. ... 'The folk psychology of the Greeks is essentially the folk psychology we use today, and we are negligibly better at explaining human behavior in its terms than was Sophocles.' And, *thirdly*, 'its categories appear (so far) to be incommensurable with or orthogonal to the categories of the background physical science whose long-term claim to explain human behaviour seems undeniable.' In other words, he thinks that the beliefs, desires, fears, hopes, intentions, hunches, anxieties and so on that we use to explain behaviour cannot be expected to correspond to the categories that

*To be fair to Pinker, although he welcomes McGinn's hypothesis he wants philosophical, computational, neurological and evolutionary work on consciousness to continue unabated.

are likely to be provided by a fully worked out account of the physical events going on in our brains.

One way of answering the criticism that folk psychology is such a bad theory is to say that it isn't a theory at all – that beliefs, desires, hopes, fears, and so on are simply things that exist; things that we are aware of in ourselves and others and that themselves need explanation. But this won't wash. The fact is that we use combinations of such concepts to predict and explain each other's behaviour; indeed, as Jerry Fodor has pointed out, for many kinds of behaviour they are indispensable for this purpose, for we have no other way of going about it.[27] But if this is the only way we have, and we seem to manage pretty well, can folk psychology be so bad as an explanatory theory?

What then about Churchland's specific criticisms? In response to the first, I shall only say what I think my grandfather would have said in the same situation: 'The solution to that shopping-list of problems will have to await not one psychological theory but the coming of the Messiah.' A possible response to the second would be to say that survival for twenty-five centuries is hardly a ground for rejecting a theory, and that, anyway, it is not quite true that 'we are negligibly better at explaining human behaviour in [folk psychological] terms than was Sophocles'. We don't accept everything that Freud tells us, but one of his teachings that modern folk do accept is the importance of unconscious motivation. Sophocles never thought in those terms. Both in the myth and in Sophocles' play based on the myth, Oedipus killed his father and married his mother unwittingly, doomed by the gods, not by the complex that Freud named after him – despite Jocasta's remark that 'in dreams too, as well as oracles, many a man has lain with his own mother'.[28]

Churchland's third criticism needs closer attention. The eliminative materialists compare our present concepts of folk psychology with the concepts of 'caloric', 'phlogiston' and the 'essences' of the alchemists, as these concepts were viewed before we learnt to understand the nature of heat, of oxidation, and of elemental chemistry. All were useful and had some explanatory and predictive value but, because they happened to be cockeyed ways of looking at the phenomena they aimed to explain, they were displaced by later unrelated and much more effective theories.

This comparison, of course, begs the question, because it assumes the impossibility of the reduction from folk-psychological to physical that it is trying to prove; but Churchland gives three arguments for assuming that such a reduction is improbable. The first is that folk psychology is such a false and misleading theory that we shouldn't expect its categories to correspond to the categories of a fully worked out account of the physical events going on in our brains. But, as we have seen, the categories of folk psychology serve us rather well. The second is that folk psychology lacks 'coherence and continuity with fertile and well-established theories in adjacent and overlapping domains – with evolutionary theory, biology and neuroscience, for example…'. But, until well into the nineteenth century you might have pointed to a similar lack of coherence and continuity in the relations between the concept of temperature and Newtonian theories of motion; it was only when the relation between heat and molecular motion was discovered that the temperature of a gas was shown to 'reduce' nicely to the mean kinetic energy of its molecules. Churchland's third argument is peculiar: 'The new theory', he says, 'must entail a set of principles and embedded concepts that mirrors very closely the specific conceptual structure to be reduced. And the fact is, that there are vastly many more ways of being an explanatorily successful neuroscience while *not* mirroring the structure of folk psychology, than there are ways of being an explanatorily successful neuroscience while also *mirroring* the very specific structure of folk psychology. Accordingly, the *a priori* probability of eliminative materialism is not lower but substantially *higher* than that of…its competitors.' This won't work. Of course, removing a constraint may allow a far greater number of theories to compete, but you can't argue from that that the constraint is unlikely to exist, since you can't assume that each theory has an equal chance of being the correct one. If the constraint does in fact exist, the chance of any of the extra theories being successful is zero, however many there are.

All this, as well as 'the sword of common sense', makes the views of the eliminative materialists look unpromising – which means that the task of explaining conventional mental phenomena in terms of physical events in the brain has to be faced.

PROPOSITIONAL ATTITUDES

Beliefs, desires, fears, intentions, expectations and anxieties are members of a large category of mental states that can all be expressed in the form: someone believes/desires/etc that…something. Bertrand Russell called them *propositional attitudes*. A characteristic of these mental states is that, though they constantly indicate dispositions, they only intermittently involve consciousness. You probably believe that zebras have stripes, and you may believe that 'Guinness is good for you', but though your responses to questions about zebras and Guinness would suggest that your beliefs on these matters do not fluctuate, you spend little of your time thinking about either.

A second characteristic of propositional attitudes is that, like nearly all mental states, they are *about something* – even if the something is a non-existent thing like a unicorn or phlogiston. In 1874, Franz Brentano, former Dominican priest, former professor of philosophy at Würzburg, prodigious linguist, composer of riddles, and described as looking like a Byzantine Christ,[29] put forward the thesis that this quality of *aboutness* was an exclusive and defining feature of mental phenomena, distinguishing them from physical phenomena. A feature that did that deserved a name, and he called it *intentionality* – from the Latin *intendo*, meaning 'I point' – a confusing choice for English speakers as intentionality usually has nothing to do with the *intentions* that may or may not pave the way to hell, or whose nature may distinguish between murder and manslaughter.

What matters more, Brentano's notion that intentionality distinguishes the mental from the physical doesn't quite work. In the first place, *aboutness* seems to be a feature of many things that lack minds but are produced by beings with minds. Darwin's book *Orchids* is *about* the various contrivances by which orchids are fertilized by insects; Leonardo's painting *The Last Supper* is *about* Christ and his disciples at the Passover meal just after Christ has uttered the words 'One of you shall betray me'; the computer program that allows travel agents all over the world to book vacant flights with a good chance of success is *about* the current state of seat bookings of available flights. Examples like these are usually said to derive their intentionality from the intentionality of the authors, painters or programmers that produced them, but not all examples of intentionality in mindless entities can be explained in this way. Consider the orchid genes that are responsible for the contrivances that Darwin describes. These genes provide information *about* the amino-acid sequences of all the proteins needed for the building of the various contrivances. And this information is used to build the contrivances, not by any organism with a mind but by the orchids themselves. If intentionality is simply *aboutness*, the claim that the relevant orchid genes possess it seems every bit as strong as the claim that Darwin's book possesses it, and unlike the intentionality of Darwin's book it has not been derived from the intentionality of anything with a mind. Darwin was conscious of what he was writing, and we are conscious of what we are reading, but the orchid isn't conscious of anything. As a defining feature for mental phenomena, intentionality (in the sense of aboutness)* seems to be a dead duck – even a decoy duck.

A third characteristic of propositional attitudes is that we take them to have causal powers. My belief that I need a new driving licence (itself caused by the receipt of an official letter from the agency in Swansea) causes certain behaviour (checking that I can still read a number-plate at 25 yards, filling in a form and writing a cheque), and it also causes other mental states (gloomy thoughts about the passage of time). Leaving aside the question of consciousness, can we explain in principle how physical events in the brain could account for the characteristics of propositional attitudes, including the tendency of the behaviour caused by them to be effective?

*'Intentionality' is used in more than one way and some philosophers are unhappy at the separation of intentionality from consciousness. Searle suggests that the – perhaps unconscious – motivation for the 'urge to separate intentionality from consciousness is that we do not know how to explain consciousness, and we would like to get a theory of mind that will not be discredited by the fact that it lacks a theory of consciousness'. (Searle, J.R., 1992, *The Rediscovery of the Mind*, p. 153, MIT Press, Cambridge Mass.)

If mental states are represented as – the identity theorist would say 'exist as' – patterns of synaptic strengths in a real neural network, or in any other physical way that the neurophysiologists might prefer, and if these physical representations incorporate information about the outside world provided currently or in the past by sense organs, it is clear that the semantic properties of the mental states – their *contents* or *what in the outside world they are about* – are represented in entities that, being physical, have causal properties. In other words, as Fodor puts it 'there are mental states with both semantic contents and causal roles'.[30] The representation of one mental state (say, a desire to kiss Jenny) could therefore, in principle, interact with the representation of another (say, a belief that Jenny was alone and not otherwise occupied) to produce some effect. For the interaction to be useful, and the effect to be appropriate, the machinery would have to be such that the outcome was logically related to the input, but, as Fodor points out, this is precisely the condition that has to be met, and is met, by computers:

> Computers are a solution to the problem of mediating between the causal properties of symbols and their semantic properties. So *if* the mind is a sort of computer, we begin to see how you can have a theory of mental processes that succeeds where – literally – all previous attempts have abjectly failed; a theory which explains how there could be nonarbitrary content relations among causally related thoughts.

Computers and their programs have designers. Presumably it is natural selection, acting over a long period of our evolutionary history, that has ensured that, by and large, the causal relations of our mental states do match the semantic relations. The survival value of rationality is not to be under-estimated. A theory along these lines could account for complex and apparently purposive behaviour, but it still leaves one puzzle – the subjective aspects of consciousness.

23 The 'Mind–Body Problem' – Consciousness and Qualia

Other people bring me problems; he brings me solutions.
— Prime Minister Margaret Thatcher, speaking of Lord Young,
a minister in her government

For major problems to do with consciousness there seems to be no Lord Young. The best anyone has done is to clarify the problems. Dictionary definitions are almost ludicrously circular: consciousness is the state of being conscious; to be conscious is to be aware, with mental facilities awake; aware means conscious or knowing, awake means not asleep, and asleep means being in a state of which one of the characteristics is that consciousness is nearly suspended. No wonder Dr Johnson described lexicographers as harmless drudges. If, instead of ordinary dictionaries, you use dictionaries of philosophy, you are led not in circles but along innumerable paths that tend to be paved not with statements of fact but with difficult questions. If, though, forsaking all dictionaries, you stick to Wittgenstein's advice: 'the meaning of a word is its use in the language',[1] you find a fair consensus about the everyday use of the words conscious and consciousness, but rather sharp disagreement among philosophers about the nature and existence of different kinds of consciousness. I want to say something about both the consensus and the disagreements.

It is, of course, because of the consensus that lay people, along with physiologists, psychologists, physicians, surgeons and, in certain contexts, even philosophers, manage to talk about consciousness without further definition and without confusion. Everyone recognizes the distinction between the use of the word conscious in: 'a few minutes after the tamping iron had passed through his head, Phineas Gage was conscious and talking to his colleagues'; and its use in: 'picking up the glass from which the dead man had drunk, the detective was immediately conscious *of* a smell of almonds'. There is also a well recognized concept of self-consciousness – consciousness of one's own identity, and reflection on one's thoughts or feelings or (particularly in everyday use) the way in which one's behaviour is likely to be viewed by others. Difficulties arise when attempts are made to subdivide consciousness further, or to say just what it is, how we know about it, what function it serves, or how it can be explained.

When we smell freshly ground coffee, or hear an oboe playing, or see the intense blue of a Mediterranean sky, or have toothache, we are having experiences that it is impossible to describe except by pointing to similar experiences on other occasions. We can know what it is like to have these experiences only by having them or having had them. There are all sorts of other things we may be told about these experiences – what in the outside world is causing us to have them; what can be deduced about us or the outside world from the fact that we are having them; what effect they are having or likely to have on our behaviour; what is going on in our brains while we are having them – but none of these things tells us about the subjective qualities of the experiences; what it is like for us to have them. We need a word to refer to these subjective qualities and we have a choice of several. *Raw feels* is the most expressive; *qualia* (singular *quale*) is the most usual, with *phenomenal properties* a close runner-up, and *qualitative states* an also-ran. 'Raw feels' and 'qualitative states' need no explanation. Philosophical dictionaries tell us that, etymologically, 'quale' is to 'quality' as 'quantum' is to quantity' – which is more elegant than helpful. 'Phenomenal' in this sort of context means 'evidenced only by the senses'.

Although qualia are most obviously associated with sensations and perceptions, they are also found in other mental states, such as beliefs, desires, hopes and fears, during the conscious episodes of these states. Just as it feels like something to smell freshly ground coffee, so it can feel like something (at least intermittently) to believe that —, or to desire that —, or to hope that —, or to fear that —, where the dashes stand for whatever it is that is believed, desired, hoped or feared.

QUEASINESS ABOUT QUALIA

Although the existence of qualia seems so obvious, the eliminative materialists are totally dismissive about them. And in an article published in 1988, Daniel Dennett (who is too selective in his eliminations to be a full-blooded eliminative materialist)[2] explains that, though he does not deny the reality of conscious experience, his intention is to make it as uncomfortable for anyone to talk of qualia – or 'raw feels' or 'phenomenal qualities' or ... 'the qualitative character' of experience – as it would be for them to appeal to the concept of *élan vital*.[3] 'Qualia', he says, 'is a philosophers' term which fosters nothing but confusion, and refers in the end to no properties or features at all.' And he follows a similar line in his 1991 book, *Consciousness Explained*.[4] Since most of us appear to spend a substantial fraction of our time engaged in activities designed to enable us to experience qualia that we find pleasant and to avoid those that we dislike, and since many major industries could be regarded as similarly engaged, Dennett's attitude to qualia is startling and needs explanation.

There is, of course, something peculiarly elusive about qualia. Many of us, as children, have asked questions such as: If you and I look at the same bit of blue sky, or the same red rose, does the blue look to you the same as it does to me? Does the red look to you the same as it does to me? How do I know that the colour the sky looks to you – not the name of the colour but how it looks – isn't the same as the colour the rose looks to me? Such questions were considered in the 17th century by the English philosopher John Locke, who imagined two men whose eyes differed at birth in such a fashion that each saw the colour of a violet as the other saw the colour of a marigold, and vice versa.[5] 'Because one man's mind could not pass into another man's body, to perceive what appearances were produced by these organs', he concluded that the colour inversion could never be known because it would have no effects. You might wonder whether this conclusion of Locke's can be reconciled with our ability to detect colour-blindness, but it can. We usually detect colour blindness by tests based on the inability of the sufferer to distinguish between colours that a normal person can distinguish between, or vice versa. If I cannot distinguish between two colours that look different to most people, then at least one of them must look differently to me from the way it looks to most people, and I have some kind of colour blindness. But I cannot know how either of the colours looks to most people, nor can they know how those colours look to me beyond the fact that they both look the same.

Two and a half centuries after Locke's conclusion about colour inversion, Wittgenstein made the same point about pain:

> Now someone tells me that *he* knows what pain is only from his own case! – Suppose everyone had a box with something in it: we call it a 'beetle'. No one can look into anyone else's box, and everyone says he knows what a beetle is only by looking at *his* beetle. – Here it would be quite possible for everyone to have something different in his box. One might even imagine such a thing constantly changing.[6]

Wittgenstein was close to the logical positivists of the Vienna circle, and their central doctrine was the *verification principle*, which denied literal meaning to any statement that was not *either* a tautology of logic or mathematics, *or* verifiable by observation (not necessarily directly) through the senses. Since statements about the inversion of qualia (such as Locke's statement about the inversion of colours seen by his two imagined men) seemed to fit in neither category, such statements were suspected of being meaningless, and the whole notion of qualia was thrown into doubt. Later, as difficulties in deciding what counted as verification grew, the verification principle lost, or at least loosened, its grip; but difficulties with qualia remained.

Wittgenstein pointed out that just as there are no criteria to tell us whether sensations experienced in similar situations and spoken of in identical ways by different people are identical, there are also no sure criteria to tell us whether sensations described as identical by the same person at different times are identical. 'Let us imagine the following case', he says:

I want to keep a diary about the recurrence of a certain sensation. To this end I associate it with the sign 'S' and write this sign in a calendar for every day on which I have the sensation. – I will remark first of all that a definition of the sign cannot be formulated. – But still I can give myself an ostensive definition [as in teaching a child the word cat by pointing at one]. – How? Can I point to the sensation? Not in the ordinary sense. But I speak, or write the sign down, and at the same time I concentrate my attention on the sensation – and so, as it were, point to it inwardly. – But what is this ceremony for? for that is all it seems to be! A definition surely serves to establish the meaning of a sign. – Well, that is done precisely by the concentrating of my attention; for in this way I impress on myself the connexion between the sign and the sensation. – But 'I impress it on myself' can only mean: this process brings it about that I remember the connexion *right* in the future. But in the present case I have no criterion of correctness. One would like to say: whatever is going to seem right to me is right. And that only means that here we can't talk about 'right'.[7]

This may sound contrived, but Dennett discusses several imaginary or real situations which suggest that there *are* problems in relating present and past qualia.[8] One of the imaginary situations concerns two professional tasters, Mr Chase and Mr Sanborn, employed by Maxwell House to ensure that the flavour of the famous coffee is preserved. After six years in the job both are unhappy because, though they used to like Maxwell House coffee, they find they no longer like it. But they interpret this change differently. Mr Chase says that the taste of the coffee hasn't changed; it's simply that his tastes have changed. He has become a more sophisticated coffee drinker and no longer likes the Maxwell House taste. Mr Sanborn says that he accepts that the coffee hasn't changed – all the other tasters agree about that – but his tasting machinery must have changed for the coffee tastes different to him. If it tasted like it used to he would still like it.

To show the difficulties inherent in qualia, Dennett employs qualia-speak to analyse the situation. For Mr Chase, there seem to be three possibilities. First is his own view, that the taste-qualia he gets by drinking Maxwell House coffee haven't changed but his reactive attitude to those qualia (reflecting his preferences) has. An alternative possibility is that he is wrong about the constancy of his qualia, and he is in the same state as Mr Sanborn but without Sanborn's insight. Or the explanation might lie somewhere in between. For Mr Sanborn, too, there are three possibilities. His own view may be right. Or he may be in the same state as Mr Chase but without Chase's insight. Or the truth might lie somewhere in between.

The crucial point is that neither Mr Chase nor Mr Sanborn can decide between these possibilities despite having what is usually (if metaphorically) described as privileged access to knowledge of their own qualia. Experimental tests of the men's ability to discriminate between different coffees or to recognize individual varieties after a time interval could make one or other possibility seem more likely, and so, in principle, could neurophysiological investigations; but, as

Dennett puts it: ' The idea that one should consult an outside expert, and perform elaborate behavioural tests on oneself to confirm what qualia one had, surely takes us too far away from our original idea of qualia as properties with which we have a particularly intimate acquaintance.'

Arguments of this kind lead Dennett to regard qualia as an unsatisfactory concept, and he retreats in the direction of behaviourism.* 'I want', he explains, 'to demonstrate some of the positive benefits of...the "reductionist" path of *identifying* "the way it is with me" with the sum total of all the idiosyncratic reactive dispositions inherent in my nervous system as a result of my being confronted by a certain pattern of stimulation'.[9] And a couple of pages later: 'When you say "*This* is my quale," what you are singling out, or referring to, *whether you realize it or not*, is your idiosyncratic complex of dispositions. You *seem* to be referring to a private, ineffable something-or-other in your minds eye, a private shade of homogeneous pink, but this is just how it seems to you, not how it is.'

If this notion is not to suffer from the defects of classical behaviourism, we must (*pace* Ryle – p. 370) regard possessing a set of dispositions to react in particular ways as evidence that the possessor is in a particular physical or functional state, a state that is, among other things, responsible for the dispositions to behave in those ways. That presents no problem. But the question we want answered (and which Dennett is not even asking) is: why should that state also be responsible for a 'raw feel' – a private shade of homogeneous pink in Dennett's example? For however insecure the arguments of Wittgenstein and Dennett may make us feel about our identifications of current qualia with past qualia, they provide no reason for pretending that we are not experiencing the current qualia. Dennett seems to anticipate this point for he mentions that a common response to his arguments is to say: 'But after all is said and done, there is still something I know in a special way: I know *how it is with me right now*.'[10] And he then goes on: 'But if absolutely nothing follows from this presumed knowledge...what is the point of asserting that one has it? Perhaps people just want to reaffirm their sense of proprietorship over their own conscious states.' The answer to Dennett's question, I suspect, has less to do with proprietorship than with the difficulty of being a behaviourist when you are suffering from toothache. We all find it irritating to be told that what is salient in our consciousness at any moment is actually a disposition to behave in certain ways. Indeed, John Searle accuses Dennett of claiming to embrace consciousness as just another material property, but doing 'this by so redefining "consciousness" as to deny...its subjective quality'.[11]

---

*After his resignation from Yale, and while he was working for the J. Walter Thompson agency, J.B. Watson, the founder of behaviourism, ran a very successful advertising campaign for Maxwell House coffee. Does this account for Dennett's choice of Maxwell House as the setting for his story of Mr Chase and Mr Sanborn?

SEARLE'S 'BIOLOGICAL NATURALISM'

If Dennett sometimes seems to be trying to avoid what Searle has called the materialist's 'terror of consciousness' by redefining consciousness, Searle himself sometimes seems to be trying to make consciousness less terrifying by domesticating it – like calling a tiger a big pussy-cat. In the opening paragraph of his book *The Rediscovery of the Mind* he writes:

> The famous mind-body problem, the source of so much controversy over the past two millennia, has a simple solution. This solution has been available to any educated person since serious work began on the brain nearly a century ago, and, in a sense, we all know it to be true. Here it is: Mental phenomena are caused by neurophysiological processes in the brain and are themselves features of the brain. To distinguish this view from the many others in the field, I call it 'biological naturalism.' Mental events are as much part of our biological natural history as digestion, mitosis, meiosis, or enzyme secretion.[12]

And a few pages later:

> Consciousness is a higher-level or emergent property of the brain in the utterly harmless sense of 'higher-level' or 'emergent' in which solidity is a higher-level emergent property of H_2O molecules when they are in a lattice structure (ice) and liquidity is similarly a higher-level emergent property of H_2O molecules when they are, roughly speaking, rolling around on each other (water).

Searle is certainly right in saying that mental phenomena are caused by neurophysiological processes in the brain, that they are themselves features of the brain, and that they are as much part of our biological natural history as digestion etc; but it is not clear how this solves the mind–body problem. To say, as he does, that 'the *mental* state of consciousness is just an ordinary biological, that is, *physical*, feature of the brain' is an extraordinary use of 'ordinary'. Calling a tiger a big pussy-cat doesn't blunt its claws. And though it is true that consciousness is a higher-level or emergent property of the brain in the same sense as that in which solidity and liquidity are higher-level emergent properties of H_2O molecules in different states, that true statement conceals a formidable difference. We can understand why solidity should emerge from water molecules that are fixed in a lattice, and liquidity from water molecules that roll round one another; but we haven't much more idea why consciousness should emerge from the events in the brain than Thomas Henry Huxley had when he wrote:

> How it is that anything so remarkable as a state of consciousness comes about as a result of irritating nervous tissue, is just as unaccountable as the appearance of Djin when Aladdin rubbed his lamp.[13]

Searle is, of course, aware of this difference. Later in his book he points out that 'not all explanations in science have the kind of necessity that we found in the relation between molecule movement and liquidity'. And he gives the example of the inverse square law, which tells us how the gravitational attraction between two bodies varies with the distance between them, but 'does not show why bodies *have to have* gravitational attraction'.[14] The solution to the gravitational problem will doubtless have to await rather fundamental advances in physics, and it may be, as Roger Penrose and Graham Cairns-Smith have suggested, that the solution to the problem of consciousness will also depend on advances at that level.[15] Meanwhile one can only continue to investigate on the basis of what is understood.

CONSCIOUSNESS AND CREATIVITY

An interesting claim made by Searle is that consciousness gives us much greater powers of discrimination, and greater flexibility and creativity than unconscious mechanisms could have, and that this accounts for the evolution of consciousness.[16] In support of his claim, Searle refers to studies by Penfield of epileptic patients who had *petit mal* seizures while walking, driving or playing the piano, and who continued these activities in a routine mechanical (though still goal-directed) way despite what was described as a total loss of consciousness. He argues that the association of the loss of consciousness with loss of flexibility and creativity of behaviour suggests that a function of consciousness is to promote flexibility and creativity. Others have noted that patients with blindsight, despite their frequent successes when obliged to guess, are unable to describe what lies in the blind part of their visual field, or to use information from that part of the field as a basis for reasoning, or to guide behaviour. They have therefore argued that consciousness is essential if information represented in the brain is to be used for reporting or reasoning or for rationally guided action.[17] And it is, of course, true that humans – the animals who show the greatest discrimination, flexibility and creativity in their behaviour – also (so far as we can tell) have the most highly developed consciousnesses. On the other hand, forced to play chess with the formidable 'Deep Blue', most of us would find that the discrimination, flexibility and creativity the machine displayed – albeit in a very limited field – far exceeded our own, but we would not attribute the machine's superiority to its consciousness. And, as Tony Marcel has remarked, 'most theories of cognition make no call at all on consciousness'.[18]

The arguments based on Penfield's *petit mal* patients and on the patients with blindsight have been attacked by Ned Block, of the Massachusetts Institute of Technology, in an article in which he attempts to distinguish between two concepts of consciousness: *phenomenal-consciousness* – which is what we normally think of as consciousness, and which is particularly but not exclusively associated with sensations — and what he calls *access*-consciousness.[19] A mental state is a state of access-consciousness, he says, if a representation of its content is available for use in reasoning and in rationally guided action (including, in humans, rationally guided speech). Although these two kinds of consciousness may interact — a change in how one interprets what one is seeing may alter 'what it feels like' to see it — Block regards the two kinds of consciousness as distinct, and he claims that it is the failure to distinguish between them has led to misleading arguments about the functions of consciousness.

There are, it seems to me, and to several of Block's critics whose comments are published with his article,[20] difficulties with Block's concept of access consciousness, but I do not think that they affect his conclusion that the *petit mal* and blindsight findings do not prove that (phenomenal) consciousness confers greater discrimination and flexibility and creativity on behaviour. For reaching this conclusion, the crux of his argument, if I understand it correctly, is that, in both clinical conditions, the reduction in discrimination, flexibility and creativeness could be (and is more likely to be) the result of damage to the machinery responsible for complex information-processing tasks — tasks that would be needed for sophisticated discrimination and for guiding rational behaviour even in a robot — rather than damage to the machinery essential for phenomenal consciousness. This sounds convincing. After all, in the various agnosias we see selective losses of particular kinds of discrimination without any loss of phenomenal consciousness.

Of course, the validity of Block's criticisms doesn't prove that the machinery responsible for consciousness *does not* facilitate discrimination, flexibility and creativity, or that this facilitation did not play a part in the evolution of consciousness. The fact is that both discriminative, flexible and creative behaviour and highly developed consciousness depend on the ability to acquire, represent and store information about the external world, and to use that information, together with information about the internal states of the individual, in extremely complex information processing tasks — tasks whose ultimate role is to control behaviour.

We know enough about the machinery responsible to feel reasonably confident that there is no impenetrable barrier to gaining a much fuller understanding of the objective processes that are involved. In contrast, we understand very little about the production of conscious states. The machinery responsible seems to be at least partly the same as the machinery responsible for the objective processes but, as we have seen, we don't know yet what physical events are necessary and sufficient to cause particular conscious states — let alone why such events should have these causal powers. That — as the surgeon whom I consulted in Sienna said of my inability to speak Italian — is *una grande lacuna*. But there is no point in trying to persuade ourselves that the lacuna isn't there, or that filling it is going to be straightforward, or that because filling it is impossible we might as well give up. So long as it remains unfilled, by the way, we cannot even begin to see what would be needed to make a machine conscious; nor, presented with an allegedly conscious machine, could we tell whether it was really conscious or just a robot clever enough to behave like a conscious being – which is how Descartes thought of animals. Perhaps we are like our 17th-century forebears trying to understand nerve conduction before electricity was understood. Fortunately, we don't need to understand conscious states in order to experience them.

24   Free Will and Morality

Others apart sat on a hill retir'd
In thoughts more elevate, and reason'd high
Of providence, foreknowledge, will and fate,
Fix'd fate, free will, foreknowledge absolute,
And found no end, in wand'ring mazes lost.
— Milton, *Paradise Lost*, bk ii

Sir, We know *our will is free, and* there's *an end on't.*
— Dr Johnson (from Boswell's *Life*)

There is an old paradox that, in modern dress, goes something like this. At this moment the state of my brain, with its 10^{14} synapses, is determined by the genes I acquired in the womb and by all the things that have subsequently happened to me or are currently happening to me. If any decision I take is determined by physical events in my brain – which seems likely – then that decision must be determined by my genes, my history, and my present circumstances. Whether I am to buy my coffee at Tesco's or Sainsbury's seems to be decided for me. I seem to have no choice. Yet I have the strong impression that I can choose; indeed, even if, as happens to be the case, I don't care a bean which of the two stores I buy my coffee from, if I want to buy it from either I have to choose – if only by choosing to spin a coin. How can the determination of brain states be reconciled with the impression of free will?

'Hold on!' you may be saying (if you have a nodding acquaintance with modern physics), 'That's old hat! We don't live in a deterministic world any more. What about Heisenberg and his uncertainty principle?' But Heisenberg won't help. It follows from his principle that at subatomic levels the result of an action can be expressed only in terms of the probability that a certain effect will occur. You can't, for example, predict when a given atom of a radioactive element will disintegrate, though you can say what the probability is that it will have disintegrated within a certain period. But if you have a sample of radioactive material, the vast number of atoms present makes it possible to say rather precisely how long it will take for half of the atoms to have disintegrated. At the levels we are concerned with – people and objects – Heisenberg uncertainty can be almost entirely ignored. And to the extent that it cannot be it provides no comfort. If there is one thing more disconcerting than believing that, in any given situation, all the decisions we think we are making are predetermined by our genetic make up and previous experience, it is to believe that that is not quite true because there is, in addition, an element of chance. The unpredictability that chaos theory tells us is likely in complex deterministic systems is equally unhelpful.

So the problem has to be faced. And it's a problem with serious implications, because if determinism and free will are incompatible, and we are obliged to give up the idea that we have free will, can we be responsible – or be held responsible – for our actions? Should we ever be praised or blamed; rewarded or punished?

I have presented the problem in modern dress and in secular terms, but for two thousand years people have wrestled with a very similar problem in religious garb. If God is omniscient and foresees all things, we can only choose what he has foreseen so how can we have free will? This problem is never raised in the Bible but, starting at the latest in the first century with Rabbi Akiba – 'Everything is foreseen, yet freedom of choice is given' – it has worried Jews, Christians and Muslims ever since.[1] Augustine, in the fourth century, attempted to argue that the solution to the problem was that God's foreknowledge, though infallible, did not *cause* what was going to happen. 'God compels no man to sin, though he sees beforehand those who are going to sin by their own will...Just as you apply no compulsion to past events by having them in your memory, so God by his fore-knowledge does not use compulsion in case of future events.'[2] A similar solution was advocated by the Mutakallimun, a school of heterodox Islamic philosophers founded in the eighth century, and later by Saadia Gaon, the head of the talmud-ical academy at Sura in Babylonia in the tenth century. (It is a spurious solution which, Louis Jacobs tells us, received its *coup de grâce* in 1754 from the New England Congregationalist minister Jonathan Edwards – later, briefly, president of what was to become Princeton.[3] Edwards pointed out that even if it is not God's foreknowledge that *causes* an event, that foreknowledge, being infallible, proves the *necessity* of the event.) Maimonides, the great Jewish teacher, philoso-pher and physician of the 12th century, recognized the magnitude of the problem – its solution, he said, was larger than the earth and wider than the sea – but all he could offer by way of solution was to say (though at greater length) that there was no doubt either about God's foreknowledge or 'that man's deeds are in his own hands', but that God's knowledge was not like human knowledge, and was beyond our comprehension. I doubt if 'cop-out' was part of the vocabulary of 12th-century rabbinical Hebrew, but Maimonides was rebuked by his contem-porary Abraham ben David of Posquières for conduct unbecoming to a sage, in embarking on a problem that he was not capable of seeing through to the end. In the 13th century, Thomas Aquinas and in the 16th Samuel ben Isaac of Uceda, a rabbi in Safed, suggested that the solution to the problem lay in the fact that God exists outside time. 'When a man sees someone else doing something,' wrote

Samuel ben Isaac, 'the fact that he sees it exercises no compulsion on the thing that is done. In exactly the same way the fact that God sees a man doing the act exercises no compulsion over him to do it. For before God there is no early or late since he is not governed by time.' This is more ingenious than convincing.

But, leaving God aside, what is the solution to the secular paradox? How can the determination of brain states be reconciled with the impression of free will? And – a separate though closely related problem – can we be responsible for our actions?

Look at the first problem like this. Since thoughts are effective in the physical world they must be correlated with or identical with, physical – presumably neural – events. But these events, being physical, must themselves have physical causes, which must themselves have physical causes…and so on. Hence our thoughts, including our desires and decisions and any actions associated with them, must be determined by our genes, our history, and our current circumstances. Even if, at some point in the chain, Heisenberg uncertainty comes in, that only adds an element of randomness. If all this makes you uncomfortable, what alternative can you suggest? If our thoughts are not determined in this way, in what way are they determined? What would you prefer to believe? – not, presumably, that our thoughts are entirely random. You say, perhaps, that you would like to believe that, in thinking, you are free to create your own thoughts and make your own decisions, unconstrained by your genes or your past history. But how would you do that? Certainly you can have thoughts about thoughts, and thoughts about thoughts about thoughts. The problem is not that you may be heading for an infinite regression – you can stop when you like – but that all these thoughts must have neural correlates (or be identical with neural events), so you don't escape from determinism. There is no coherent alternative to determinism. There is no 'homuncular you' controlling the machinery, no independent 'sanctuary within the citadel' (William James),* no 'transcendental ego' (Kant).** But if that is right, why do we all – not just Dr Johnson – feel so confident that our will is free?

There seem to be three reasons. In the first place, in any given situation, what ultimately determines the neural events that are correlated with (or identical with) our thoughts, including our decisions and desires, is our own genes and our own history; not anybody else's. In the limited sense that, in any given situation, it is our own characteristics that determine what we think, we *are* free.

---

*James used the phrase to refer to the 'spiritual something' in us which 'whatever qualities a man's feeling may possess, or whatever content his thoughts may include…welcomes or rejects' them.

**Kant used the phrase to refer to an ego which is supposed to be distinct from the empirical ego associated with ordinary self consciousness, and to account for the unity of our experience.

Secondly, the neural events are invisible to us; we can't introspect them; we can only introspect thoughts and feelings. What we observe is what William James observed – a 'stream of thought', not the machinery that controls the flow or contents of that stream. Our thoughts (even if, as Dennett supposes, they depend on multiple drafts)[4] appear to flow freely so why should we not feel that our will is free? Putting the matter another way, and thinking specifically of desires: If we can only want what the neural events determined by our genes, our history and our current circumstances make us want, we may be frustrated by not being able to get what we want – either because it's not achievable or because it's incompatible with attending to other, more pressing, wants – but we won't feel prevented from wanting it. What the neural events don't make us want we don't want. (We may, of course, feel guilty at wanting to possess our neighbour's goods, because we also want to obey the commandment not to covet our neighbour's goods, but though simultaneous incompatible wants present a problem, it is not the problem of free will.)

The third reason is, perhaps, the most important. It is that we feel ourselves to be part of the system that affects the future (at any rate, locally), and that our decisions, predetermined though they may be, do affect the future. And we are justified in feeling this. We can't opt out. We can't climb out of James's stream of thought and rest on the bank, arguing that since everything is predetermined nothing will be altered by our inertness. Our inertness would itself have effects. To reach a decision to act in a particular way we have to deliberate, and until we have deliberated we won't know which way the predetermined cat will jump. But note that, though our decision – the behaviour of the notional cat – is predetermined because the mental/neural events that cause it are predetermined, the predetermination of those events does not make them any less effective. And since we are mostly unaware of the predetermining machinery – and if we are aware of its existence we have no idea in which direction it is heading – we feel able to make decisions both freely and effectively. In this sense – and it is this sense rather than that given in the dictionary* that most matters to us – we have free will.

These considerations lead to a comforting corollary. One of the things that makes people regard determinism with such misgiving is the feeling that it inevitably leads to fatalism. If all our decisions and actions are predetermined, why try? Why not just sit back and let whatever happens happen? But, as we have seen, because we are part of the system, inertness has effects, just as any other behaviour has effects. There is no opting-out of the causal chain. Fatalism is not a coherent philosophy, and we can accept determinism without worrying about the bogeyman thought to follow in its train.

If this view of free will is right, where does that leave questions about

---

*The Shorter Oxford (1993) still includes Dr Johnson's definition: the power of directing our own actions without constraint by necessity or fate.

responsibility? Responsibility, of course, has many meanings but what we are immediately concerned with here is moral and legal responsibility for our actions. If these actions are wholly determined by our genes, our history, and our current situation, in what circumstances, if any, do we deserve praise or blame, gratitude or resentment, reward or punishment?

Laws – man-made laws – may seem more arbitrary than morals, but because they have to be designed to make clear-cut distinctions, it is helpful to see how laws distinguish between circumstances in which we are or are not responsible – or, since laws are usually concerned with blame rather than praise, circumstances in which we are or are not *liable*.

Although we are liable for speeding just by exceeding the speed limit, and we commit a parking offence just by parking in a prohibited area, with more serious offences liability depends not only on our doing something illegal but also on our intending to do it – or (sometimes) being so negligent that we fail to avoid doing it. To be guilty of murder we must not only have killed but intended to kill; and that intention must be uncoerced. But the question of coercion is tricky. The policeman who shoots the criminal who is pointing a gun at him will not be liable; but the criminal who has shot someone in the course of a crime and who claims that, as a result of his genes and his deprived childhood his character is such that, in the position he was in, he had no choice but to shoot, can only hope at best to persuade the court to regard his personal problems as mitigating circumstances. On the other hand, the repeatedly battered wife who, 'provoked beyond endurance', as they say, eventually stabs her husband with the carving knife, might well be found 'not liable'.

The suggested outcome of these three cases probably seems reasonable to most people, however deterministic their views. But why should a convinced determinist be reluctant to accept the criminal's defence, when what he says in his defence is true and accounts for his unfortunate behaviour? One answer to that question is, presumably, that the verdict and punishment become part of the history of the criminal and of everyone in his community, and so influence the future behaviour of the criminal and the whole community. Both would be damaged by accepting as a valid defence what may well be a valid explanation. Acquitting the battered wife would not have similar undesirable effects; it might even encourage battery-prone husbands to behave better. Since both laws and the moral codes they reflect are thought to have originated in social behaviour that evolved by natural selection, it is not surprising that they tend to judge individuals on current behaviour, without making allowances for individual histories in a way that would encourage further departures from the code. In other words they usually ignore the predetermined nature of our acts, and it is usually to our advantage, as a community, that they do.

In saying that both laws and the moral codes they reflect are thought to have originated in social behaviour that evolved by natural selection, I am not suggesting that the origin of our moral sense was a cold-blooded rational assessment of the effects – and hence the desirability – of different kinds of behaviour. It is true that Darwin thought it highly probable 'that any animal whatever, endowed with well-marked social instincts...would inevitably acquire a moral sense or conscience, as soon as its intellectual powers had become as well, or nearly as well developed, as in man';[5] and he pointed out that the Roman Emperor Marcus Aurelius had also regarded social instincts as the basis of morals. But it is clear that what Darwin had in mind was not a dispassionate weighing up of pros and cons. In fact, he took exception to the utilitarian approach of Jeremy Bentham and John Stuart Mill – the approach which held that 'actions are right in proportion as they tend to promote happiness' – because he felt that individuals acted rightly *instinctively*, without a preliminary rational calculation. (Darwin uses the word 'instinct' to mean an inherited pattern of behaviour.) Conscience arose, he believed, when animals with well developed social instincts became intelligent enough to remember the discomfort, 'or even misery', of occasions in which they had failed to satisfy their social instincts because of the temporarily more urgent pressure of some other instinct such as hunger. After language had been acquired, 'the common opinion how each member ought to act for the public good, would naturally become in a paramount degree the guide to action', provided that there was sympathy between those expressing the opinion and those acting on it. And, finally, 'habit in the individual would ultimately play a very important part in guiding the conduct of each member; for the social instinct, together with sympathy, is, like any other instinct, greatly strengthened by habit.' (And Darwin still believed that habits could be genetically inherited.)

When Darwin published these views in *The Descent of Man*, the theory of evolution by natural selection was already widely accepted, but the notion that the basis of morals was natural rather than divine caused alarm and dismay.[6] *The Times* felt that, in dealing with moral questions, Darwin was 'quite out of his element', the *Athenaeum* that he was at his 'feeblest', and *The Quarterly* that 'Darwin's power of reasoning seems to be in an inverse ratio to his power of observation.' But when you compare the plausible reasoning of Darwin's suggested explanation of the origin of human morality and law with the Just-So stories of 'covenants' or 'compacts' or 'social contracts' entered into by individuals in a 'state of nature' that were proposed as explanations by the card-carrying philosophers Hobbes and Locke in the seventeenth century and Rousseau in the eighteenth, this kind of attack seems misplaced.*

In fact, Darwin's theory was subtle in several ways.

---

*On the other hand, unlike Hobbes, Darwin was never threatened with being burnt for heresy by a committee of bishops in the House of Lords.

People who do not believe that our ethical codes are God-given are often tempted to believe that they are simply a matter of learnt personal beliefs; we believe differently because we have been taught differently. In a lecture to the Leicester Philosophical Society in 1921, the Cambridge philosopher G. E. Moore amused himself by pointing out that if that were true, when he judged of a particular action that it was wrong, and someone else judged of the same action it was right, there would be no more disagreement between them than if the one had said 'I came from Cambridge today' and the other had said 'I did not come from Cambridge today', both statements being true.[7] Anyway, this is not how we feel. If we feel that something – fox-hunting, abortion, euthanasia – is wicked or cruel and others don't (or vice versa), we are tempted to regard them as perverse, wrong-headed, beyond the pale. We have a strong gut feeling that many kinds of behaviour are intrinsically good or intrinsically bad, and we need to be able to reconcile the existence of that feeling with the regrettable fact that there is not universal agreement about which category a particular piece of behaviour belongs to. By suggesting that our ethical feelings arise from an instinct, Darwin provides some explanation of their deep and spontaneous nature – deep and spontaneous feelings accompany instinctive behaviour connected with nutrition or sex or parenthood or self-defence – and by making the instinct the social instinct, an instinct which (like the language instinct) is associated with a capacity for highly complex behaviour of many kinds, he provides a possible explanation for at least some of the variability that is found in our ethical attitudes.

A crucial feature of Darwin's theory of the evolution of conscience and moral behaviour was that such behaviour would have a selective advantage. Although the Manchester essayist William Greg had pointed out that moral behaviour might weaken a tribe by helping the unfit to survive,[8] the evolutionists argued that, by facilitating cooperation between individuals, moral behaviour could greatly help a tribe in competing with other tribes. If the helpful effect of moral behaviour predominated, Darwin's theory offered an escape from the moral relativism that followed from the assumption that our ethical code is determined arbitrarily by what we have been taught. Teaching is, of course, important in imparting an ethical code but, for our ancestors to have survived and flourished, the code that they were taught must have been influenced by the kind of behaviour that favoured their survival.

Despite explaining so much so plausibly, and providing an escape from moral relativism into the bargain, the theory that natural selection could account for the origin of morality has had a very chequered history over the century and a half since Darwin first thought of it. Apart from its threat to religion, a major cause of its unpopularity was a reaction to the extravagances of the so-called social Darwinists – a heterogeneous collection of people who attempted to apply Darwin's theory of natural selection to explain the social relations of individuals, groups, nations and races.[9] Chief among these was Herbert Spencer – rarely read now but very influential both in the United Kingdom and the United States in the 1870s and 1880s. Even before Darwin and Wallace had given sketches of their theories of natural selection to the Linnaean Society, Spencer had suggested that the scarcity of resources had improved the human race by ensuring that it was the best in each generation that tended to survive. He welcomed with enthusiasm Darwin's notion that natural selection could explain the evolution of animals and plants, and he suggested the phrase 'survival of the fittest' to describe the crux of the theory. But his aim was to produce a grand philosophy that would account for all evolution – of planets, of animals and plants, of embryos, and of societies – in terms of a progression from homogeneity to heterogeneity; this would be based on both physics (especially the law of conservation of energy, which for some reason he called the law of persistence of force) and the laws of biology (including not only natural selection but also Lamarckian inheritance of acquired characteristics). As Richard Hofstadter has put it, referring particularly to the USA, all this 'made Spencer the metaphysician of the home-made intellectual, and the prophet of the cracker-barrel agnostic'.[10] But his non-conformist background, his confidence that individuals were endowed with a moral sense, and his recognition of an inscrutable power behind phenomena, which he called 'the unknowable', ensured that his views were also acceptable to many theists.

What made Spencer's views matter politically was his extreme *laissez-faire* attitude, his antipathy to any action by the state beyond preserving the right of each man to do as he pleases (subject only to his not interfering with the rights of others), and a profound and wholly unjustified optimism – extreme even for a Victorian – that, without undue interference, the evolution of the ideal society was inevitable:

> Thus the ultimate development of the ideal man is logically certain – as certain as any conclusion in which we place the most implicit faith; for instance that all men will die...Progress, therefore is not an accident but a necessity. Instead of civilization being artificial, it is part of nature; all of a piece with the development of the embryo or the unfolding of a flower.[11]

With such a desirable consummation to be expected, it was important not to interfere with the machinery bringing it about. Spencer therefore opposed the poor laws, free education, and the setting of minimal standards for housing. Of the unfit poor he felt: '... the whole effort of nature is to get rid of such to clear the world of them, and make room for better.' ... 'If they are sufficiently complete to live, they *do* live, and it is well they should live. If they are not sufficiently complete to live, they die, and it is best they should die.'[12]

As Hofstadter has pointed out, in the USA just after the War of Independence, sentiments such as these fell on fertile soil; 'American Society saw its own image in the tooth-and-claw version of natural selection', and the phrase 'survival of the fittest' was seized on by the likes of John D. Rockefeller and Andrew Carnegie both to describe and to justify their ruthless business practices.[13] But by making Spencer the *bête noire* of later sociologists they also ensured that, in that field, not only Spencer but the whole subject of evolutionary ethics was viewed with distaste. Towards the end of the 19th century and in the first part of the 20th, further aversion therapy was provided by the naïve and unprincipled use of Darwinian theory to try to justify the most appalling racial prejudice and even genocide. Spencer's views have been out of fashion for a very long time, but, with rare exceptions,* it is only in the last few decades that sociologists and social psychologists have been ready to think again in terms of evolutionary theory.[14]

There are, of course, more serious grounds for criticizing some aspects of Darwin's theory of the evolution of moral behaviour. His belief that habits could be inherited became increasingly untenable as time went by without producing any convincing evidence of the inheritance of acquired characters. The final blow was the demonstration by August Weismann, in Freiburg, around the time of Darwin's death, that the genetic material destined for the egg or sperm produced by an individual is separated off very early in the life of that individual. That separation makes it very difficult to see how the necessary modification of the genetic material in the egg or sperm could be brought about by changes in the rest of the body.

A more serious difficulty was the uncertainty about whether group selection could account for the spread of moral behaviour in a population. We saw (Chapter 3) that kin selection can account for altruism among blood relations, and that reciprocal altruism can account for altruism among unrelated members of a population small enough to ensure that failure to reciprocate can be recognized and punished by exclusion from further favours. For larger populations, it has generally been thought that the ability of individuals to cheat — that is, to get the advantages of the altruism of others without contributing to the cost — would tend to lead to an increase in the ratio of cheats to altruists and so prevent altruism from spreading in the population.

Since our primate ancestors probably lived in small groups largely made up of relatives, kin selection and reciprocal altruism could explain the development of altruistic behaviour in them. But we, alone among animals, show highly cooperative and sometimes altruistic behaviour while living in groups containing large numbers of individuals, each unrelated to the great majority of the

*A striking exception was Wilfred Trotter, whose *Instincts of the Herd in Peace and War* was published in 1916 and reprinted thirteen times by 1942. But Trotter was only a sociologist *de facto*. *De jure* he was an extremely distinguished surgeon who is said to have refused a knighthood from King George V, one of his patients, later remarking to a friend 'I have no time for such buffoonery'.

group. How can this have happened?

Roughly speaking, for an individual facing competition within a large group, it pays to cheat; for a group facing competition with other groups, it is helpful if its members behave altruistically. A gene that encourages altruistic behaviour will therefore handicap the individual in competition *within* the group, even where it helps the group in its competition with other groups. For many years, the prevailing view has been that, unless several unlikely (but possible) conditions are satisfied, the disadvantage to the individual outweighs the advantage to the group, and the gene encouraging altruistic behaviour will fail to spread.[15] Recently though, computer simulations have shown that cooperative altruistic behaviour is more likely to spread and become stable if reciprocation is not all-or-nothing but variable, so that rewards can be matched to the benefit received. In this situation, a strategy encouraging spread (and resistant to cheats) is to offer little at the first encounter with a new partner, and gradually to 'raise the stakes' if the partner reciprocates satisfactorily.[16]

In humans, much behaviour is determined by our culture rather than our genes, so could our cooperative behaviour be, at least in part, the result of selection of behaviour patterns that are culturally rather than genetically determined? In other words: does human behaviour represent 'a compromise between genetically inherited selfish impulses and more cooperative culturally acquired values'?[17] Richard Boyd and Peter Richerson, in California, have analysed the behaviour to be expected on this hypothesis, and have shown that the substitution of cultural for genetic inheritance will not by itself alter the likelihood that behaviour costly to the individual though beneficial to the group will spread through the group. But they also show that the situation changes dramatically if there is what they call 'conformist cultural transmission'. By this they mean that, where there are alternative behaviour patterns, children are *disproportionately* likely to acquire the variant that is most common in their adult role models. (In the extreme case, all children would adopt the pattern shown by the majority of their adult role models.) Their calculations show that when conformist transmission is strong, it can create and maintain differences among groups, and, as a result, group selection can cause cooperation to predominate even if groups are large.

But why should a tendency to conformism have evolved? They point out that in a heterogeneous environment the rule 'imitate the common type' (or, 'when in Rome do as the Romans do') provides a simple way of acquiring the best behaviour without costly individual learning. People moving from the centre of a tropical savanna to temperate woodland at its margin would be well advised to follow the example of the majority of the woodlanders rather than their own traditional practices. If the 'conformist tendency' is genetically inheritable, the relevant genes will be favoured by natural selection.

Whether a mixture of cultural evolution and conformism played a major part in the early development of our cooperative and sometimes altruistic behaviour is uncertain, but we are, at all sorts of levels, notoriously conformist, and it is a plausible theory. There can, of course, be no doubt about the importance of cultural evolution and conformism in our more recent history. Just think of the profound effect of religious beliefs, and of the devastating changes in what was regarded as moral behaviour that occurred in China during the Cultural Revolution.

Epilogue

Near the end of *Northanger Abbey*, Jane Austen notes that her readers 'will see in the tell-tale compression of the pages before them, that we are all hastening together to perfect felicity'. In this book, the paraphernalia of notes, references and index conceals the tell-tale compression; and perfect felicity at the end was neither promised nor, I hope, expected. But, if not perfect felicity, what have we achieved?

We have cleared away much dross, and examined a great variety of approaches by many people in many places over a long time. It is clear that, incomplete though our knowledge is, there is no reason to look for some deep mystery in the origin of life; and clear too that, despite the contrary impression given by vocal creationists, the evidence that we are the products of evolution by natural selection – already strong where Darwin left it – has become overwhelming as the result of more recent work by geneticists, biochemists, molecular biologists and anthropologists.

We have seen how nerve cells can carry information about the body, how they can interact with one another to form control devices capable of making decisions and of remembering, and how specialized cells in our sense organs make it possible to encode information about the outside world in streams of nervous impulses. We have looked at the evolution and structure of the human brain, with its mixture of parallel and hierarchical organization, and at the striking localization of function within it. Using this as a basis, we have seen how observations made by investigators in many different scientific fields (and working with healthy people, patients with selective brain damage, animals, and computer analogs) have revealed much about the organization of the neural machinery inseparable from our minds and about its detailed working at different levels of organization. This information has enabled us to follow a good deal of the complex information processing involved in seeing, communicating, thinking and feeling. Although there is complexity that we do not understand, we can, to an encouraging extent, break down these processes into their logical parts, and say where particular kinds of analysis are occurring and often how they are occurring.

All this not only clarifies what you might call the robotic aspects of our brains, but also gives us a picture, admittedly very incomplete though sometimes detailed, of the relations between particular patterns of neural events and our sensations, thoughts and feelings. What it has so far failed to do is to account for the existence of those sensations, thoughts and feelings. The common-sense view of the reciprocal relations between mental and physical events is not tenable, and the available alternatives are not wholly satisfactory. Functional-state identity theory *is* tenable – it can be rescued from Kripke's criticisms, and may even be true – but it is unsatisfying because it fails to account for the subjective features of the mental states and events that it identifies with neural states and events.

As to the future, it is clear that many of the approaches we have looked at will continue, and in many areas one can see the sort of direction in which progress is likely, though the unpredictability of the appearance of new techniques and new concepts makes forecasting unreliable. It is safe to say, though, that as the human genome project progresses we shall learn about tens of thousands of new genes, and it would be very surprising if many of the proteins they code for did not have important roles in our nervous systems. Discovering what these roles are will be a task in what it is fashionable to call 'reverse engineering' – finding roles for devices rather than inventing devices for roles. Given the number of new genes likely to be discovered, the task will be a mammoth one, but not insuperable given also the availability of techniques that can knock out selected single genes. Nor should it be too difficult to discover the distribution in nervous tissues of the proteins coded for by the new genes.*

With all this new information, we should be in a position to get a much better picture of the way in which the colossal network of nerve cells is connected up in the adult brain, and the way in which that network is formed as the individual brain develops. Advances in all these areas should help us to understand the interactions between nerve cells that are responsible for the complex information processing that goes on in normal brains, and what has gone wrong in diseased brains. Since this may sound like the vague optimism of the politician's jam tomorrow, let me give one example of the help provided by the discovery of a gene – though not, as it happens, as a result of the genome project. One variety of Alzheimer's disease has a strong familial tendency, and has been found to be associated with a mutation of a particular gene. In October 1998, Zhuohua Zhang, at Harvard, and colleagues in the USA, Switzerland, The Netherlands, Germany and Belgium, reported that the role of the protein produced by the normal gene is to bind to and stabilize members of a family of signalling proteins, and that the protein produced by the mutant form of the gene is much less good at doing this. The result is increased breakdown of the signalling protein and death of the affected cells – a plausible cause of Alzheimer's disease in this group of patients.[1]

---

*They could be detected by using tailor-made labelled antibodies; alternatively, the messenger RNAs that contain the instructions for making the proteins could be detected by using tailor-made labelled complementary stretches of RNA.

The area in which we can be least confident of progress is, alas, the missing explanation of our sensations and thoughts and feelings. It may be true, as some believe, that sufficiently complicated neural networks of the right kind – which is, presumably, what our nervous systems are – would necessarily be conscious, but we have no idea how, if it is true, the consciousness could be explained. Whether better understanding of the neural correlates of consciousness will help here is uncertain, but their study would seem to be an essential first step. After all, if consciousness is uniquely associated with very large networks of nerve cells connected in complicated ways and functioning in complicated ways, it seems reasonable to want to know not only how they are connected and how they function, but also how that functioning is related to the current contents of consciousness.

Though I hope that what I have said clarifies some of these problems, I cannot escape the charge that Abraham ben David made against Maimonides – of forgetting that 'no one should embark on a discussion of a problem unless he is capable of seeing it through to the end'. But then, I am not a sage.

Notes

I CLEARING THE GROUND

1 What this book is about

1 Short, R. V. (1994) *Lancet*, 343, 528–9.

2 The failure of the common-sense view

1 Warnock, G. J. (1979) In a review of *The Self and Its Brain: an Argument for Interactionism*, by K. R. Popper & J. C. Eccles (1977), *Brain*, 102, 225–32.

2 Descartes, R. (1649) *The Passions of the Soul*, 1, 31–2. (1664) *Treatise on Man*. For both refs see *The Philosophical Writings of Descartes*, vol. I, trans. by J. Cottingham, R. Stoothoff & D. Murdoch (1985), Cambridge University Press, Cambridge.

3 Adam, Ch. & Tannery, P. (1897–1913) *Oeuvres de Descartes*, vol. 3, p. 660. J. Vrin, Paris.

4 Descartes, R. (1643) Letter to Princess Elizabeth, 21 May 1643. In *The Philosophical Writings of Descartes*, trans. by J. Cottingham, R. Stoothoff, D. Murdoch & A. Kenny (1991), vol. III, pp. 217–18. Cambridge University Press, Cambridge.

5 Hodgson, S. H. (1870) *The Theory of Practice*, vol. I, ch. III, s. 57. Longmans, Green, Reader and Dyer, London.

6 Schiff, J. M. (1858) *Physiologie des Menschen*, vol. I, Schauenburg & Co., Lahr.

7 James, W. (1890) *The Principles of Psychology*, ch. 5, Henry Holt, New York.

8 Huxley, T. H. (1874) *Fortnightly Review* (n.s.), 16, 555–80.

9 James, W. (1879) *Mind*, 4, 1–22. The same difficulty was pointed out a little later by Romanes, G. (1882) *Nineteenth Century*, 871–88.

10 Eldred, E., Granit, R. & Merton, P. A. (1953) *J. Physiol.*, 122, 498–523.

3 Evolution by natural selection

1 Keynes, R. D. (1998) In *Exobiology: Matter, Energy, and Information in the Origin and Evolution of Life in the Universe* (eds J. Chela-Flores & F. Raulin) pp. 35–49, Kluwer, Netherlands.

2 Freud, S. (1933) *Introductory Lectures on Psycho-Analysis*, 2nd edn, 18th lecture, Allen & Unwin, London.

3 Disraeli, B. *Tancred*, ch. 9.

4 Sedgwick, A. (1845) *Edinburgh Review*, 82, 1–86.

5 Darwin, C. (1844?) Letter to J. D. Hooker quoted in Darwin, F. (1887) *Life and Letters of Charles Darwin*, vol. i, p. 333, John Murray, London.

6 Huxley, T. H. (1887) In *Life and Letters of Charles Darwin* (ed. F. Darwin) vol. 2, pp. 188–9, John Murray, London.

7 Barlow, N. (ed.) (1963) *Bull. Br. Mus. Nat. Hist. (Hist. Ser.)* 2, 203–78.

8 *Life and Letters of Charles Darwin* (ed. Francis Darwin) vol. i, p. 276; John Murray, London, 1887.

9 Barlow, N. (ed.) (1958) *The Autobiography of Charles Darwin, 1809–1882*, Collins, London.

10 For discussion and references, see (i) Himmelfarb, G. (1959) *Darwin and the Darwinian Revolution*, Norton, New York; (ii) Hodge, M. J. S. (1983) in *Evolution from Molecules to Men* (ed. Bendall, D. S.) pp. 43–62, Cambridge University Press, Cambridge; (iii) Mayr, E. (1991) *One Long Argument. Charles Darwin and the Genesis of Modern Evolutionary Thought*, Harvard University Press, Cambridge, Mass.; (iv) Browne, J. (1995) *Charles Darwin: Voyaging*, Jonathan Cape, London; (v) Keynes, R. D. (1997) *J. Theor. Biol.*, 187, 461–71.

11 The personal and scientific causes of this prolonged gestation are discussed by Janet Browne (see ref. 10). See also Porter, D. M. (1993) *J. Hist. Biol.*, 26, 1–38.

12 Wallace, A. R. (1905) *My Life*, vol. i, pp. 362–3.

13 Darwin, F. (1887) *Life and Letters of Charles Darwin*, vol. ii, pp. 116–17, John Murray, London.

14 See Desmond, A. and Moore, J. (1991) *Darwin*, ch. 35, Michael Joseph, London.

15 Huxley, L. (1900) *Life and Letters of Thomas Henry Huxley*, vol. i, p. 170, Macmillan, London. Interesting accounts of Huxley's anti-evolutionary views before the publication of the *Origin* are given in Desmond, A. and Moore, J. (1991) *Darwin*, ch. 29, Michael Joseph, London; and in Desmond, A. (1994) *Huxley: the Devil's Disciple*, Michael Joseph, London.

16 Indeed, Ernst Mayr has pointed out that for much of the period between the publication of the *Origin* and the 1930s evolution by common descent was more widely accepted than the role of natural selection in guiding that evolution – see Mayr, E. (1991) *One Long Argument*, ch. 7, Harvard University Press, Cambridge, Mass.

17 Sulloway, F. J. (1982) *J. Hist. Biol.*, 15, 1–53.

18 For references see pp. 333–4 of Diamond, J. (1991) *The Rise and Fall of the Third Chimpanzee*, Radius, London.

19 Diamond, J. (1991) *The Rise and Fall of the Third Chimpanzee*, Radius, London.

20 Von Baer's story is quoted in later editions of the *Origin*. In the first edition, the same story, but without direct quotation, is attributed to Agassiz.

21 Following the original discovery, five further specimens of *Archaeopteryx* were found, all of them in the Solnhofen limestone. In 1985, a group of people including the British Astronomer Frederick Hoyle declared that the British Museum specimen was a fake, but detailed re-examination failed to support that view (see Charig, A. J., Greenaway, F., Milner, A. C., Walker, C. A. & Whybrow, P. J. (1986) *Science*, 232, 622–6). The timing of the discovery of the individual specimens itself makes the notion of forgery difficult to sustain. (See Wellnhofer, P. (1990) *Scientific American*, 262, May, 42–9.)

22 Barnes, R. S. K., Calow, P., and Olive, P. J. W. (1988) *The Invertebrates: a New Synthesis*, ch. 16, s. 7, Blackwell, Oxford; Nilsson, D.-E. & Pelger, S. (1994) *Proc. Roy. Soc. (Lond.), B*, 250, 53–8.

23 Dawkins, R. (1986) *The Blind Watchmaker*, ch. 4, Longman, Harlow; (1995) *River Out of Eden*, ch. 3, Weidenfeld & Nicolson, London; (1996) *Climbing Mount Improbable*, ch. 5, Viking, London.

24 Hoyle, F. & Wickramasinghe, N. C. (1981) *Evolution from Space*, Dent: London.

25 See, for example, Alberts, B., Bray, D., Lewis, J., Raff, M., Roberts, K. and Watson, J. D. (1994) *Molecular Biology of the Cell*, 3rd edn, pp. 387–8, Garland, New York.

26 Dawkins, R. (1976) *The Selfish Gene*, Oxford University Press, Oxford.

27 Fisher, R. A. (1914) *Eugenics Review*, 5, 309–19; (1930) *The Genetical Theory of Natural Selection*, Clarendon Press, Oxford. Haldane, J. B. S. (1927) *Possible Worlds and Other Essays*, Chatto & Windus, London; (1955) *New Biology*, 18, 34–51.

28 Hamilton, W. D. (1964) *J. Theoretical Biology*, 7, 1–52. Trivers, R. (1985) *Social Evolution*, p. 47, Benjamin/Cummings: Menlo Park, Calif. See also Williams, G. C. (1966) *Adaptation and Natural Selection*, Princeton University Press, Princeton, N.J.

29 Trivers, R. (1985) *Social Evolution*, Benjamin/Cummings: Menlo Park, California.

30 G. S. Wilkinson (1990) *Scientific American*, 262 (Feb.), 64–70.

31 On the first page of his *Genetical Theory of Natural Selection*, Fisher points out that 'the need for an alternative theory to blending inheritance was certainly felt by Darwin, though probably he never worked out a distinct idea of a particulate theory'.

32 Mendel, G. J. (1865) *Verhandlungen des naturforschenden Vereines in Brünn*, 4, 3–47.

33 The population is assumed to be of a fixed size, with random mating. An important further assumption is that the product of the population size and the mutation rate (for the particular mutation being considered) is adequate. The point is that a mutation, even if favourable, is not likely to establish itself if it occurs only once. Fisher calculated that a mutation occurring only once, and conferring an advantage of 1 per cent in survival, has a chance of about one in fifty of establishing itself and spreading through the species; there would therefore need to be 25 occurrences of such a mutation for it to have an even chance of success. With 250 occurrences the chances of success would be more than 99 per cent. These occurrences could, of course, be spread over many generations. (See Fisher, R. A. (1930), *The Genetical Theory of Natural Selection*, pp. 73–8, Clarendon Press, Oxford.)

34 Pilbeam, D. (1985) In *Ancestors: the Hard Evidence* (ed. E. Delson), Alan R. Liss: New York, pp. 51–9.

35 This was first shown by E. Zuckerkandl and L. Pauling, and by E. Margoliash and E. L. Smith, in studies of evolutionary changes in haemoglobins and in cytochrome c. See V. Bryson and H. J. Vogel (eds) (1965) *Evolving Genes and Proteins*, pp. 97–166 and 221–42, Academic Press, New York.

36 Kimura, M. (1983) *The Neutral Theory of Molecular Evolution*, Cambridge University Press: Cambridge; (1986) *Phil. Trans. Roy. Soc. Lond. B*, 312, 343–54.

37 For references and discussion see Wilson, A. C., Carlson, S. S. & White, T. J. (1977) *Ann. Rev. Bioch.*, 46, 573–639; Kimura, M. (1983) *The Neutral Theory of*

Molecular Evolution, chs. 4 and 5, Cambridge University Press, Cambridge; Sibley, C. G. & Ahlquist, J. E. (1987) *J. Molec. Evolution*, 26, 99–121; Miyamoto, M. M., Koop, B. F., Slightom, J. L., Goodman, M. & Tennant, M. R. (1988) *Proc. Natl. Acad. Sci. USA*, 85, 7627–31.

38 Margulis, L. (1970) *Origin of Eukaryotic Cells*, Yale University Press, New Haven, Conn; (1981) *Symbiosis in Cell Evolution: Life and Its Environment on the Early Earth*, W. H. Freeman, New York.

4 'The Descent of Man'

1 Letter to Lyell, 11 Oct. 1859.

2 Wallace, A. R. (1869) (in a review of the 10th edition of Lyell's *Principles of Geology*), *Quarterly Review*, 126, 359–94.

3 Gould, S. J. & Lewontin, R. C. (1979) *Proc. Roy. Soc. B* 205, 581–98.

4 *The Descent of Man*, ch. 6.

5 Descriptions of fossils mentioned in this chapter are mainly based on (i) articles in *The Cambridge Encyclopaedia of Human Evolution*, 1992, (eds S. Jones, R. Martin & D. Pilbeam, Cambridge University Press); (ii) Stringer, C. & Gamble, C. (1993) *In Search of the Neanderthals*, Thames and Hudson, London; (iii) Leakey, R. (1994) *The Origin of Humankind*, Weidenfeld & Nicolson, London; (iv) Foley, R. (1995) *Humans Before Humanity*, Blackwell, Oxford.

6 See Wills, C. (1993) *The Runaway Brain*, ch. 4, Basic Books, USA.

7 Dart, R. A. (1925) *Nature*, 115, 195–9.

8 *Australopithecus boisei*.

9 *Laetoli: a Pliocene Site in Northern Tanzania* (eds M. D. Leakey and J. M. Harris), Clarendon Press, Oxford, 1987.

10 Brunet, M., Beauvilain, A., Coppens, Y., Heintz, E., Moutaye, A. H. E. & Pilbeam, D. (1995) *Nature* 378, 273–5.

11 Wood, B. (1994) *Nature*, 371, 280–1; White, T. D., Suwa, G. & Asfaw, B. (1994) *Nature*, 371, 306–12.

12 Rodman, P. S. & McHenry, H. M. (1980) *Amer. J. Physical Anthropology*, 52, 103–6.

13 Lovejoy, C. O. (1988) *Scientific American*, 259, Nov., 82–9; (1981) *Science*, 211, 341–50; and related correspondence (1982) *Science*, 217, 295–306.

14 Foley, R. (1995) *Humans Before Humanity*, ch. 7, Blackwell, Oxford.

15 Wheeler, P. E. (1991) *J. Human Evolution*, 21, 117–36.

16 For a discussion of dates of *Homo erectus* fossils see Lewin, R. (1998) *Principles of Human Evolution*, pp. 332–4, Blackwell Science, Malden, Mass.

17 Dzaparidze, V., Bosinski, G. *et al.* (1989) *Jahrbuch des Römisch-Germanischen Zentralmuseums Mainz*, 36, 67–116.

18 Swisher, C. C. III, Curtis, G. H., Jacob, T., Getty, A. G., Suprijo, A. & Widiasmoro (1994), *Science*, 263, 1118–21.

19 Huang Wanpo *et al.* (1995) *Nature*, 378, 275–8; Wood, B. & Turner, A. (1995) *Nature*, 378, 239–40.

20 Shipman, P. *American Anthropologist*, 88, 27–43 (1986).

21 Toth, N. (1987) *Scientific American*, 256, Apr. 104–113.

22 Alexander, R. D. (1974) *Annual Review of Ecology and Systematics*, 5, 325–83. Humphrey, N. K. (1976) In *Growing Points in Ethology* (eds P. P. G. Bateson & R. A.

Hinde), pp. 303–17, Cambridge University Press, Cambridge. Humphrey, N. (1983) *Consciousness Regained: Chapters in the Development of Mind*, ch. 2, Oxford University Press, Oxford.

23 Ridley, M. (1993) *The Red Queen: Sex and the Evolution of Human Nature*, ch. 10, Viking, London.

24 Baron-Cohen, S., Leslie, A. & Frith, U. (1985) *Cognition*, 21, 37–46. Baron-Cohen, S. (1995) *Mindblindness: an Essay on Autism and Theory of Mind*, MIT Press, Cambridge Mass.

25 Articles in Byrne, R. W. & Whiten, A. (eds) (1988) *Machiavellian Intelligence*, Oxford University Press, Oxford. Byrne, R. W. & Whiten, A. (1992) *Man* (N.S.) 27, 609–27. Baron-Cohen, S. (1995) *Mindbliness: an Essay on Autism and Theory of Mind*, ch. 8, MIT Press, Cambridge Mass.

26 Trivers, R. (1985) *Social Evolution*, ch. 15, Benjamin/Cummings, Menlo Park, California.

27 Articles in Byrne, R. W. & Whiten, A. (eds) (1988) *Machiavellian Intelligence*, Oxford University Press, Oxford. Byrne, R. W. & Whiten, A. (1992) *Man* (N.S.) 27, 609–27.

28 Cosmides, L. & Tooby, J. (1992) In *The Adapted Mind* (eds J. H. Barkow, L. Cosmides & J. Tooby), pp. 163–228, Oxford University Press, New York.

29 Petralona, in northern Greece; Arago, in the French Pyrenees; Burgos, in northern Spain; Mauer, near Heidelberg; and Steinheim, near Stuttgart.

30 Elandsfontein (Saldanha) in South Africa; Broken Hill (Kabwe) in Zambia; Lake Ndutu in Tanzania; Omo in Ethiopia; and in China at Dali, about 60 miles northeast of Xian, and at Jingnuishan, about 300 miles east-northeast of Beijing.

31 Remains of an inner ear with Neanderthal features have been found at a site that may be more recent – see Hublin, J.-J., Spoor, F., Braun, M., Zonnefeld, F. and Condemi, S. (1996) *Nature*, 381, 224–6.

32 Thorne, A. G. & Wolpoff, M. H. (1992) *Scientific American*, 266, Apr. 28–33.

33 Tobias, P. V. (1991) *Olduvai gorge IV, The Skulls, Endocasts and Teeth of Homo habilis*, pp. 614, 794, Cambridge University Press; Bräuer, G. and Smith, F. H. (eds) (1992) *Continuity or Replacement*, Balkema, Rotterdam.

34 Brown, W. M., George, M. Jr. & Wilson, A. C. (1979) *Proc. Natl. Acad. Sci. USA* 76, 1967–71; Brown, W. M. (1980) *Proc. Natl. Acad. Sci. USA*, 77, 3605–9; Cann, R. L., Stoneking, M. & Wilson, A. C. (1987) *Nature*, 325, 31–6; Wilson, A. C. & Cann, R. L. (1992) *Scientific American*, 266, Apr., 22–7.

35 Krings, M., Stone, A., Schmitz, R. W., Krainitzki, H., Stoneking, M. & Pääbo, S. (1997) *Cell*, 90, 19–30. Ward, R. & Stringer, C. (1997) *Nature*, 388, 225–6.

36 Stringer, C. and Gamble, C. (1993) *In Search of the Neanderthals*, ch. 6 and plates 60–8 Thames and Hudson, London. Bräuer, G. (1992) In Bräuer, G. and Smith, F. H. (eds) 1992, *Continuity or Replacement*, pp. 83–98, Balkema, Rotterdam.

37 Smith, F. H. (1992) *Phil. Trans. Roy. Soc. B.*, 337, 243–50.

38 Schwarcz, H. P. & Grün, R. (1992) *Phil. Trans. Roy. Soc. B.*, 337, 145–8. Stringer, C. & Gamble, C. (1993) *In Search of the Neanderthals*, ch. 6, Thames and Hudson, London. Leakey, R. (1994) *The Origin of Humankind*, ch. 5, Weidenfeld & Nicholson, London.

39 Stringer, C. and Gamble, C. (1993) *In Search of the Neanderthals*, ch. 5 and

plates 52–9, Thames and Hudson, London.

40 Aitken, M. J. & Valladas, H. (1992) *Phil. Trans. Roy. Soc. B.*, 337, 139–44; Schwarcz, H. P. & Grün, R. (1992) *Phil. Trans. Roy. Soc. B.*, 337, 145–8; Stringer, C. & Gamble, C. (1993) *In Search of the Neanderthals*, ch. 5, Thames and Hudson, London.

41 Coppens, Y. (1994) *Scientific American*, 270, May, 62–9. The phrase as well as the theory is Coppens's.

42 Coppens, Y. (1994) *Scientific American*, 270, May, 62–69.

5 The origin of life

1 In a letter to J. D. Hooker dated 29 March 1863.

2 Joyce, G. F. & Orgel, L. E. (1993) In *The RNA World* (eds R. F. Gesteland & J. F. Atkins) Cold Spring Harbor Laboratory Press, pp. 1–25. Lazcano, A. & Miller, S. L. (1996) *Cell*, 85, 793–8.

3 For reviews see (i) Schopf, J. W. (ed.) (1983) *Earth's Earliest Biosphere: Its Origin and Evolution*, Princeton, New Jersey; (ii) Schopf, J. W. and Klein, C. (eds) (1992) *The Proterozoic Biosphere*, Cambridge University Press, New York; (iii) Pace, N. R. (1991) *Cell*, 65, 531–3; (iv) Lazcano, A. & Miller, S. L. (1996) *Cell*, 85, 793–8.

4 Schopf, J. W. (1993) *Science*, 260, 640–6.

5 Maher, K. A. & Stevenson, D. J. (1988) *Nature*, 331, 612–14. Sleep, N. H., Zahnle, K. J., Kasting, J. F. & Morowitz, H. J. (1989) *Nature*, 342, 139–42. Chyba, C. & Sagan, C. (1992) *Nature*, 355, 125–32.

6 Kasting, J. F. (1993) *Science*, 259, 920–5.

7 Safronov, V. S. (1972) *NASA Reports*, TT F-677.

8 Gold, T. (1987) *Power from the Earth*, Dent, London.

9 Chyba, C. & Sagan, S. (1992) *Nature*, 355, 125–32.

10 Corliss, J. B., Baross, J. A. & Hoffman, S. E. (1981) *Oceanologica Acta*, No. SP, 59–69. Jannasch, H. W. & Mottl, M. J. (1985) *Science*, 229, 717–25.

11 See review by Miller, S. L. (1987) *Cold Spring Harbor Symp. Quant. Biol.*, 52, 17–27.

12 Harada, K. & Fox, S. W. (1965) In *The Origins of Prebiological Systems and Their Molecular Matrices* (ed. S. W. Fox), p. 190, Academic Press, New York.

13 Chyba, C. & Sagan, C. (1992) *Nature*, 355, 125–32.

14 Bar-Nun, A. & Shaviv, A. (1975) *Icarus*, 24, 197–210.

15 Müller, D., Pitsch, S., Kittaka, A., Wagner, E., Wintner, C. E. & Eschenmoser, A. (1990) *Helv. Chim. Acta*, 73, 1410–68. Joyce, G. F. & Orgel, L. E. (1993) In *The RNA World* (eds R. F. Gesteland and J. F. Atkins), pp. 1–25, Cold Spring Harbor Laboratory Press.

16 Nielsen, P. E. (1993) *Orig. Life Evol. Biosph.*, 23, 323–7.

17 See Sanchez, R. A., Ferris, J. P. & Orgel, L. E. (1967) *J. Mol. Biol.*, 30, 223–53. Miller, S. L. (1987) *Cold Spring Harbor Symp. Quant. Biol.*, 52, 17–27. Ferris, J. P. (1987) *Cold Spring Harbor Symp. Quant. Biol.*, 52, 29–35. Ponnamperuma, C., Sagan, C. & Mariner, R. (1963) *Nature*, 199, 222–6. Ferris, J. P. & Hagen, W. J. (1984) *Tetrahedron*, 40, 1093–120.

18 Orò, J. (1961) *Nature*, 191, 1193–4.

19 Ferris, J. P., Sanchez, R. A. & Orgel, L. E. (1968) *J. Mol. Biol.*, 33, 693–704. Robertson, M. P. & Miller, S. L. (1995) *Nature*, 375, 772–3 and 377, 257.

20 Kerridge, J. F. & Matthews, M. S. (1988) *Meteorites and the Early Solar System*,

University of Arizona Press, Tucson.

21 See article on Svante Arrhenius in the 11th edition of the *Encyclopaedia Britannica*.

22 McKay, D. S., Gibson, E. K., Thomas-Keprta, K. L. *et al.*, (1996) *Science*, 273, 924–30. Bada, J. L., Glavin, D. P., McDonald, G. D. & Becker, L. (1998) *Science*, 279, 362–5. Jull, A. J. T., Courtney, C., Jeffrey, D. A. & Beck, J. W. (1998) *Science*, 279, 366–9.

23 Woese, C. R. (1967) *The Genetic Code*, pp. 179–95. Harper & Row, New York. Crick, F. H. C. (1968) *J. Mol. Biol.*, 38, 367–79. Orgel, L. E. (1968) *J. Mol. Biol.*, 38, 381–93.

24 Lee, D. H., Granja, J. R., Martinez, J. A., Severin, K. and Ghadiri, M. R. (1996) *Nature*, 382, 525–8. Orgel, L. E. (1987) *Cold Spring Harbor Symp. Quant. Biol.*, 52, 9–16.

25 Cairns-Smith, A. G. (1982) *Genetic Takeover and the Mineral Origins of Life*, Cambridge University Press, Cambridge.

26 Inoue, T. & Orgel, L. E. (1983) *Science*, 219, 859–62. Orgel, L. E. (1987) *Cold Spring Harbor Symp. Quant. Biol.*, 52, 9–16.

27 For an earlier though much less striking result of the same kind see Naylor, R. & Gilham, P. T. (1966) *Biochemistry*, 5, 2722–28.

28 Cech, T. R., Zaug, A. J. & Grabowsky, P. J. (1981) *Cell*, 27, 487–96; Kruger, K., Grabowski, P. J., Zaug, A. J., Sands, J., Gottschling, D. E. & Cech, T. R. (1982) *Cell*, 31, 147–57. Guerrier-Takada, C., Gardiner, K., Marsh, T. *et al.* (1983) *Cell*, 35, 849–57. Cech, T. R. (1986) *Proc. Natl. Acad. Sci. USA*, 83, 4360–3. Westheimer, F. H. (1986) *Nature*, 319, 534–6. Articles in *The RNA World* (1993) (eds R. F. Gesteland & J. F. Atkins) Cold Spring Harbor Laboratory Press.

29 Doudner, J. A., Couture, S. & Szostak, J. W. (1991) *Science*, 251, 1605–8.

30 The phrase was first used by W. Gilbert (1986) *Nature*, 319, 618.

31 Szostak, J. W. & Ellington, A. D. (1993) In *The RNA World* (eds R. F. Gesteland & J. F. Atkins), pp. 511–33, Cold Spring Harbor Laboratory Press.

32 Moore, M. J. (1995) *Nature*, 374, 766–7. Wilson, C. & Szostak, J. W. (1995) *Nature*, 374, 777–82. Ekland, E. H., Szostak, J. W. & Bartel, D. P. (1995) *Science*, 269, 364–70.

33 For a contrary view see Miller, S. L. & Bada, J. L. *Nature*, 334, 609–11, and a reply by Corliss, J. B. (1990) *Nature*, 347, 624.

34 Written in 1871. See Darwin, F. (1887) *Life and Letters of Charles Darwin*, vol. III, p. 18, fn.

35 Blobel, G. (1980) *Proc. Natl. Acad. Sci. USA*, 77, 1496–500. See also Cavalier-Smith, T. (1987) *Cold Spring Harbor Symp. on Quant. Biol.*, 52, 805–23.

36 Oparin, A. I. (1924) *The Origin of Life* (Russ. *Proiskhozdenic Zhizny*), Moskovski Rabochii, Moscow.

37 Haldane, J. B. S. (1928) *Rationalist Annual*, 148, 3.

II NERVES AND NERVOUS SYSTEMS

6 *The nature of nerves*

1 Where specific references are not given, the account of Galen's life and work draws on (i) Castiglione, A. (1947) *A History of Medicine*, 2nd American Edition

(trans. and ed. E. B. Krumbhaar), Knopf, New York; (ii) R. H. Major (1954) *History of Medicine*, Blackwell, Oxford; (iii) Singer, C. & Underwood, E. A. (1962) *A Short History of Medicine*, 2nd edn. Clarendon Press, Oxford; (iv) Sarton, G. (1954) *Galen of Pergamon*, University of Kansas Press, Lawrence, Kansas; (v) Siegel, R. E. (1968) *Galen's System of Physiology and Medicine*, (1970) *Galen on Sense Perception*, (1973) *Galen on Psychology, Psychopathology, and Function and Diseases of the Nervous System*, Karger: Basel; (vi) May, M. T. (1968) *Galen on the Usefulness of the Parts of the Body*, 2 vols, Cornell, Ithaca.

2 *De Animae Passionibus*, 1.8.41, p. 31 Marquardt.

3 In this he was anticipated by Herophilos of Chalcedon (about 300 BC).

4 The word is used by Plato for the Divine Craftsman.

5 See, for example Castiglione, A. *A History of Medicine*, 2nd American edn, 1947 (trans. and ed. E. B. Krumbhaar), p. 220, Knopf: New York.

6 Walzer, R. (1949) *Galen on Jews and Christians*, Oxford University Press: London.

7 Galen: *On the Usefulness of the Parts of the Body* (book xi, section 14) translated by M. T. May (1968) vol. 2, p. 553, Cornell, Ithaca.

8 My colleague Gisela Striker points out that the harsh view of the human body taken by Plato in the *Phaedo* has to be balanced by his belief (expressed in the *Timaeus*) that the *demiourgos* makes the human body as good as is possible. She suggests that Galen's creator-god is most likely to have come from Plato's *Timaeus*, on which Galen wrote commentaries.

9 See Duckworth, W. L. H. (1962) *Galen on Anatomical Procedures: the Later Books*, English translation, Cambridge University Press, Cambridge.

10 Foster, M. (1901) *Lectures on the History of Physiology During the Sixteenth, Seventeenth and Eighteenth Centuries* (repr. 1924), p. 257, Cambridge University Press, London.

11 Foster, M. (1901) *Lectures on the History of Physiology During the Sixteenth, Seventeenth and Eighteenth Centuries* (repr. 1924), p. 15, Cambridge University Press, London.

12 All quotations from Descartes are taken from *The Philosophical Writings of Descartes* (Transl. by J. Cottingham, R. Stoothoff, D. Murdoch, and, for vol. 3, A. Kenny; 3 vols., 1985–1991, Cambridge University Press, Cambridge.

13 *The Passions of the Soul*, I, 6.

14 *The Passions of the Soul*, I, 31, 32.

15 Quotations in this and the next paragraph are from the *Treatise on Man*.

16 The notion that fire consists of minute particles in violent motion is Descartes's. The argument for it is given in *The World*.

17 Letter to the Cambridge philosopher and poet Henry More, 5 Feb. 1649.

18 Letter to Plempius for Fromondus, 3 Oct. 1637.

19 Replies to Fourth Set of Objections to Descartes's *Meditations on First Philosophy*. The phrase in single quotation marks towards the end of the passage had been used by Arnauld.

20 Letter to the Marquess of Newcastle, 23 Nov. 1646.

21 See Part Five of *Discourse on the Method*.

22 Letter to Henry More, 5 Feb. 1649.

23 Letter to the Marquess of Newcastle, 23 Nov. 1646.

24 Letter to Henry More, 5 Feb. 1649.

25 This and the following quotations from Borelli are taken from Sir Michael Foster's *Lectures on the History of Physiology During the Sixteenth, Seventeenth and Eighteenth Centuries*, 1901 (repr. 1924), pp. 61–82 and 279–81, Cambridge University Press, London.

26 Swammerdam, J. *Biblia Naturae*, (Published in Leyden by Boerhaeve in 1737–8, long after Swammerdam's death).

27 Goddard's experiment is usually associated with Francis Glisson, who described it in his *De ventriculo et intestinis*, 1677, but it seems to have been first demonstrated by Goddard in 1669 before the Royal Society, then in its seventh year – see Birch, T. (1756–7) *The History of The Royal Society*, vol. 2, pp. 411–2, Millar, London.

28 Pliny described the paralysis of the muscles experienced by someone who touches the torpedo with a spear; Scribonius, a first century physician who dedicated a book of prescriptions to Claudius, recommended 'a living black torpedo ... placed on the spot where the pain is' as a treatment for headache; a freedman of Tiberius, was said to have been cured of gout by shocks from the torpedo; such shocks were also said to have been helpful in childbirth and in the treatment of epilepsy. (See Thompson, D'Arcy W. *A Glossary of Greek Fishes*, p. 169, 1947, Oxford University Press: London.)

29 In a book whose title, translated into English, was *On the Magnet, Magnetic Bodies, and the Great Magnet, the Earth.*

30 Priestley, J. (1775) *History and Present State of Electricity*, 3rd edn, vol. 1, pp. 35–40, London.

31 Quoted from Joseph Priestley's *History and Present State of Electricity*, 3rd edn, vol. 1, p. 58, London, 1775.

32 Later experiments showed that the direction in which charge is transferred on rubbing depends not only on the composition of the object being rubbed but also on that of the rubber and on the character of the surfaces of each.

33 In June 1752, Franklin, accompanied by his son, stood in the doorway of a hut holding by a dry silk thread the string of a kite being flow in a thunderstorm. A metal key was attached to the end of the string, and by presenting his knuckles to it Franklin obtained sparks, which became more copious as the string got wetter.

34 For clarity I have used the modern term 'electric charge'; Franklin used the phrase 'electrical fire'.

35 Letter from Benjamin Franklin in Philadelphia to Peter Collinson in London, 11 July 1747. In *Benjamin Franklin: the Autobiography and Other Writings* (ed. L. Jesse Lemisch), 1961, pp. 228–34, New American Library, New York.

36 The description of this and the following experiments by Galvani is based on (i) the English translations of his *De Viribus Electricitatis in Motu Musculari* (1791), by M. G. Foley, 1953, Burndy Library, Norwalk, Conn.; and (ii) a detailed account of Galvani's work, and English translations of extracts of his *Giornale* and other writings in *The Ambiguous Frog* (1992) by M. Pera, Princeton University Press.

37 The quotation is from the second part of Galvani's *De Viribus Electricitatis in Motu Musculari* – see endnote 36.

38 Volta, A. (1800) *Phil. Trans. Roy. Soc.*, 90, 403–31 (in French).

39 Carlisle, A. and Nicholson, W. (1801) *Journal of Natural Philosophy, Chemistry and the Arts*, 4, 179–87.

40 The way Volta's pile works is this. At the surface of the zinc in contact with the salt-water, atoms of zinc give up negative charges (electrons) and pass into solution as (positively charged) zinc ions; here they combine with (negatively charged) hydroxyl ions from the water to form a precipitate of zinc hydroxide. The electrons acquired by the zinc disc pass to the silver, and at the surface of the silver in contact with the salt-water they combine with (positively charged) hydrogen ions from the water to form hydrogen gas. The net result is the reaction of zinc and water to form zinc hydroxide and hydrogen, and a flow of electrons through the pile.

41 Cavendish, H. (1776) *Phil. Trans. Roy. Soc.*, 66, 196–225. Cavendish pointed out that it was possible to experience an unpleasant electric shock from a partially charged Leyden jar – a form of electrical condenser (capacitor) invented in 1775 – when the voltage was too low to cause a spark.

42 Walker, W. C. (1937) *Annals of Sci.*, 2, 84–113.

43 Matteucci, C. (1840) *Essai sur les phénomènes électrique des animaux*. Paris.

44 (i) Matteucci, C. (1842) *Annales de chimie et de physique* 3rd series, vol. 6, pp. 301–39; (ii) *On Animal Electricity: Being an Abstract of the Discoveries of E. Du Bois-Reymond* (ed. H. Bence Jones, 1852), John Churchill, London. Similar experiments seem to have been done by Galvani, by his nephew Aldini, and by Humboldt, in the last decade of the eighteenth century – see Rowbottom, M. and Susskind, C. (1984), *Electricity and Medicine: History of Their Interaction*, ch. 3, San Francisco Press.

45 Matteucci, C. (1842) *Annales de chimie et de physique* 3rd series, vol. 6, pp. 301–39.

46 (i) Du Bois-Reymond, E. (1848–84) *Untersuchungen über thierische Elektricität*, Reiner, Berlin; (ii) Bence Jones, H. ed. (1852) *On Animal Electricity: Being an Abstract of the Discoveries of Emil Du Bois-Reymond*, John Churchill, London.

47 Quoted by Mary Brazier in her article 'The historical development of neurophysiology' in vol. 1 of the American Handbook of Physiology, 1st edn, 1959, American Physiological Society, Washington, DC.

48 Morse had invented his electric telegraph in 1838, and the 1840s were pioneering years for telegraphy.

49 von Helmholtz, H. (1850) *Arch. Anat. Physiol. wiss. Med.*, pp. 71–73.

50 Bernstein, J. (1868) *Pflügers Archiv*, 1, 173–207.

7 The nerve impulse

1 Adrian, E. D. (1928) *The Basis of Sensation*, Christophers, London.

2 Lucas, K. (1909) *J. Physiology*, 38, 113–33.

3 Bernstein, J. (1902) *Pflügers Archiv*, 92, 521–62; (1912) *Elektrobiologie*, Braunschweig, Vieweg.

4. Hermann, L. (1899) *Pflügers Archiv*, 75, 574–90.

5 Overton had shown that muscles were inexcitable in solutions lacking sodium ions. It was also known that muscles lost some potassium and gained some sodium each time they contracted, but it was not clear whether these losses and gains were related to the excitation process or the contractile process. See Overton, E. (1902)

Pflügers Archiv, 92, 346–86.

6 The story is told in Ch. 10 of A. L. Hodgkin's *Chance and Design*, 1992, Cambridge University Press, Cambridge.

7 Hodgkin, A. L. (1964) *The Conduction of the Nervous Impulse*, Liverpool University Press, Liverpool.

8 For a readable account, see Hodgkin, A. L. (1964) *The Conduction of the Nervous Impulse*, Liverpool University Press, Liverpool.

9 The potential difference across the membrane swung from a resting level of about –70 millivolts (i.e. the inside of the fibres was about 70 millivolts negative to the outside) to about +40 millivolts (i.e. the inside of the fibre was about 40 millivolts positive relative to the outside), and then swung back to its original value, the whole process taking about one millisecond.

10 Keynes, R. D. (1951) *J. Physiol. (London)*, 114, 119–50.

11 You might, of course, argue that with a wider nerve a greater area of membrane would be involved, and so more current would be needed. This is true, but the area of membrane per unit length of nerve fibre increases in proportion to the diameter, whereas the resistance per unit length of fluid contained in the fibre decreases with the square of the diameter.

12 Tasaki, I. & Takeuchi, T. (1941) *Pflügers Archiv.*, 244, 696–71; (1942) *Pflügers Archiv.*, 245, 764–82. Huxley, A. F. & Stämpfli, R. (1949) *J. Physiol. (London)*, 108, 315–39.

13 Noda, M. *et al.* (1984) *Nature*, 312, 121–7.

14 For a brief account of the possible evolution of the sodium pump see Glynn, I. M. (1993) *J. Physiol. (London)*, 462, 1–30.

8 Encoding the message

1 Descartes' *Principles of Philosophy*, part 4, s. 198.

2 The main mechanism for pitch discrimination in the ear makes use of labelled line coding: the basilar membrane in the cochlea vibrates maximally at different points depending on the pitch, and those nerve fibres are stimulated most vigorously that originate in the bit of membrane that is vibrating most vigorously. With sounds lower than the G below middle C, however, this mechanism is rather ineffective, yet we can discriminate surprisingly well. The explanation is thought to be that, with sounds of low pitch, the action potentials in the nerve fibres from the cochlea are in phase with the sound vibrations, so that action potentials will arrive at the brain with a frequency identical with the frequency of the sound. Information about pitch is therefore available to the brain. (A single nerve fibre may not be able to conduct action potentials at the frequency of the sound, but if the action potentials in all fibres are in phase with the sound, cohorts of action potentials will arrive with a frequency equal to the frequency of the sound.)

9 Interactions between nerve cells

1 Loewi, O. (1921) *Pflügers Archiv.* 189, 239–42.

2 Dale, H. H., Feldberg, W. & Vogt, M. (1936) *J. Physiol. (London)*, 86, 353–80.

3 Brown, G. L., Dale, H. H. & Feldberg, W. (1936) *J. Physiol. (London)*, 87, 394–424.

4 This is because only sodium and potassium ions are present in high concentration, and the sodium ions are driven inwards both by the electrical potential difference across the membrane – outside positive to inside – and by the high concentration of sodium outside the muscle fibre; for potassium ions the inward electrical driving force is more than offset by the high potassium concentration inside the fibre.

5 Unwin, N. (1989) *Neuron*, 3, 665–76. Kandel, E. R. & Siegelbaum, S. A. (1991) in *Principles of Neural Science*, 3rd edn (eds. E. R. Kandel, J. H. Schwartz & T. M. Jessell), ch. 10, Appleton and Lange, East Norwalk, Conn. Changeux, J-P. (1993) *Scientific American*, 269 (5) 30–7.

6 Fatt, P. & Katz, B. (1952) *J. Physiol. (London)*, 117, 109–28.

7 Del Castillo, J. & Katz, B. (1954) *J. Physiol. (London)*, 124, 560–73.

8 Heuser, J. E. & Reese, T. S. (1981) *J. Cell Biology*, 88, 564–80.

9 Brock, L. G., Coombs, J. S. & Eccles, J. C. (1952) *J. Physiol. (London)*, 117, 431–60.

10 The decrease in excitability has two causes: (i) the increase in the potential difference across the membrane means that a greater reduction in this potential difference is necessary to reach the threshold for excitation; (ii) the increase in chloride or potassium permeability makes it more difficult to reduce the potential difference across the membrane. (Any reduction will upset the balance between the effects on ion movements of the membrane potential and of the transmembrane concentration differences and will therefore tend to be offset by a rapid entry of negatively charged chloride ions or a rapid loss of positively charged potassium ions.)

11 Bowyer, J. F,. Spuhler, K. P. & Weiner, N. (1984) *J. Pharmacol. Exp. Ther.*, 229, 671–80.

12 The outward-facing portion of the receptor having bound the neuromodulator, the portion that projects into the cell reacts with a so-called G-protein attached to the inner surface of the cell membrane, altering it so that it can in turn activate another membrane enzyme that then generates the second messenger.

13 Furshpan, E. J. and Potter, D. D. (1959) *J. Physiol. (London)*, 145, 289–325.

14 This is because the current caused by the action potential at the terminal of the presynaptic cell will depolarize the postsynaptic cell. There is a single known example of an inhibitory electrical synapse, which depends on special anatomical features.

15 The analogy was made by Stephen Kuffler and John Nicholls.

10 'The doors of perception'

1 Nagel, T. (1974) *The Philosophical Review*, 83, 435–50.

2 Partridge, B. L. and Pitcher, T. J. (1980) *J. Comp. Physiol.*, 135, 315–25.

3 Fettiplace, R. (1987) *Trends in Neuroscience Research*, 10, 421–25.

4 Ashmore, J. F. (1987) *J. Physiol. (London)*, 388, 323–47. Yates, G. K., Johnstone, B. M., Patuzzi, R. B. & Robertson, D. *Trends in Neurosciences*, 15, 57–61.

5 For a discussion of timbre perception, and of many other aspects of hearing, see Moore, B. C. J. (1997) *An Introduction to the Psychology of Hearing*, 4th edn, Academic Press, London.

6 This is because a vibrating tuning fork behaves as a perfectly elastic body. At any moment the restoring force on the prong, and therefore its acceleration, is proportional to the prong's displacement from its equilibrium position. As the beetle goes round the rim of the clockface its acceleration towards the 3 o'clock–9 o'clock line is also proportional to its distance from that line.

7 See Wald, G. (1968) *Nature*, 219, 800–807.

8 In bright light much of the rhodopsin will be bleached and therefore out of action. The changes in calcium concentration occur because the sodium channels are not quite specific for sodium. In the dark, more sodium channels are open and more calcium enters the cell.

9 Hecht, S., Schlaer, S. & Pirenne, M. H. (1942) *J. Gen. Physiol.*, 25, 819–40.

10 Thomas Young's 1801 Bakerian Lecture, *Phil. Trans. Roy. Soc. for 1802*, pp. 12–48.

11 Le Blon, J. C. (c. 1725) *Il Coloritto*, W. & J. Innys, London. There still exist, in Dresden and in Paris, prints in red, yellow and blue, of the three plates which, together with a black plate, Le Blon used to reproduce Van Dyke's self portrait. A fourth print shows the result of combining the yellow plate and the black plate. See Franklin, C. (1977) *Early Colour Printing*, C. and C. Franklin, Culham, Oxford.

12 All forms of red-green colour blindness are almost exclusively male because the genes for the red-sensitive and green-sensitive cone pigments are carried on the X chromosome.

13 See article on George Palmer (alias Giros von Gentilly) by J. D. Mollon, in the *Missing Persons* volume of the *Dictionary of National Biography*.

14 Rushton, W. A. H. (1963) *J. Physiol. (London)*, 168, 345–59; (1975) *Scientific American*, 232(3), 64–74.

15 Brown, P. K. & Wald, G. (1963) *Nature* 200, 37–43. Marks, W. B., Dobelle, W. H. & MacNichol, E. F. (1964) *Science* 143, 1181–3.

16 For recent and surprising information about Dalton's colour blindness see Mollon, J. D. (1995) *Science* 267, 933–1064.

17 The Duke of Marlborough referred to was the third Duke, not the famous soldier. See *Diary and Letters of Madame D'Arblay*, entry for 16th December, 1785. Vol ii, p. 328, of Macmillan's 1904 edition.

18 Nathans, J., Thomas, D. & Hogness, D. S. (1986) *Science*, 232, 193–202.

19 The position in New World monkeys is complicated – see Mollon, J. D. (1995) In *Colour: Art and Science*, (eds T. Lamb and J. Bourriau), pp. 127–50, Cambridge University Press, Cambridge.

20 See endnote 19.

21 Kuffler, S. W. (1953) *J. Neurophysiol.* 16, 37–68.

22 Whether a particular ganglion cell is an *on-centre cell* or an *off-centre cell* depends not on anything intrinsic to the ganglion cell but on the nature of the bipolar cells that transmit information to the ganglion cell from the different regions of its receptive field. The transmitter released by the photoreceptor is always the same but, as explained on page 149, the synapses between photoreceptors and bipolar cells may be stimulatory or inhibitory.

23 Barlow, H. B. (1953) *J. Physiol. (London)*, 119, 69–88.

24 Maturana, H. R., Lettvin, J. Y., McCulloch, W. S. and Pitts, W. H. (1960)

J. Gen. Physiol., 43, Suppl. 2, 129–75.

25 Barlow, H. B., Hill, R. M. and Levick, W. R. (1964) *J. Physiol. (London)*, 173, 377–407; Barlow, H. B. and Levick, W. R. (1965) *J. Physiol. (London)*, 178, 477–504.

26 Maturana, H. R. and Frenk, S. (1963) *Science*, 142, 977–9.

27 Barlow, H. B. and Levick, W. R. (1965) *J. Physiol. (London)*, 178, 477–504.

28 This does not imply that there are only five types of receptor, but the number is not thought to be very large. See Lindemann, B. (1996) *Physiological Reviews*, 76, 719–66.

29 The mechanisms involved in the tasting of salt and sour things is simpler. Tasting salt depends on the entry of sodium ions through sodium-selective channels; tasting sour things depends on the blocking by hydrogen ions of potassium-selective channels.

30 Alberts, B., Bray, D., Lewis, J., Raff, M., Roberts, K. & Watson, J. D. (1994) *Molecular Biology of the Cell*, 3rd edn, pp. 721–2.

31 Reed, R. R. (1990) *Cell*, 60, 1–2. Buck, L. & Axel, R. (1991) *Cell*, 65, 175–87. Chess, A., Simon, I., Cedar, H. & Axel, R. (1994) *Cell*, 78, 823–34. Axel, R. (1995) *Scientific American*, 273, Oct. 130–7.

32 For references see Lancet, D. (1986) *Ann. Rev. Neurosci.*, 9, 329–55.

33 Carpenter, R. H. S. (1996) *Neurophysiology*, 3rd edn, p. 176, Arnold, London.

34 From time to time there are suggestions that molecular vibrations could provide the basis for a continuously variable quantity associated with smells. For a recent suggestion see Turin, L. (1996) *Chemical Senses*, 21, 773–91.

11 *A Cook's Tour of the brain*

1 See (i) Singer, W. (1994) *Nature*, 369, 444–5; (ii) Crick, F. (1984) *Proc. Natl. Acad. Sci. USA*, 81, 4586–90; (iii) Sherman, S. M. and Koch, C. (1986) *Exptl. Brain Res.*, 63, 1–20; (iv) Sillito, A. M., Jones, H. E., Gerstein, G. L. and West, D. C. (1994) *Nature*, 369, 479–82.

2 The Edwin Smith Surgical Papyrus, in the New York Academy of Medicine. It is thought to be a copy of a much older document.

3 See accounts in ch. VI of *Garrison's History of Neurology* (revised and enlarged) ed. McHenry, L. C. Jr (1969), Thomas, Springfield, Ill.

4 See article on 'Phrenology' by A. Macalister in the 11th edition of the *Encyclopaedia Britannica*.

5 Crick, F. H. C. (1994) *The Astonishing Hypothesis*, p. 85, Charles Scribner's Sons, New York.

6 See (i) *Garrison's History of Neurology*, revised and enlarged (1969) ed. McHenry, L. C. Jr, Thomas, Springfield Illinois. (ii) numerous entries in Schiller, F. (1979) *Paul Broca*, University of California Press, Berkeley and Los Angeles.

7 Schiller, F. (1979) *Paul Broca*, University of California Press, Berkeley and Los Angeles.

8 See *Garrison's History of Neurology*, revised and enlarged (1969) ed. McHenry, L. C. Jr, Thomas, Springfield, Illinois.

9 Fritsch, G. and Hitzig, E. (1870) *Arch. f. Anat. Physiol. u. wiss. Med.*, 37, 300–32. See also (i) *Garrison's History of Neurology*, revised and enlarged (1969)

ed. McHenry, L. C. Jr, Thomas, Springfield, Illinois. (ii) Article by Kuntz, A. in *The Founders of Neurology*, ed. W. Haymaker (1953), Thomas, Springfield, Illinois.

10 See (i) Ferrier, D. (1876) *The Function of the Brain*, pp. 144 and 173–5, Smith and Elder, London; (ii) *Garrison's History of Neurology*, revised and enlarged (1969) ed. McHenry, L. C. Jr, Thomas, Springfield, Illinois; (iii) Article by David McK. Rioch in *The Founders of Neurology*, ed. W. Haymaker, (1953) Thomas, Springfield, Illinois; (iv) Fulton, J. F. (1966) *Selected Readings in the History of Physiology*, 2nd edn, Thomas, Springfield, Illinois.

11 Wernicke, C. (1874) *Der aphasische Symptomenkomplex*, Cohn & Weigert, Breslau.

12 Accounts of Wernicke's life and of his contribution to the study of aphasia may be found in chs 4 and 12 of Norman Geschwind's *Selected Papers on Language and the Brain* (vol. 16 of *Boston Studies in the Philosophy of Science*), eds R. S. Cohen and M. W. Wartofsky, 1974, Reidel, Dordrecht.

13 See (i) N. W. Winkelman (1953) in *The Founders of Neurology* (ed. W. Haymaker), Thomas, Springfield, Illinois, (ii) Zeki, S. (1993) *A Vision of the Brain*, Blackwell, Oxford.

14 See (i) Bartholow, R. (1874) *Am. J. of the Medical Sciences*, Apr., pp. 305–13; (ii) entry in *Dictionary of American Biography*.

15 *Garrison's History of Medicine* (revised and enlarged) 1969 (ed. L. C. McHenry Jr) p. 336, Thomas, Springfield, Illinois.

16 Cushing, H. (1909) *Brain*, 32, 44–53. See also Fulton, J. F. (1946) *Harvey Cushing: a Biography*, Thomas, Springfield, Illinois.

17 Penfield, W. and Rasmussen, T. (1950) *The Cerebral Cortex of Man*, Macmillan, New York; Penfield, W. (1958) *The Excitable Cortex in Conscious Man*, Liverpool University Press.

18 Penfield, W. (1938) *Arch. Neurol. & Psychiat.*, 40, 417–42. Partly reproduced in ch. 9 of Penfield, W. and Rasmussen, T. (1950) *The Cerebral Cortex of Man*, Macmillan, New York.

19 Penfield, W. and Rasmussen, T. (1950) *The Cerebral Cortex of Man*, pp. 180–1, Macmillan, New York.

20 The quotation is from an article by the anatomist J. Szentágothai at a symposium in Pavia to celebrate Golgi's discovery – see Szentágothai, J. (1975) in *Golgi Centennial Symposium* (ed. M. Santini), pp. 1–12, Raven Press, New York.

21 Ramón y Cajal, S. (1937) *Recollections of My Life*, (translated by E. Horne Craigie with the assistance of J. Cano) MIT Press, Cambridge, Mass. First published as vol. 8 of *Memoirs of the American Philosophical Society*.

22 See legend to Figure 8.3 in Zeki, S. (1993) *A Vision of the Brain*, Blackwell, Oxford.

23 Zeki, S. (1993) *A Vision of the Brain*, ch. 28, Blackwell, Oxford.

24 See Lashley, K. S. (1931) *Science*, 73, 245–54. For an informative and amusing account of Lashley's attitude to localization of function in the association areas, see Zeki, S. (1993) *A Vision of the Brain*, ch. 28, Blackwell, Oxford.

III LOOKING AT SEEING

Introduction

1 Marr, D. (1982) *Vision*, W. H. Freeman, New York.

12 Illusions

1 Marr, D. (1982) *Vision*, p. 16, W. H. Freeman, New York.

2 Helmholtz, H. von (i) (1856–76) *Treatise on Physiological Optics*, vol. 3, ch. 26; Voss, Hamburg and Leipzig; English trans. 1925, reprinted Dover, New York, 1962. (ii) (1894) *Zeitschrift für Psychologie*, 7, 81–96.

3 Ramachandran, V. S. (1988) *Nature*, 331, 163–6.

4 Ramachandran, V. S. (1992) *Scientific American*, 266(5), 44–9.

5 The new approach was started at the University of Frankfurt by Max Wertheimer and his colleagues, Kurt Koffka and Wolfgang Köhler.

13 Disordered seeing with normal eyes

1 Verrey, L. (1888) *Archs. Ophtalmol. (Paris)*, 8, 289–301.

2 Munk, H. (1878) *Arch. Anat. Physiol.*, 2, 162–78; (1890) *Ueber die Functionen der Grosshirnrinde* (17 papers 1877–89), Hirschwald, Berlin.

3 Wilbrand, H. (1887) *Die Seelenblindheit als Herderscheinung und ihre Beziehung zur Alexie und Agraphie*, Bergmann, Wiesbaden. For an English translation of Wilbrand's case-report, and discussion of its significance, see Solms, M., Kaplan-Solms, K. & Brown, J. W. (1996) in *Classic Cases in Neuropsychology* (ed. C. Code, C-W. Wallesch, Y. Joanette & A. R. Lecours), pp. 89–110, Psychology Press, Hove, Sussex.

4 Lissauer, H. (1890) *Archiv für Psychiatrie und Nervenkrankheiten*, 21, 222–70. For a slightly abbreviated English translation see Shallice, T. & Jackson, M. (1988) *Cognitive Neuropsychology*, 5 (2), 153–92.

5 See, for example, McCarthy, R. A. & Warrington, E. K. (1990) *Cognitive Neuropsychology*, Academic Press, San Diego.

6 The first case described as visual form agnosia was also caused by carbon monoxide poisoning in a shower – see Benson, D. F. & Greenberg, J. P. (1969) *Archives of Neurology, Chicago*, 20, 82–9.

7 Goodale, M. A., Milner, A. D., Jakobson, L. S. & Carey, D. P. (1991) *Nature*, 349, 154–6. A fuller account is given in Milner, A. D. *et al.* (1991) *Brain*, 114, 405–28. See also Milner, A. D. & Goodale, M. A. (1995) *The Visual Brain in Action*, ch. 5, Oxford University Press, Oxford.

8 For discussion and references see ch. 1 of Shallice, T. (1988) *From Neuropsychology to Mental Structure*, Cambridge University Press, Cambridge.

9 Rubens, A. B. & Benson, D. F. (1971) *Archives of Neurology* (Chicago), 24, 305–16.

10 McCrae, D. & Trolle, E. (1956) *Brain*, 79, 94–110.

11 Hécaen, H. & Ajuriaguerra, J. de (1956) *Revue Neurologique*, 94, 222–3.

12 McCarthy, R. A. & Warrington, E. K. (1986) *Journal of Neurology, Neurosurgery, and Psychiatry*, 49, 1233–40.

13 Bornstein, B., Sroka, M. & Munitz, H. (1969) *Cortex*, 5, 164–9. Lhermitte, F., Chain, F., Escourolle, R., Ducarne, B. & Pillon, B. (1972) *Rev. Neurol.*, 126, 329–46. Newcombe, F. (1979) In *Research in Psychology and Medicine* (eds. D. J. Oborne, M. M. Gruneberg, & J. R. Eiser), vol. 1, pp. 315–22, Academic Press, London.

14 Damasio, A. R., Tranel, D. & Damasio, H. (1990) *Annu. Rev. Neurosci.*, 13, 89–109.

15 Wigan, A. L. (1844) *The Duality of the Mind*, Longman, London.

16 Goren, C. C., Sarty, M. & Wu, P. Y. K. (1975) *Pediatrics*, 56, 544–9. For further references see Carey, S. (1992) *Phil. Trans. Roy. Soc. Lond.*, B 335, 95–103; Ellis, H. D. (1992) *Phil. Trans. Roy. Soc. Lond.*, B 335, 105–11.

17 Bushnell, I. W. R., Sai, F. & Mullin, J. T. (1989) *Br. J. Dev. Psychol.* 7, 3–15; Walton, G. E. & Bower, T. G. R. (1991) quoted in Carey, S. (1992) see endnote 16.

18 Wigan, A. L. (1844) *The Duality of the Mind*, pp. 170–1, Longman, London.

19 Bodamer, J. (1947) *Archiv für Psychiatrie und Nervenkrankheiten*, 179, 6–54; an English translation is given by Ellis, H. D. & Florence, M. (1990) *Cognitive Neuropsychology*, 7, 81–105.

20 De Renzi, E. (1986) *Neuropsychologia* 24, 385–9.

21 Etkoff, N. K. L. (1984) *Neuropsychologia*, 22, 281–95; Adolphs, R., Tranel, D., Damasio, H. & Damasio, A. (1994) *Nature*, 372, 669–72.

22 Campbell, R., Landis, T. & Regard, M. (1986) *Brain*, 109, 509–21; Campbell, R. (1992) In *Processing the Facial Image* (ed. V. Bruce, A. Cowey, A. W. Ellis & D. I. Perrett), pp. 39–45, Clarendon Press, Oxford.

23 For further references see McCarthy, R. A. & Warrington, E. K. (1990) *Cognitive Neuropsychology*, ch. 3, Academic Press, San Diego.

24 See, for example, Bruce, V. & Young, A. (1986) *British J. of Psychology* 77, 305–27.

25 Sergent, J., Ohta, S. & MacDonald, B. (1992) *Brain*, 115, 15–36; Sergent, J. & Signoret, J-L. (1992) In *Processing the Facial Image* (ed. Bruce, V., Cowey, A., Ellis, A. W. & Perrett, D. I.), pp. 55–62, Clarendon Press, Oxford.

26 Sergent, J. & Signoret, J-L. (1992) see endnote 25; Meadows, J. C. (1974) *Journal of Neurology, Neurosurgery, and Psychiatry* 37, 489–501; Damasio, A. R., Tranel, D. & Damasio, H. (1990) *Annu. Rev. Neurosci.*, 13, 89–109; McCarthy, R. A. & Warrington, E. K. (1990) *Cognitive Neuropsychology*, pp. 63–5, Academic Press, San Diego.

27 Bruyer, R. *et al.* (1983) *Brain & Cognition* 2, 257–84; Bauer, R. M. (1984) *Neuropsychologia* 22, 457–69; Tranel, D. & Damasio, A. R. (1985) *Science*, 228, 1453–4.

28 de Haan, E. H. F., Young, A. & Newcombe, F. (1987) *Cognitive Neuropsychology*, 4, 385–415. See also de Haan, E. H. F. & Newcombe, F. (1991), *New Scientist*, 9, Feb. pp. 49–52.

29 See Ellis, A. W. & Young, A. W. (1988) *Human Cognitive Neuropsychology*, Erlbaum, Hove.

30 Sergent, J. & Poncet, M. (1990) *Brain*, 113, 989–1004.

31 Zihl, J., von Cramon, D. & Mai, N. (1983) *Brain*, 106, 313–40.

32 Riddoch, G. (1917) *Brain*, 40, 15–57.

33 Barlow, H. B., Blakemore, C. & Pettigrew, J. D. (1967) *J. Physiol. (Lond.)*, 193, 327–42.

34 Hubel, D. H., Wiesel, T. N. & LeVay, S. (1977) *Phil. Trans. Roy. Soc. B.*, 278, 377–409. Atkinson, J. & Braddick, O. J. (1989) 'Development of Basic Visual Functions.' In *Infant Development* (eds Slater, A. & Bremner, G.), Erlbaum, Hove.

35 Carmon, A. & Bechtoldt, H. P. (1969) *Neuropsychologia*, 7, 20–39.

36 Riddoch, G. (1917) *Brain*, 40, 15–57.

37 Holmes, G. & Horrax, G. (1919) *Archives of Neurology and Psychiatry*, 1, 385–407.

38 For a fuller discussion see Shallice, T. (1988) *From Neuropsychology to Mental Structure*, Cambridge University Press, Cambridge, p. 234.

39 Meadows, J. C. (1974) *Brain*, 97, 615–32.

40 Pöppel, E., Held, R. & Frost, D. (1973) *Nature*, 243, 295–6.

41 Weiskrantz, L., Warrington, E. K., Sanders, M. D. & Marshall, J. (1974) *Brain*, 97, 709–28.

42 Weiskrantz, L. (1997) *Consciousness Lost and Found*, Oxford University Press, Oxford.

43 Stoerig, P. & Cowey, A. (1992) *Brain*, 115, 425–44.

44 Kolb, F. C. & Braun, J. (1995) *Nature*, 377, 336–8. See also Cowey, A. (1995) *Nature*, 377, 290–1.

14 Opening the black box

1 Hubel, D.H. (1988) *Eye, Brain, and Vision*, Scientific American Library, New York.

2 Recent work on macaque monkeys suggests that there are two colour pathways, an evolutionary older pathway carrying information from ganglion cells that compare short wavelength (blue) and medium wavelength (green) light in their receptive fields, and a newer pathway with ganglion cells that compare medium wavelength (green) light and long wavelength (red) light. See Hendry, S.H.C. & Yoshioka, T. (1994) *Science*, 264, 575–7; Martin, P.R. *et al.* (1997) *European J. Neurosci.*, 9, 1536–41; Mollon, J. (1999) *Proc. Ntl. Acad. Sci. USA*, 96, 4743–5.

3 The story is told on pp. 69 and 70 of Hubel's *Eye, Brain and Vision* – see endnote 1.

4 Such dyes were developed by Larry Cohen at Yale in the 1960s.

5 Bonhoeffer, T. & Grinvald, A. (1991) *Nature*, 353, 429–31.

6 Barlow, H.B., Blakemore, C. & Pettigrew, J.D. (1967) *J. Physiol. (London)*, 193, 327–42.

7 Wiesel, T.N. & Hubel, D.H. (1963) *J. Neurophysiol*, 26, 1003–1017.

8 Blakemore, C. & Van Sluyters, R.C. (1974) *J. Physiol. (London)*, 237, 195–216.

9 Blakemore, C. & Cooper, G. (1970) *Nature*, 228, 477–8.

10 Livingstone, M.S. & Hubel, D.H. (1984) *J. Neurosci.* 4, 309–56.

11 Zeki, S. (1993) *A Vision of the Brain*, ch. 14, Blackwell, Oxford.

12 An area containing, predominantly, cells sensitive to motion in a particular direction had been noted earlier in the prestriate cortex of the cat. Hubel, D.H. & Wiesel, T.N. (1969) *J. Physiol. (London)*, 201, 251–60.

13 Zeki, S. *et al.* (1991) *J. Neurosci.*, 11, 641–9.

14 Barlow, H.B. (1995) In *The Cognitive Neurosciences*, (ed. Gazzaniga, M.S.), pp. 415–35. MIT Press, Cambridge, Mass.

15 For references see Gross, C.G. (1992) In *Processing the Facial Image*, (ed. V. Bruce, A. Cowey, A.W. Ellis, & D.I. Perrett), pp. 3–10, Clarendon Press, Oxford.

16 Gross, C.G., Bender, D.B. & Rocha-Miranda, C.E. (1969) *Science*, 166, 1303–6.

Gross, C.G., Rocha-Miranda, C.E. & Bender, D.B. (1972) *J. Neurophysiol.*, 35, 96–111.

17 For references see (i) articles by C.G. Gross, by E.T. Rolls, by D.I. Perrett *et al.*, and by C.A. Heywood & A. Cowey (1992) in *Processing the Facial Image* (eds V. Bruce, A. Cowey, A.W. Ellis & D.I. Perrett), Clarendon Press, Oxford; (ii) review by Logothetis, N.K. & Sheinberg, D.L. (1996) *Annu. Rev. Neurosci.*, 19, 577–621.

18 The neighbourhood of the superior temporal sulcus, and the lowest part of the prefrontal cortex.

19 Perrett, D.I., Hietanen, J.K., Oram, M.W. & Benson, P.J. (1992) in *Processing the Facial Image*, (eds V. Bruce, A. Cowey, A.W. Ellis & D.I. Perrett), pp. 23–30, Clarendon Press, Oxford.

20 Desimone, R. (1991) *J. Cognitive Neuroscience*, 3, 1–8. Hasselmo, M.E., Rolls, E.T. & Bayliss, G.C. (1989) *Behavioral Brain Research*, 32, 203–18.

21 For references see (i) endnote 21 of Ch. 13; (ii) Heywood, C.A. & Cowey, A. (1992) in *Processing the Facial Image*, (eds V. Bruce, A. Cowey, A.W. Ellis & D.I. Perrett), pp. 31–38. Clarendon Press, Oxford.

22 Sergent, J. & Signoret, J.-L. (1992) in *Processing the Facial Image* (eds V. Bruce, A. Cowey, A.W. Ellis & D.I. Perrett), pp. 55–62, Clarendon Press, Oxford.

23 Logothetis, N.K. & Sheinberg, D.L. (1996) *Annu. Rev. Neurosci.*, 19, 577–621.

24 Newsome, W.T. & Paré, E.B. (1988) *J. Neurosci.*, 8, 2201–11.

25 Salzman, C.D., Britten, K.H. & Newsome, W.T. (1990) *Nature*, 346, 174–7.

26 Zeki, S. (1993) *A Vision of the Brain*, p. 280 and Plate 21, Blackwell, Oxford. Zeki, S. & Lamb, M. (1994) *Brain*, 117, 607–36.

27 For references see Snowden, R.J. (1994) in *Visual Detection of Motion* (eds A.T. Smith, & R.J. Snowden), pp. 51–83, Academic Press, London.

28 Gaspard Monge (1789) *Annales de chimie*, 3, 131-47. For a very readable account, see Mollon, J. (1995) In *Colour: Art and Science* (eds T. Lamb & J. Bourriau), pp. 127–50, Cambridge University Press, Cambridge.

29 Land, E.H. (1959) *Proc. Natl. Acad. Sci. USA*, 45, 115–29.

30 Land, E.H. (1986) *Proc. Natl. Acad. Sci. USA*, 83, 3078–3080.

31 Land, E.H. (1983) *Proc. Natl. Acad. Sci. USA*, 80, 5163–5169.

32 Livingstone, M.S. & Hubel, D.H. (1984) *J. Neurosci.*, 4, 309–56.

33 Zeki, S. (1983) *Neuroscience*, 9, 741–65 and 767–81.

15 Natural computers and artificial brains

1 Barlow, H.B. (1972) *Perception*, 1, 371–94.

2 Rumelhart, D.E., McLelland, J.L. & the PDP Research Group (1986) *Parallel Distributed Processing: Explorations in the Microstructure of Cognition*, vol. 1, p. 122, MIT Press, Cambridge, Mass.

3 Reichardt, W. & Poggio, T. (1976) *Quart. Rev. Biophys.*, 9, 311–75.

4 Rumelhart, D.E., McLelland, J.L. & the PDP Research Group (1986) *Parallel Distributed Processing: Explorations in the Microstructure of Cognition*, vol. 1, MIT Press, Cambridge, Mass.

5 Cottrell, G. (1991) In *Connectionist Models: Proceedings of the 1990 Summer School* (eds D. Touretzky, J. Elman, T. Sejnowski & G. Hinton), Morgan Kaufmann, San Mateo, CA. A more readily accessible, and very readable, account of this work is in ch. 3 of Churchland, P.M. (1995) *The Engine of Reason, the Seat of the Soul*, MIT

Press, Cambridge, Mass.

6 Rumelhart, D.E., Hinton, G.E. & Williams, R.J. (1986) In *Parallel Distributed Processing: Explorations in the Microstructure of Cognition* (eds D.E. Rumelhart, J.L. McLelland & the PDP Research Group), vol. 1 ch. 8, MIT Press, Cambridge, Mass.

7 Lehky, S.R. & Sejnowski, T. J. (1990) *Proc. Roy. Soc. B.*, 240, 251–78.

8 Hinton, G.E., McLelland, J.L. & Rumelhart, D.E. (1986) In *Parallel Distributed Processing: Explorations in the Microstructure of Cognition*, (eds Rumelhart, D.E., McLelland, J.L. & the PDP Research Group) vol. 1, ch. 11, MIT Press, Cambridge, Mass.

9 Marr, D. & Poggio, T. (1976) *Science*, 194, 283–7.

10 The algorithm used by Marr and Poggio was not conceived as a neural net algorithm.

11 McLelland, J.L., Rumelhart, D.E. & Hinton, G.E. (1986) In *Parallel Distributed Processing: Explorations in the Microstructure of Cognition* (eds D.E. Rumelhart, J.L. McLelland & the PDP Research Group), vol. 1, ch. 1, pp. 33–7, MIT Press, Cambridge, Mass.

12 Hinton, G.E., Dayan, P., Frey, B.J. & Neal, R.M. (1995) *Science*, 268, 1158–61. Dayan, P., Hinton, G.E., Neal, R.M. & Zemel, R.S. (1995), *Neural Computation*, 7, 889–904.

IV TALKING ABOUT TALKING

16 In the steps of the 'diagram makers'

1 Gardner, H. (1977) *The Shattered Mind*, Routledge, London.

2 Lichtheim, L. (1885) *Brain*, 7, 433–84.

3 Dejerine, J. (1891) *Memoires-Société Biologie*, 3, 197–201; (1892) *Memoires-Société Biologie*, 4, 61–90.

4 Trescher, J.H. & Ford, F.R. (1937) *Archives of Neurology and Psychiatry*, 37, 959–73.

5 See Shallice, T. (1988) *From Neuropsychology to Mental Structure*, ch. 1, Cambridge University Press, Cambridge.

6 Cole, M.F. & Cole, M. (eds) (1971) Pierre Marie's papers on speech disorders, Hafner, New York.

7 Castaigne, P., Lhermitte, F., Signoret, J.-L. & Abelanet, R. (1980) *Revue Neurologique*, 136, 563–83. For a slightly later report, in English, see Signoret, J.-L., Castaigne, P., Lhermitte, F., Abelanet, R. & Lavorel, P. (1984) *Brain and Language*, 22, 303–19.

8 For references see McCarthy, R.A. & Warrington, E.K. (1990) *Cognitive Neuropsychology*, pp. 174–6, Academic Press, San Diego. See also Berndt, R.S. & Caramazza, A. (1980) *Appl. Psycholinguistics*, 1, 225–78.

9 Smuts, J.C. (1926) *Holism and Evolution*, Macmillan, London.

10 For example, Kurt Goldstein in the USA, Aleksandr Luria in Russia.

11 Geschwind, N. (1965) *Brain*, 88, 237–94 and 585–644.

12 Geschwind, N. & Kaplan, E. (1962) *Neurology*, 12, 675–85.

13 Geschwind, N., Quadfasel, F.A. & Segarra, J.M. (1968) *Neuropsychologia*, 6, 327–40.

14 Goldstein, K. (1917) *Die Transkortikalen Aphasien*, Fischer, Jena. Stengel, E. (1936) *Z. ges. Neurol. Psychiat*, 154, 778–82; (1947) *J. Ment. Sci.*, 93, 598–612.

15 Kosslyn, S.M. & Intriligator, J.M. (1992) *J. Cog. Neurosci.*, 4, 96–106.

16 Caramazza, A. (1992) *J. Cog. Neurosci.*, 4, 80–95.

17 For a discussion on whether such behaviour could be produced by non-modular systems see Shallice, T. (1988) *From Neuropsychology to Mental Structure*, ch. 11, Cambridge University Press, Cambridge.

18 For references see McCarthy, R.A. & Warrington, E.K. (1990) *Cognitive Neuropsychology*, ch. 6, Academic Press, San Diego.

19 For references see McCarthy, R.A. & Warrington, E.K. (1990) *Cognitive Neuropsychology*, ch. 7, Academic Press, San Diego.

20 Hart, J., Berndt, R.S. & Caramazza, A. (1985) *Nature*, 316, 439–40.

21 Lhermitte, F. & Beauvois, M.-F. (1973) *Brain*, 96, 695–714.

22 Hier, D.B. & Mohr, J.P. (1977) *Brain and Language*, 4, 115–26.

23 For reviews see: McCarthy, R.A. & Warrington, E.K. (1990) *Cognitive Neuropsychology*, ch. 10, Academic Press, San Diego. Shallice, T. (1988) *From Neuropsychology to Mental Structure*, ch. 5, Cambridge University Press, Cambridge. Benson, D.A. & Ardila, A. (1996) *Aphasia*, ch. 11, Oxford University Press, Oxford.

24 See, for example, Seidenberg, M.S. & McClelland, J.L. (1989) *Psychological Review*, 96, 523–68.

25 Schwartz, M.F., Saffran, E.M. & Marin, O.S.M. (1980) In *Deep Dyslexia* (eds M. Coltheart, K.E. Patterson & J.C.Marshall), pp. 259–69, Routledge, London.

26 Bramwell, B. (1897) *Lancet*, i, 1256–9. Reproduced with commentary in Ellis, A.W. (1984) *Cognitive Neuropsychology*, 1, 245–58. Allport, D.A. & Funnell, E. (1981) *Phil. Trans. Roy. Soc. Lond.*, B., 295, 397–410.

27 Marshall, J.C. & Newcombe, F. (1973) *J. Psycholinguistic Research*, 2, 175–99. Papers in Coltheart, M., Patterson, K. & Marshall, J.C. (eds) (1980) *Deep Dyslexia*, Routledge, London.

28 Luria, A.S. (1970) *Traumatic Aphasia*, Mouton, The Hague.

29 Beauvois, M.-F. & Dérouesné, J. (1981) *Brain*, 104, 21–49. Shallice, T. (1981) *Brain*, 104, 413–29. Roeltgen, D.P. & Heilman, K.M. (1984) *Brain*, 107, 811–27.

30 Miura, K. (1901) *Iji-Shimbun (J. Med.)*, 584, 249–56.

31 Sasanuma, S. (1980) In *Deep Dyslexia* (eds M. Coltheart, K. Patterson & J.C. Marshall), pp. 48–90, Routledge, London. Iwata, M. (1984) *Trends in Neurosciences*, Aug., 290–3. Soma, Y., Sugushita, M., Kitamura, K., Maruyama, S. & Imanaga, H. (1989) *Brain*, 112, 1549–61. Benson, D.A. & Ardila, A. (1996) *Aphasia*, pp. 209–11, Oxford University Press, Oxford.

32 Beauvois, M.-F. & Dérouesné, J. (1979) *J. Neurology, Neurosurgery and Psychiatry*, 42, 1115–24; (1981) *Brain*, 104, 21–49.

33 Shallice, T. (1988) *From Neuropsychology to Mental Structure*, p. 159, Cambridge University Press, Cambridge.

34 Allport, D.A. & Funnell, E. (1981) *Phil. Trans. Roy. Soc. Lond.*, B, 295, 397–410.

35 Rasmussen, T. & Milner, B. (1977) *Ann. N.Y. Acad. Sci.*, 299, 355–69.

36 McCarthy, R.A. & Warrington, E.K. (1990) *Cognitive Neuropsychology*, chs 2 & 4, Academic Press, San Diego. Milner, B. (1962). In *Interhemispheric Relations and Cerebral Dominance* (ed, V. B. Mountcastle), pp. 177–95, Johns Hopkins Press,

Baltimore. Tramo, M., Bharucha, J. & Musiek, F. (1990) *J. Cognitive Neurosci.*, 2, 195–212.

37 Jackson, J.H. (1915) *Brain*, 38, 107–74.

38 Heilman, K.M., Scholes, R. & Watson, R.T. (1975) *J. Neurology, Neurosurgery, and Psychiatry*, 38, 69–72.

39 Ross, E.D. & Mesulam, M.–M. (1979) *Archives of Neurology*, 36, 144–8. Ross, E.D. (1981) *Archives of Neurology*, 38, 561–9.

40 Monrad-Krohn, G.H. (1947) *Brain*, 70, 405–15.

41 Poizner, H., Bellugi, U. & Iragui, V. (1984) *Amer. J. Physiol.*, 246, R868–83.

42 Ross, E.J., Seibert, E.G. & Chan, J.-L. (1992) *J. Phonetics*, 20, 441–56. Packard, J. (1986) *Brain and Language*, 29, 212–23.

43 Sperry, R.W. (1961) *Science*, 133, 1749–57.

44 Gazzaniga, M.S. & LeDoux, J. (1978) *The Integrated Mind*, Plenum, New York.

45 Gazzaniga, M.S. (1983) *American Psychologist*, 38, 525–37. See also Gazzaniga, M.S. (1985) *The Social Brain*, pp. 88–91, Basic Books, New York.

46 Sermon preached by F.A. Simpson on 6 November 1932. *The Cambridge Review*, 1932, reprinted in *The Cambridge Review*, 1954.

47 Howard, D. *et al.* (1992) *Brain*, 115, 1769–82.

48 Baddeley, A.E. (1986) *Working Memory*, ch. 5, Clarendon Press, Oxford; (1996) *Your Memory*, 3rd Edn, ch. 3, Prion, London.

49 Paulesu, E., Frith, C.D. & Frackowiak, R.S.J. (1993) *Nature*, 362, 342–5.

50 Warrington, E.K., Logue, V. & Pratt, R.T.C. (1971) *Neuropsychologia*, 9, 377–87.

51 Eden, G.F. *et al.* (1996) *Nature*, 382, 66–9. See also Frith, C. & Frith, U. (1996) *Nature*, 382, 19–20.

17 Chomsky and after

1 Chomsky, N. (1957) *Syntactic Structures*, Mouton, the Hague.

2 Darwin, C. (1871) *The Descent of Man*, ch. 3, Murray, London.

3 Chomsky, N. (1965) *Aspects of the Theory of Syntax*, pp. 51–3, MIT Press, Cambridge, Mass.; (1986) *Knowledge of Language*, Praeger, New York.

4 Brown, R. (1958) *Words and Things*, pp. 3–7 and 186–93, The Free Press, Macmillan, New York. Malson, L. (1972) *Wolf Children and the Problem of Human Nature*, Monthly Review Press, New York. Curtiss, S. (1977) *Genie: A Psycholinguistic Study of a Modern-day 'Wild Child'*, Academic Press, New York.

5 Chomsky, N. (1957) *Syntactic Structures*, Mouton, the Hague.

6 Pinker, S. (1994) *The Language Instinct*, Penguin, London.

7 Pinker, S. (1994) *The Language Instinct*, pp. 150–1, Penguin, London.

8 Gopnik, M. (1990) *Nature*, 344, 715. See also Gopnik, M., Dalalakis, J., Fukuda, S.E., Fukuda, S. & Kehayia, E. (1996) *Proceedings of the British Academy*, 88, 223–49.

9 Bickerton, D. (1983) *Scientific American* 249(1), 108–15; (1984) *Behavioral and Brain Sciences*, 7, 173–221; (1990) *Language and Species*, University of Chicago Press, Chicago; (1995) *Language and Human Behavior*, University of Washington Press.

10 Roberts, J. (1995) *Journal of Pidgin and Creole Languages*, 10.1. See also Bickerton, D. (1995) *Language and Human Behavior*, pp. 38–9, University of Washington Press.

11 Bickerton, D. (1984) *Behavioral and Brain Sciences*, 7, 173–221; (1990) *Language*

and Species, University of Chicago Press, Chicago.

12 I am grateful to my colleagues Sidney Allen and Simon Keynes for telling me about Alfred's writing and for checking the Anglo-Saxon original.

13 Bickerton, D. (1984), p. 218 in *Behavioral and Brain Sciences, 7,* 173–221; (1990) *Language and Species*, University of Chicago Press, Chicago.

14 Marler, P. (1977) In *Interaction, Conversation and the Development of Language.* Eds. M. Lewis & L.A. Rosenblum, Wiley, New York, pp. 95–114.

18 Monkey puzzles

1 Geschwind, N. & Levitsky, W. (1968) *Science,* 161, 186–7.

2 Lemay, M. & Culebras, A. (1972) *New England J. Med.* 287, 168–70.

3 Tobias, P.V. (1981) *Phil. Trans, Roy. Soc. B. Lond.,* 292, 43–56. Holloway, R. (1983) *Human Neurobiology,* 2, 105–14.

4 Deacon, T.W. (1986) 'Human Brain Evolution: 1. Evolution of Language Circuits' In *Intelligence and Evolutionary Biology* (eds. Jerison, H.J. & Jerison, I.) pp. 363–81. Springer-Verlag, Berlin.

5 Lieberman, P. (1984) *The Biology and Evolution of Language*, Harvard University Press, Cambridge, Mass.

6 Laitman, J.T. & Heimbuch, R.C. (1982) *Amer. J. Physical Anthropology,* 59, 323–44.

7 Leakey, R. (1994) *The Origin of Humankind*, ch. 7, Weidenfeld & Nicolson, London.

8 Pinker, S. (1994) *The Language Instinct*, p. 354, Penguin, London.

9 Struhsaker, T.T. (1967) In *Social Communication Among Primates* (ed. S.A. Altmann), pp. 281–324, University of Chicago Press, Chicago.

10 Cheney, D.L. & Seyfarth, R.M. (1990) *How Monkeys See the World*, University of Chicago Press, Chicago.

11 See Dawkins, R. (1995) *River Out of Eden*, chs 3 & 4, Weidenfeld & Nicolson, London.

12 Thornhill, R. (1979) *Science,* 205, 412–14.

13 Goodall, J. (1986) *The Chimpanzees of Gombe: Patterns of Behaviour*, ch. 6, Harvard University Press, Cambridge, Mass.

14 For a readable and critical review of most of these attempts see Wallman, J. (1992) *Aping Language*, Cambridge University Press, Cambridge.

15 Hayes, K.J. & Hayes C.H. (1951) *Proceedings of the American Philosophical Society,* 95, 105–9.

16 Warden, C.J. & Warner, L.H. (1928) *Quarterly Review of Biology,* 3, 1–28.

17 Gardner, R.A. & Gardner, B.T. (1969) *Science,* 165, 664–72; (1984) *J. Comp. Psychol.* 98, 381–404.

18 Terrace, H.S., Petitto, L.A., Sanders, R.J. & Bever, T.G. (1980) In *Children's Language*, vol. 2, (ed. K.E. Nelson), pp. 371–495, Gardner, New York.

19 Patterson, F.G. (1978) *Brain and Language,* 5, 72–97; (1980) In *Children's Language*, vol. 2 (ed. K.E. Nelson), pp. 497–561, Gardner, New York; (1987) *Koko's Story*, Scholastic, New York. Terrace, H.S. (1979) *Nim*, Knopf, New York. Terrace, H.S., Petitto, L.A., Sanders, R.J. & Bever, T.G. (1979) *Science,* 206, 891–902; (1981) *Science,* 211, 87–8; (1980) In *Children's Language*, vol. 2, (ed. K.E. Nelson), pp. 371–495, Gardner, New York.

20 Premack, D. (1971) *Science*, 172, 808–22.

21 Gill, T.V. (1977) In *Language Learning by a Chimpanzee* (ed. D.M. Rumbaugh), Academic, New York.

22 Straub, R.O., Seidenberg, M.S., Bever, T.G. & Terrace, H.S. (1979) *Journal of Experimental Analysis of Behavior*, 32, 137–48.

23 For critical discussion see (i) Terrace, H.S. (1979) *J. Experimental Analysis of Behavior*, 31, 161–75; (ii) Wallman, J. (1992) *Aping Language*, chs 3 and 4, Cambridge University Press, Cambridge.

24 Savage-Rumbaugh, S., McDonald, K., Sevcik, R.A., Hopkins, W.D. & Rupert, E. (1986) *J. Experimental Psychology: General*, 115, 211–35. Savage-Rumbaugh, E.S. (1986) *Ape Language: From Conditioned Response to Symbol*, Columbia University Press, New York; (1987) *J. Experimental Psychology: General*, 116, 288–92.

25 Savage-Rumbaugh, E.S. (1988) In *Comparative Perspectives in Modern Psychology*, (*Nebraska Symposium on Motivation 1987*), (ed. D.W. Leger), pp. 201–255, University of Nebraska Press, Lincoln, Nebraska.

26 Wallman, J. (1992) *Aping Language*, pp. 102–6, Cambridge University Press, Cambridge.

27 Humboldt, Wilhelm von (1836) *Linguistic Variability and Intellectual Development*, Trans. 1971 by G.C. Buck & F.A. Raven, University of Miami Press, Coral Gables, Florida.

28 Chomsky, N. (1988) *Language and Problems of Knowledge: The Managua Lectures*, p. 167, MIT Press, Cambridge, Mass; (1980) Discussion of Putnam's comments, In *Language and Learning: The Debate Between Jean Piaget and Noam Chomsky* (ed. M. Piattelli-Palmarini), p. 321, Harvard University Press.

29 Pinker, S. & Bloom, P. (1990) *Behavioral and Brain Sciences*, 13, 707–84.

30 Bickerton, D. (1995) *Language and Human Behavior*, University of Washington Press.

31 Pinker, S. (1997) *How the Mind Works*, p. 367, Penguin, London.

32 The concept of storing information in the part of the environment that we use as inputs is discussed by Tooby, J. & Cosmides, L. (1990) *Journal of Personality*, 58, 17–67.

V THINKING ABOUT THINKING

19 Memory

1 Scoville, W.B. & Milner, B. (1957) *J. Neurol. Neurosurg. Psychiat.*, 20, 11–21.

2 Specifically, the parahippocampal gyrus and the uncus.

3 Scoville, W.B. & Milner, B. (1957) *J. Neurol. Neurosurg. Psychiat.*, 20, 11–21.

4 Quoted in Milner, B. (1966) In *Amnesia* (eds C.W.M.Whitty, & O.L. Zangwill), pp. 109–33, Butterworth, London.

5 Scoville, W.B. & Milner, B. (1957) *J. Neurol. Neurosurg. Psychiat.*, 20, 11–21.

6 Zola-Morgan, S., Squire, L.R. & Amaral, D.G. (1986) *J. Neuroscience*, 6, 2950–67.

7 The medial thalamic nuclei and the mamillary bodies.

8 Polster, M.R. (1993) *Psychological Bulletin*, 114, 477–93.

9 Rang, H.P., Dale, M.M. & Ritter, J.M. (1995) *Pharmacology*, 3rd edn, p. 556,

Churchill Livingstone, Edinburgh.

10 Milner, B. (1966) In *Amnesia*, (eds C.W.M. Whitty & O.L. Zangwill), pp. 109–33, Butterworth, London.

11 Milner, B., Corkin, S. & Teuber, H.-L. (1968) *Neuropsychologia*, 6, 215–34.

12 Claparède, E. (1911) *Arch. de Psychol. Geneva*, 11, 79–90. English trans. (1951). In *Organisation and Pathology of Thought*, (ed. D. Rapaport), pp. 58–75, Columbia University Press, New York.

13 Corkin, S. (1968) *Neuropsychologia*, 6, 255–65.

14 Tulving, E. (1972) In *Organisation of Memory* (eds E. Tulving & W. Donaldson), pp. 381–403, Academic Press, New York; (1983) *Elements of Episodic Memory*, Oxford University Press, Oxford.

15 Lancaster, O. (1947) *Classical Landscape*, John Murray, London.

16 Vargha-Khadem, F., Gadian, D.G., Watkins, K.E., Connelly, A., Van Paesschen, W. & Mishkin, M. (1997) *Science*, 277, 376–80.

17 For references see O'Keefe, J. & Nadel, L. (1978) *The Hippocampus as a Cognitive Map*, ch. 4, Oxford University Press, Oxford.

18 For references see Maguire, E.A., Frackowiak, R.S.J. & Frith, C.D. (1997) *J. Neuroscience* 17, 7103–10.

19 Maguire, E.A., Frackowiak, R.S.J. & Frith, C.D. (1997) *J. Neuroscience*, 17, 7103–10.

20 Maguire, E.A., Burgess, N., Donnett, J.G., Frackowiak, R.S.J., Frith, C.D. & O'Keefe, J. (1998) *Science*, 280, 921–4.

21 For references see Shallice, T. (1988) *From Neuropsychology to Mental Structure*, ch. 3, Cambridge University Press, Cambridge.

22 Warrington, E.K. & Rabin, P. (1971) *Quart. J. Exp. Psychol.*, 23, 423–31. De Renzi, E. & Nichelli, P. (1975) *Cortex*, 11, 341–54. Squire, L.R., Knowlton, B. & Musen, G. (1993) *Annual Review of Psychology*, 44, 453–95.

23 Baddeley, A.D., Papagno, C. & Vallar, G. (1988) *J. Memory and Language*, 27, 586–95.

24 See Baddeley, A. (1997) *Human Memory: Theory and Practice*, rev. edn, ch. 4, Psychology Press, East Sussex.

25 Jacobsen, C.F. (1935) *Archives of Neurology and Psychiatry*, 33, 559–69.

26 Fuster, J.M. (1997) *The Prefrontal Cortex*, 3rd edn, ch. 5, Lippincott-Raven, Philadelphia.

27 Funahashi, S., Bruce, C.J. & Goldman-Rakic, P.S. (1989) *J. Neurophysiol.*, 61, 331–49.

28 Wilson, F.A.W., O Scalaidhe, S.P. & Goldman-Rakic, P.S. (1993) *Science*, 260, 1955–8.

29 See review by Kandel, E.R. & Schwartz, J.H. (1982) *Science*, 218, 433–43.

30 Bliss, T.V.P. & Lømo, T. (1973) *J. Physiol. (Lond.)* 232, 331–56.

31 Frey, U. & Morris, R.G.M. (1997) *Nature*, 385, 533–6; Bear, M.F. (1997) *Nature*, 385, 481–2.

32 Bonhoeffer, T., Staiger, V. & Aertsen, A. (1989) *Proc. Natl. Acad. Sci. USA*, 86, 8113–7. Engert, F. & Bonhoeffer, T. (1997) *Nature*, 388, 279–84.

33 Schuman, E.M. & Madison, D.V. (1994) *Science*, 263, 532–6. See also review by Hölscher, C. (1997) *Trends in Neurosciences*, 20, 298–303.

34 Morris, R.G.M., Davis, S. & Butcher, S.P. (1990) *Phil. Trans. Roy. Soc. Lond. B.*

329, 187–204.

35 Tsien, J.Z. *et al.* (1996) *Cell*, 87, 1317–26. Tsien, J.Z. Huerta, P.T. & Tonegawa, S. (1996) *Cell*, 87, 1327–38. Morris, R.G.M. & Morris R.J. (1997) *Nature*, 385, 680–1.

36 Barondes, S.H. & Cohen, H.D. (1968) *Science*, 160, 556–7.

37 Chapman, P.F. *et al.* (1992) *NeuroReport* 3, 567–70.

38 Kendrick K.M. *et al.* (1997) *Nature*, 388, 670–4.

39 Butters, N. & Cermak, L.S. (1986) In *Autobiographical Memory*, (ed. D.C. Rubin), pp. 253–72, Cambridge University Press, Cambridge.

40 The name is still used though the long-established notion that these parts of the cortex are evolutionarily newer than the areas adjacent to the hippocampus is now questioned – see Northcutt, R.G. & Kaas, J.H. (1995) *Trends in Neurosciences*, 18, 373–9.

41 Milner, B. (1965) In *Cognitive Processes and the Brain* (eds P.M. Milner & S.E. Glickman), pp. 97–111, Van Nostrand, Princeton, N.J. Milner, P.M. (1989) *Neuropsychologia*, 27, 23–30. Alvarez, P. & Squire, L.R. (1994) *Proc. Natl. Acad. Sci. USA*, 91, 7041–5. Squire, L.R. & Alvarez, P. (1995) *Current Opinion in Neurobiology*, 5, 169–77.

42 Marr, D. (1971) *Phil. Trans. Roy. Soc. Lond. B.*, 262, 23–81.

43 Chrobak, J.J. & Buzsaki, G. (1994) *J. Neurosci.*, 14, 6160–70.

44 Wilson, M.A. & McNaughton, B.L. (1994) *Science*, 265, 676–9. Qin, Y-L., McNaughton, B.L., Skaggs, W.E. & Barnes, C.A. (1997) *Phil. Trans. Roy. Soc. Lond. B.*, 352, 1525–33.

45 McClelland, J.L., McNaughton, B.L. & O'Reilly, R.C. (1995) *Psychological Review*, 102, 419–57.

46 McClelland, J.L., McNaughton, B.L. & O'Reilly, R.C. (1995) *Psychological Review*, 102, 419–57.

47 Shadmehr, R. & Holcombe, H.H. (1997) *Science*, 277, 821–5.

48 The account which follows is taken from the English translation of Luria's 1965 book – see Luria, A.R. (1968) *The Mind of a Mnemonist*, trans. by L. Solotaroff, Basic Books, New York.

49 Baron-Cohen, S. & Harrison, J.E. (eds) (1997) *Synaesthesia*, Blackwell, Oxford.

50 Gregory, R.L. (1988) In *Consciousness in Contemporary Science* (eds Marcel, A.J. & Bisiach, E.), pp. 257–72, Oxford University Press, Oxford.

51 See the chapter on prodigies in Sacks, O. (1995) *An Anthropologist on Mars*, Knopf, New York.

20 *The emotions*

1 James, W. (1884) *Mind*, 9, 188–205. The article was described as notorious by James Ward (1911) in his article on Psychology in the 11th edition of the *Encyclopaedia Britannica*.

2 Schacter, S. & Singer, J. (1962) *Psychological Review*, 69, 379–99.

3 Goltz, F. (1892) *Pflügers Arch.*, 51, 570–614. Dusser de Barenne, J.G. (1920) *Arch. Néerland. de Physiol.*, 4, 31–123. Cannon, W.B. & Britton, S. W. (1925) *Am. J. Physiol.*, 72, 283–94.

4 Bard, P. (1928) *Am. J. Physiol.* 84, 490–515. Hess, W.R. & Brügger, M. (1943) *Helv. Physiol. Acta*, 1, 33–52. Hess, W.R. (1954) *Diencephalon: Automatic and Extrapyramidal Functions*, Heineman, London.

5 Klüver, H. & Bucy, P.C. (1937) *Am. J. Physiol.*, 119, 352–3; (1939) *Archives of Neurology and Psychiatry*, 42, 979–1000.

6 Schreiner, L. & Kling, A. (1953) *J. Neurophysiol.* 16, 643–59. Weiskrantz, L. (1956) *J. Comp. Physiol. Psychol.*, 49, 381–91. Aggleton, J.P. & Passingham, R.E. (1981) *J. Comp. Physiol. Psychol.*, 95, 961–77.

7 Downer, J.L. deC. (1961) *Nature*, 191, 50–1.

8 Weiskrantz, L. (1956) *J. Comp. Physiol. Psychol.*, 49, 381–91. Horvath, F.E. (1963) *J. Comp. Physiol. Psychol.*, 56, 380–9. Blanchard, D.C. & Blanchard, R.J. (1972) *J. Comp. Physiol. Psychol.*, 81, 281–90. Nachman, M. & Ashe, J.H. (1974) *J. Comp. Physiol. Psychol.*, 87, 622–43. Kapp, B.S., Frysinger, R.C., Gallagher, M. & Haselton, J. R. (1979) *Physiology and Behaviour*, 23, 1109–17.

9 LeDoux, J. (1996) *The Emotional Brain*, ch. 6, Simon & Schuster, New York.

10 Jarrell, T.W., Gentile, C.G., Romanski, L.M., McCabe, P.M. & Schneiderman, N. (1987) *Brain Research*, 412, 285–94.

11 Bordi, F. & LeDoux, J. (1994) *Experimental Brain Research*, 98, 261–74 and 275–86.

12 LeDoux, J. (1996) *The Emotional Brain*, ch. 6, Simon & Schuster, New York.

13 Seldon, N.R.W., Everitt, B.J., Jarrard, L.E. & Robbins, T.W. (1991) *Neuroscience*, 42, 335–50. Phillips, R.R.G.& LeDoux, J. (1992) *Behavioral Neuroscience*, 106, 274–85. Kim.J.J. & Fanselow, M.S. (1992) *Science*, 256, 675–7.

14 Rescorla, R.A. & Heth, C.D. (1975) *J. Exp. Psychol.: Animal Behavior Processes*, 1, 88–96. Jacobs, W.J. & Nadel, L. (1985) *Psychol. Rev.* 92, 512–31.

15 Falls, W.A., Miserendino, M.J.D. & Davis, M. (1992) *J. Neuroscience*, 12, 854–63.

16 Jones, B. & Mishkin, M. (1972) *Experimental Neurology*, 36, 362–77. See also Gaffan, D. (1992) In *The Amygdala: Neurobiological Aspects of Emotion, Memory and Mental Dysfunction* (ed. J.P. Aggleton), pp. 471–483, Wiley-Liss, New York.

17 For references and discussion see Rolls, E.T. (1992) In *The Amygdala: Neurobiological Aspects of Emotion, Memory and Mental Dysfunction*, (ed. J.P. Aggleton), pp. 143–65, Wiley-Liss, New York.

18 Rolls, E.T. (1992) In *The Amygdala: Neurobiological Aspects of Emotion, Memory and Mental Dysfunction* (ed. J.P. Aggleton), pp. 143–65. Wiley-Liss, New York.

19 Terzian, H. & Dalle Ore, G. (1955) *Neurology*, 5, 373–80.

20 For references and discussion see Aggelton, J.P. (1992) In *The Amygdala: Neurobiological Aspects of Emotion, Memory and Mental Dysfunction*, (ed. J.P. Aggleton), pp. 485–503, Wiley-Liss, New York.

21 Jacobson, R. (1986) *Psychological Medicine*, 16, 439–50.

22 The account of this visit is taken from a fascinating account of the history of prefrontal leucotomy/lobotomy by Valenstein, E.S. (1986) *Great and Desperate Cures*, Basic Books, New York. The title is a quotation from John Bunyan.

23 Tooth, G.C. & Newton, M.P. (1961) *Leucotomy in England and Wales*, 1942–54. London. Ministry of Health Reports on Public Health and Medical Subjects, 104.

24 Bechara, A., Tranel, D., Damasio, H., Adolphs, R., Rockland, C. & Damasio, A.R. (1995) *Science*, 269, 1115–8.

25 Tranel, D. & Hyman, B.T. (1990) *Archives of Neurology*, 47, 349–55.

26 Adolphs, R., Tranel, D., Damasio, H. & Damasio, A.R. (1995) *Nature*, 379, 497; (1995) *J. Neuroscience*, 15, 5879–91.

27 Young, A.W., Aggleton, J.P., Hellawell, D.J., Johnson, M., Broks, P. & Hanley, J.R. (1995) *Brain*, 118, 15–24. Scott, S.K., Young, A.W., Calder, A.J., Hellawell, D.J., Aggleton, J.P. & Johnson, M. (1997) *Nature*, 385, 254–7.

28 Cahill, L., Babinsky, R. Markowitsch, H.J. & McGaugh, J.L. (1995) *Nature*, 377, 295–6.

29 Heath, R.G. (1963) *Am. J. Psychiat.*, 120, 571–7.

30 Reviews by Halgren, E., and by Gloor, P. (1992) in *The Amygdala: Neurobiological Aspects of Emotion, Memory and Mental Dysfunction*, (ed. J.P. Aggleton), pp. 191–228 and 505–38, Wiley-Liss, New York. Gloor, P. (1986) In *The Limbic System: Functional Organizations and Clinical Disorders*, (eds. P.K. Doane & K. E. Livingston), pp. 159–169, Raven Press, New York. Rolls, E.T. (1975) *Brain and Reward*, pp. 29–33, Pergamon, Oxford.

31 Olds, J. & Milner, P. (1954) *J. Comp. Physiol. Psychol.* 47, 419–27. Olds, J. (1977) *Drives and Reinforcements*, Raven, New York. Rolls, E.T. (1975) *The Brain and Reward*, Pergamon, Oxford.

32 Routtenberg, A. & Lindy, J. (1965) *J. Comp. Physiol. Psychol.*, 60, 158–161.

33 Balagura, S. & Hoebel, B.G. (1967) *Physiol. Behav.*, 2, 337–40.

34 Gallistel, C.R. & Beagley, G. (1971) *J. Comp. Physiol. Psychol.*, 76, 199–205.

35 Caggiula, A.R. (1970) *J. Comp. Physiol. Psychol.*, 70, 399–412.

36 For references see Rolls, E.T. (1975) *Brain and Reward*, pp. 29–33, Pergamon, Oxford.

37 Wise, R.A. (1996) *Annu. Rev. Neurosci.*, 19, 319–40.

38 For references see Wise, R.A. (1996) *Annu. Rev. Neurosci.*, 19, 319–40.

39 Wise, R.A. (1996) *Annu. Rev. Neurosci.*, 19, 319–40.

40 Lyness, W.H., Friedle, N.M. & Moore, K.E. (1979) *Pharmacology Biochemistry & Behavior*, 11, 553–6.

41 Connell, P.H. (1958) *Amphetamine Psychosis*, Maudsley Monograph No. 5, Chapman & Hall, London.

42 Seeman, P., Lee, T., Chau-Wong, M. & Wong, K. (1976) *Nature*, 261, 717–19.

43 Baldesarrini, R.J. & Frankenburg, F.R. (1991) *New England Journal of Medicine*, 324, 746–54.

44 See Cooper, J.R., Bloom, F.E. & Roth, R.H. (1996) *The Biochemical Basis of Neuropharmacology*, 7th edn, ch. 9.

45 Seeman, S., Guan, H-C. & Van Tol, H.H.M. (1993) *Nature*, 365, 441–5.

46 For references see Lidow, M.S., Williams, G.V. & Goldman-Rakic, P.S. (1998) *Trends in Pharmacological Sciences*, 19, 136–40.

47 Northsen, M. *et al.* (1994) *Human Molecular Genetics*, 3, 2207–12. Seeman, P. *et al.* (1994) *American Journal of Molecular Genetics*, 54, 384–90.

48 Seeman, P., Chau-Wong, M., Tedesco, J. & Wong, K. (1975) *Proc. Natl. Acad. Sci. USA*, 72, 4376–80.

49 Lidow, M. & Goldman-Rakic, P.S. (1997) *Journal of Pharmacology and Experimental Therapeutics*, 283, 939–46.

50 Pilowsky, L.S., Mulligan, R.S., Acton, P.D., Ell, P.J., Costa, D.C. & Kerwin, R.W. (1997) *Lancet*, 350, 490–1.

51 Rang, H.P., Dale, M.M. and Ritter, J.M. (1995) *Pharmacology*, 3rd edn, p. 590, Churchill Livingston, Edinburgh.

52 Rang, H.P., Dale, M.M. and Ritter, J.M. (1995) *Pharmacology*, 3rd edn, ch. 29, Churchill Livingston, Edinburgh.

53 Rang, H.P., Dale, M.M. and Ritter, J.M. (1995) *Pharmacology*, 3rd edn, p. 491, Churchill Livingston, Edinburgh.

54 Nemeroff, C.B. (1998) *Scientific American*, 278, June 28–35; (1996)*Molecular Psychiatry* 1, 336–42.

21 Planning and attention

1 Harlow, J.M. (1848) *Boston Medical and Surgical Journal*, 39, 389–93; (1849) *Boston Medical and Surgical Journal*, 39, 506–7. Bigelow, H.J. (1850) *American Journal of the Medical Sciences*, 39, 13–22. Macmillan, M. (1996) In *Classic Cases in Neuropsychology* (ed. C. Code, C-W. Wallesch, Y. Joanette & A. R. Lecours), ch. 18, Psychology Press, Hove, East Sussex.

2 Harlow, J.M. (1868) *Publications of the Massachusetts Medical Society* 2, 327–47.

3 Macmillan, M. (1996) In *Classic Cases in Neuropsychology* (eds C. Code, C-W. Wallesch, Y. Joanette & A.R. Lecours), ch. 18, Psychology Press, Hove, East Sussex.

4 Damasio, H., Grabowski, T., Frank, R., Galaburda, A.M. & Damasio, A.R. (1994) *Science*, 264, 1102–5.

5 Brickner, R.M. (1934) *Research Publications of the Association for Research in Nervous and Mental Diseases*, 13, 259–351. A fascinating account of a recent similar case is given by Damasio, A.R. (1994) *Descartes' Error: Emotion, Reason and the Human Brain*, Grosset/Putnam, New York.

6 Damasio, A.R. & Anderson, S.W. (1993) In *Clinical Neuropsychology* (eds K.H. Heilman & E. Valenstein), 3rd edn, ch. 12, Oxford University Press, New York. Fuster, J.M. (1997) *The Prefrontal Cortex*, 3rd edn, ch. 6, Lippincott-Raven, Philadelphia.

7 Milner, B. (1963) *Archives of Neurology*, 9, 90–100.

8 Duncan, J. (1995) In *The Cognitive Neurosciences* (ed. M.S. Gazzaniga), ch. 45, MIT Press, Cambridge, Mass.

9 Cummings, J.L. (1985) *Clinical Neuropsychiatry*, Grune & Stratton, Orlando. Damasio, A.R. & Anderson, S.W. (1993) In *Clinical Neuropsychology*, 3rd edn (eds K.M. Heilman & E. Valenstein), ch. 12, Oxford University Press, New York. Knight, R.T. & Grabowecky, M. (1995) In *The Cognitive Neurosciences* (ed. Gazzaniga, M.), ch. 90, MIT Press, Cambridge, Mass. Fuster, J.M. (1997) *The Prefrontal Cortex*, 3rd edn, ch. 6, Lippincott-Raven, Philadelphia.

10 See discussion by Duncan, J. & Owen, A. (1999) In *Attention and Performance*, *XVIII* (eds S. Monsell & J. Driver) MIT Press (in the press).

11 James, W. (1890) *The Principles of Psychology*, ch. 11, Henry Holt.

12 Pinker, S. (1998) *How the Mind Works*, p. 44, Penguin, London.

13 Huxley, T.H. in his 1893 Romanes Lecture – see Huxley, T.H. & Huxley, J. (1947) *Evolution and Ethics*, p. 57, Pilot Press, London.

14 Bisiach, E. & Luzzatti, C. (1978) *Cortex*, 14, 129–33.

15 Caramazza, A. & Hillis, A.E. (1990) *Nature*, 346, 267–9.

16 Silberpfennig, J. (1949) *Confinia Neurologica*, 4, 1–13. Rubens, A.B. (1985) *Neurology*, 35, 1019–24.

17 Vallar, G., Bottini, G., Rusconi, M.L. & Sterzi, R., (1993) *Brain*, 116, 71–86.

18 Vallar, G. (1997) *Phil. Trans. Roy. Soc. Lond.* B., 352, 1401–9.

19 Wurtz, R.H., Goldberg, M.E. & Robinson, D.L. (1982) *Scientific American,* 246, Dec. 100–107.

20 Moran, J. & Desimone, R. (1985) *Science,* 229, 782–4. Treue, S. & Maunsell, J. H.R. (1996) *Nature,* 382, 539–41.

21 Mangun, G.R., Hillyard, S.A. & Luck, S.J. (1993) In *Attention and Performance,* vol. 14, (eds D. Meyer & S. Kornblum), pp. 219–243, MIT Press, Cambridge, Mass.

22 Treisman, A. & Gelade, G. (1980) *Cognitive Psychology,* 12, 97–136. Treisman, A. (1996) *Current Opinion in Neurobiology,* 6, 171–8.

23 Friedman-Hill, S.R., Robertson, L.C. & Treisman, A. (1995) *Science,* 269, 853–5. Robertson, L.C., Treisman, A., Friedman-Hill, S.R. & Grabowecky, M. (1997) *J. Cognitive Neuroscience,* 9:3, 295–317.

24 Von der Malsburg, C. & Schneider, W. (1986) *Biological Cybernetics,* 54, 29–40. Singer, W. & Gray, C.M. (1995) *Annual Review Neuroscience,* 18, 555–86.

25 Singer, W. (1990) *Concepts in Neuroscience,* 1, 1–26.

26 Gray, C.M., König, P., Engel, A.K. & Singer, W. (1989) *Nature,* 338, 334–7.

27 Engel, A.K., Kreiter, A.K., König, P. & Singer, W. (1991) *Proc. Natl. Acad. Sci. USA,* 88, 6048–52.

28 Engel, A.K., König, P. & Singer, W. (1991) *Proc. Natl. Acad. Sci. USA,* 88, 9136–40.

29 Logothetis, N.K. & Schall, J.D. (1989) *Science,* 245, 761–3.

30 Leopold, D.A. & Logothetis, N.K. (1996) *Nature,* 379, 549–53.

31 Logothetis, N.K., Leopold, D.A. & Sheinberg, D.L. (1996) *Nature,* 380, 621–4.

32 Tootell, R.B.H. *et al.* (1995) *Nature,* 375, 139–41.

33 Crick, F. & Koch, C. (1990) *Seminars in the Neurosciences,* 2, 263–75.

34 Von der Malsburg, C. & Schneider, W. (1986) *Biological Cybernetics,* 54, 29–40.

35 Singer, W. & Gray, C.M. (1995) *Annual Review of Neuroscience,* 18, 555–86.

VI THE PHILOSOPHY OF MIND – OR MINDING THE PHILOSOPHERS

Introduction

1 Wittgenstein, L. (1953) *Philosophical Investigations,* (trans. G. E. M. Anscombe), I, 309, Blackwell, Oxford.

22 The 'mind–body problem' – a variety of approaches

1 Much of this chapter is taken from Glynn, I.M. (1993) *Biological Reviews of the Cambridge Philosophical Society,* 68, 599–616.

2 James, W. (1879) *Mind,* 4, 1–22.

3 But cf Jackson, F. (1982) *Philosophical Quarterly,* 32, 127–36.

4 Watson, J.B. (1913) *Psychological Review,* 20.

5 An interesting account of Watson's eventful life is given by Cohen, D. (1979) *J. B. Watson: The Founder of Behaviorism,* Routledge, London.

6 Skinner, B.F. (1974) *About Behaviorism,* Knopf, New York.

7 Ryle, G. (1949) *Concept of Mind*, Hutchinson, London.

8 Ryle, G. (1949) *Concept of Mind*, ch. 2, p. 43, Hutchinson, London.

9 Armstrong, D.M. (1968) *A Materialist Theory of the Mind*, Routledge & Kegan Paul, London; (1977) 'The causal theory of mind', In *Neue Hefte für Philosophie*, 11, 82–95, reproduced (1991) in *The Nature of Mind*, (ed. D. M. Rosenthal), pp. 181–8, Oxford University Press, Oxford. Block, N. (1980) *Readings in the Philosophy of Psychology*, vol. I, Part III, Functionalism, Methuen, London.

10 Place, U.T. (1956) *British Journal of Psychology*, 47, 44–50. Smart, J.J.C. (1959) *Philosophical Review*, 68, 141–56. Feigl, H. (1960) 'Mind-body, *not* a pseudo-problem', In *Dimensions of Mind* (ed. S. Hook), New York University Press, New York. Armstrong, D.M. (1968) *A Materialist Theory of the Mind*, Routledge & Kegan Paul, London; (1977) *Neue Hefte für Philosophie*, 11, 82–95, reproduced (1991) in *The Nature of Mind*, (ed. E.M. Rosenthal), pp. 181–88, Oxford University Press, Oxford. Armstrong, D.M. & Malcolm, N. (1984) *Consciousness and Causality*, Blackwell, Oxford.

11 Known as anomalous monism. See Davidson, D. (1970) In *Experience and Theory* (eds L. Foster & J. Swanson), pp. 79–101, University of Massachusetts Press, Amherst. Earlier versions of non-reductive materialism were discussed in the nineteenth century and in the first half of the twentieth.

12 Crane, T. & Mellor D.H. (1990) *Mind*, 99, 185–206.

13 Putnam, H. (1967) reprinted (1991) as 'The nature of mental states' in *The Nature of Mind* (ed. D.M. Rosenthal), pp. 197–203, Oxford University Press, Oxford. Jackendoff, R. (1987) *Consciousness and the Computational Mind*, MIT Press, Cambridge, Mass. Johnson-Laird, P.N. (1988) *The Computer and the Mind*, Harvard University Press, Cambridge, Mass.

14 Place, U.T. (1956) *British Journal of Psychology*, 47, 44–50.

15 Putnam, H. (1967) Reprinted (1991) as 'The nature of mental states' in *The Nature of Mind* (ed. D.M. Rosenthal), pp. 197–203, Oxford University Press, Oxford.

16 Dennett, D.C. (1978) *Brainstorms*, chs 1 and 4, Bradford, Montgomery, Vermont.

17 Johnson-Laird, P.N. (19 88) *The Computer and the Mind*, Harvard University Press, Cambridge, Mass.

18 Kripke, S. (1971) 'Identity and Necessity', in *Identity and Individuation* (ed. M.K. Munitz), pp. 135–64, New York University Press; (1972) 'Naming and Necessity', in *Semantics of Natural Language* (eds G. Harman & D. Davidson), pp. 253–355, Reidel, Dordrecht & Boston, rev. and enlarged edn, 1980, Blackwell, Oxford.

19 Slightly different genes that, because of redundancy in the genetic code, code for identical amino-acid sequences are well recognized.

20 Feigl, H. (1960) 'Mind-body, *not* a pseudo-problem', in *Dimensions of Mind*, (ed. S. Hook), New York University Press, New York.

21 McGinn, C. (1989) *Mind*, 98, no. 891 July; (1988) *Proceedings of the British Academy*, 74, 219–39. Both articles are reproduced in McGinn, C. (1991) *Problems of Consciousness*, Blackwell, Oxford.

22 Chomsky, N. (1975) *Reflections on Language*, ch. 4, pp. 155–6, Pantheon Books.

23 McGinn (1991) *Problems of Consciousness*, Blackwell, Oxford.

24 Wittgenstein, L. (1953) *Philosophical Investigations* (trans. G.E.M. Anscombe), I, 133; Blackwell, Oxford.

25 Pinker, S. (1997) *How the Mind Works*, p. 563, Penguin, London.

26 Feyerabend, P. (1963) *Review of Metaphysics*, 17, 49–67. Rorty, R. (1965) *Review of Metaphysics*, 19, 24–54; (1970) *Review of Metaphysics* 24, 112–21. Stich, S.P. (1983) *From Folk Psychology to Cognitive Science*, MIT Press, Cambridge, Mass. Churchland, P.M. (1981) *Journal of Philosophy* 78, 67–90; (1988) *Matter and Consciousness* (rev. edn), pp. 43–9, MIT Press, Cambridge, Mass. Churchland, P.S. (1986) *Neurophilosophy*, pp. 395–9, MIT Press, Cambridge, Mass.

27 Fodor, J. (1987) *Psychosemantics*, ch. 1, MIT Press, Cambridge, Mass.

28 For interesting comments on Freud, Oedipus and Sophocles see Dawe, R.D. (1996) *Sophocles: the Classical Heritage*, pp. xxii–xxv and 243–50, Garland, New York.

29 See Ellenberger, H.F. (1970) *The Discovery of the Unconscious*, p. 541, Basic Books, New York.

30 Fodor, J. (1987) *Psychosemantics*, ch. 1, MIT Press, Cambridge, Mass.

23 The 'mind–body problem' – consciousness and qualia

1 The full quotation is: 'For a *large* class of cases – though not for all – in which we employ the word "meaning" it can be defined thus: the meaning of a word is its use in the language.' See Wittgenstein. L. (1953) *Philosophical Investigations*, pt I, para 43, Blackwell, Oxford.

2 Stich, S.P. (1981) *Philosophical Topics*, 12, 39–62.

3 Dennett, D.C. (1988) In *Consciousness in Contemporary Science*, (eds A.J. Marcel & E. Bisiach), pp. 42–77, Oxford University Press, Oxford.

4 Dennett, D.C. (1991) *Consciousness Explained*, ch. 12, Little, Brown, Boston.

5 Locke, J. (1894) *An Essay Concerning Human Understanding*, Book II, ch. XXXII, para. 15.

6 Wittgenstein, L. (1953) *Philosophical Investigations*, pt I, para. 293, Blackwell, Oxford.

7 Wittgenstein, L. (1953) *Philosophical Investigations*, pt I, para. 258, Blackwell, Oxford.

8 Dennett, D.C. (1988) In *Consciousness in Contemporary Science* (eds A.J. Marcel & E. Bisiach), pp. 42–77, Oxford University Press, Oxford.

9 Dennett, D.C. (1991) *Consciousness Explained*, pp. 386–7, Little, Brown, Boston.

10 Dennett D.C. (1988) In *Consciousness in Contemporary Science* (eds A.J. Marcel & E. Bisiach), pp. 54–5, Oxford University Press, Oxford.

11 Searle, J.R. (1992) *The Rediscovery of the Mind*, p. 55, MIT Press, Cambridge, Mass.

12 Searle, J.R. (1992) *The Rediscovery of the Mind*, p. 1, MIT Press, Cambridge, Mass.

13 Huxley, T.H. (1866) *Lessons in Elementary Physiology*, ch. VIII, p. 193 Macmillan, London.

14 Searle, J.R. (1992) *The Rediscovery of the Mind*, p. 101, MIT Press, Cambridge, Mass.

15 Penrose, R. (1989) *The Emperor's New Mind*, Oxford University Press, Oxford. Cairns-Smith, A.G. (1996) *Evolving the Mind*, Cambridge University Press, Cambridge.

16 Searle, J.R. (1992) *The Rediscovery of the Mind*, pp. 107–8, MIT Press, Cambridge, Mass.

17 Barrs, B.J. (1988) *A Cognitive Theory of Consciousness,* Cambridge University Press, Cambridge. Flanagan, O.J. (1991) *The Science of the Mind,* 2nd edn, MIT Press, Cambridge, Mass; (1992) *Consciousness Reconsidered,* MIT Press, Cambridge, Mass. Marcel, A.J. (1986) *Behavioral and Brain Sciences* 9, 40–4. Van Gulick, R. (1989) *Philosophical Topics,* 17, 211–30.

18 Marcel, A.J. (1988) In *Consciousness in Contemporary Society,* (eds A.J. Marcel & E. Bisiach), pp. 121–58, Oxford University Press, Oxford.

19 Block, N. (1995) *Behavioral and Brain Sciences,* 18, 227–87.

20 I am thinking particularly of the comments by G. Graham, D. Lloyd, T. Natsoulas, and A. Revonsuo, published with Block's article.

24 Free will and morality

1 The account that follows draws heavily on Jacobs, L. (1964) *Principles of the Jewish Faith,* ch. 11, Vallentine Mitchell; and on articles on individual writers and on Arabic philosophy in *The Jewish Encyclopaedia* (1901), Funk & Wagnalls, New York, and in the 11th edition and the current edition of *The Encyclopaedia Britannica.*

2 St. Augustine, (388) *De Libero Arbitrio,* III, §1 ff., see Smart N. (1962) *Historical Selections in the Philosophy of Religion,* pp. 40–9, SCM Press, London.

3 Jacobs, L. (1964) *Principles of the Jewish Faith,* p. 339, Vallentine Mitchell, London.

4 Dennett, D.C. (1991) *Consciousness Explained,* ch. 5, Little, Brown, Boston.

5 The quotations from Darwin in this paragraph are from ch. 4 of Part I of *The Descent of Man.*

6 Richards, R.J. (1987) *Darwin and the Emergence of Evolutionary Theories of Mind and Behavior,* University of Chicago Press, Chicago, gives a critical account of the development of the views of Darwin and his contemporaries on the origin of a moral sense and of the reactions to those views.

7 Moore, G.E. (1922) *Philosophical Studies,* ch. X, Routledge, London.

8 Greg, W.R. (1868) *Fraser's Magazine,* 78, 353–62.

9 Hofstadter, R. (1945) *Social Darwinism in American Thought,* University of Pennsylvania Press, Philadelphia. Jones, G. (1980) *Social Darwinism and English Thought,* Harvester Press, Sussex. Richards, R.J. (1987) *Darwin and the Emergence of Evolutionary Theories of Mind and Behavior,* chs. 6, 7; University of Chicago Press, Chicago.

10 Hofstadter, R. (1945) *Social Darwinism in American Thought,* ch. 2, University of Pennsylvania Press, Philadelphia.

11 Spencer, H. (1866) *Social Statics,* p. 79–80, Williams and Norgate, London.

12 Spencer, H. (1866) *Social Statics,* p. 414–15, Williams and Norgate, London.

13 Hofstadter, R. (1945) *Social Darwinism in American Thought,* chs. 2, 9, 10, University of Pennsylvania Press, Philadelphia.

14 See, for example, the articles in Barkow, J.H., Cosmides, L. & Tooby, J. (eds) (1992) *The Adapted Mind,* Oxford University Press, New York and Oxford; and in Runciman, W.G., Maynard Smith, J. & Dunbar, R.I.M. (eds) (1996) *Evolution of Social Patterns in Primates and Man,* vol. 88 of the *Proceedings of the British Academy.*

15 See, for example, Maynard Smith, J. (1976) *Quarterly Review of Biology,* 51, 277–83; (1980) In *Morality as a Biological Phenomenon* (ed. G.S. Stent), pp. 21–30,

University of California Press, Berkeley.

16 Roberts, G. & Sherratt, T.N. (1998) *Nature,* 394, 175–9.

17 Boyd, R. & Richerson, P.J. (1990) In *Beyond Self-Interest* (ed. J. Mansbridge), ch. 7, University of Chicago Press, Chicago. See also Erdal, D. & Whiten, A. (1996) In *Modelling the Early Human Mind* (eds P. Mellars & K. Gibson), pp. 139–50, McDonald Institute Monographs, Cambridge. Ridley, M. (1996) *The Origins of Virtue,* ch. 9, Viking, London.

EPILOGUE

1 Zhang, Z. *et al.* (1998) *Nature,* 395, 698–702.

Extra note to page 47

The long-held belief that regular tool-making by our ancestors was begun by hominids with brains substantially larger than those of australopithecines may well be wrong. In April 1999 Tim White and his colleagues published reports suggesting that, 2.5 million years ago in the Afar region of Ethiopia, a species of Australopithecine, with a brain volume of about 450 millilitres and long legs, used stone tools to cut flesh from the carcases of antelopes and horses. See de Heinzelin, J. *et al. Science,* 284, 625–629; Asfaw, B. *et al. Science,* 284, 629–635.

Index